MEASURING THE DEMAND FOR ENVIRONMENTAL QUALITY

CONTRIBUTIONS
TO
ECONOMIC ANALYSIS

198

Honorary Editor:
J. TINBERGEN

Editors:
D. W. JORGENSON
J. WAELBROECK

NORTH-HOLLAND
AMSTERDAM • NEW YORK • OXFORD • TOKYO

MEASURING THE DEMAND FOR ENVIRONMENTAL QUALITY

Edited by

John B. BRADEN
Institute of Environmental Studies
University of Illinois
Urbana, IL, U.S.A.

Charles D. KOLSTAD
Institute of Environmental Studies
University of Illinois
Urbana, IL, U.S.A.

1991

NORTH-HOLLAND
AMSTERDAM • NEW YORK • OXFORD • TOKYO

ELSEVIER SCIENCE PUBLISHERS B.V.
Sara Burgerhartstraat 25
P.O. Box 211, 1000 AE Amsterdam, The Netherlands

Distributors for the United States and Canada:

ELSEVIER SCIENCE PUBLISHING COMPANY INC.
655 Avenue of the Americas
New York, N.Y. 10010, U.S.A.

ISBN: 0 444 88877 2

PRINTED IN THE NETHERLANDS

INTRODUCTION TO THE SERIES

This series consists of a number of hitherto unpublished studies, which are introduced by the editors in the belief that they represent fresh contributions to economic science.

The term "economic analysis" as used in the title of the series has been adopted because it covers both the activities of the theoretical economist and the research worker.

Although the analytical methods used by the various contributors are not the same, they are nevertheless conditioned by the common origin of their studies, namely theoretical problems encountered in practical research. Since for this reason, business cycle research and national accounting, research work on behalf of economic policy, and problems of planning are the main sources of the subjects dealt with, they necessarily determine the manner of approach adopted by the authors. Their methods tend to be "practical" in the sense of not being too far remote from application to actual economic conditions. In additon they are quantitative.

It is the hope of the editors that the publication of these studies will help to stimulate the exchange of scientific information and to reinforce international cooperation in the field of economics.

The Editors

PREFACE

This book stems from a unique provision of the environmental regulatory system in Illinois. Alone among states in the United States, Illinois has required economic studies of proposed environmental regulations. Virtually all of the studies have encountered great difficulty in quantifying the economic advantages — benefits — of the proposals. As a result, the Illinois Department of Energy and Natural Resources has periodically commissioned reviews of the "state of the art" in environmental benefits assessment. The department's efforts in this area previously resulted in the book, *Cost-Benefit Analysis and Environmental Regulations: Politics, Ethics, and Methods,* edited by Daniel Swartzman, Richard Liroff, and Kevin Croke and published in 1982 by the Conservation Foundation, Washington, D.C. This book results from such a commission made in 1987 to the co-editors of this book. The authors for the individual chapters were identified in early 1988 and preliminary versions of their contributions were presented in an open workshop at the Bismarck Hotel in Chicago on November 18, 1988. This book contains significantly revised versions of the papers presented at the Chicago workshop.

This study could not have been prepared without the financial support of the Illinois Department of Energy and Natural Resources, under project EA-75, and the Research Board of the University of Illinois. Additional support for Professor Braden was provided by the Illinois Agricultural Experiment Station through projects 0331 and 0334. We gratefully acknowledge these sponsors but absolve them from responsibility for the contents.

We owe a debt to Chris Leeson, Eileen McCulley, Jacqueline Smith, Stephanie Spaulding, and Sandy Waterstradt for their meticulous word processing, and to Holly Korab for her careful and professional editing. Our sincere thanks to them.

CONTENTS

Part I

Theory and Methods

Measuring the Demand for Environmental Quality
John B. Braden & Charles D. Kolstad (Editors)
© Elsevier Science Publishers B.V. (North-Holland), 1991

Chapter I

INTRODUCTION

JOHN B. BRADEN, CHARLES D. KOLSTAD, and DAVID MILTZ

University of Illinois at Urbana-Champaign and Katholieke Universiteit Leuven

Two decades ago, the National Environmental Policy Act (NEPA) set new premises for U.S. public policy toward the environment and pollution — that the environment shall be protected and pollution reduced to nonthreatening levels. These uncompromising standards echoed throughout the ambitious environmental legislation of the succeeding years; for example, they are found in the "rebuttable presumption against registration" of pesticides established in the Federal Insecticide, Fungicide, and Rodenticide Act, in the "fishable and swimmable" standards set in the Federal Water Pollution Control Act Amendments, and in the "prevention of significant deterioration" criterion established in the Clean Air Act.

The implementation of these lofty goals is often confronted with a sobering reality: a philosophy of no environmental degradation is often incompatible with everyday life. For example, electricity can be produced from many sources, such as hydropower, fossil fuels, nuclear fission, biogases, and solar energy. However, each of these sources entails some potential for harm to the environment: hydropower alters natural flowing rivers; fossil fuels generate various gaseous and solid emissions, including acid rain precursors and greenhouse gases; nuclear fission produces radioactive wastes that will endure for thousands of years; alcohol-based fuels consume huge amounts of biomass, which is already in short supply for recycling atmospheric carbon dioxide; and solar energy requires significant amounts of land and materials and is aesthetically unpleasing.

As these examples suggest, some nonzero level of environmental intrusion must often be accepted. The real question is: What level of environmental damage should people accept? Increasingly, policy makers view the trade-offs in economic terms as benefits and costs.

The costs of environmental protection are reasonably obvious in many cases — more expensive power, more costly automobiles, more costly garbage disposal, and so on. Not only are the costs obvious, but commercial interests make sure that they are well documented.

The benefits are more elusive for they are not nearly so apparent in everyday commerce. Maintenance of estuarine ecosystems, reducing the risk of contracting cancer, protecting aesthetics, and enhancing the quality of recreational experiences — such things are the consequences of protecting or improving the environment. They often have very significant "collective goods" characteristics in that individuals cannot easily be excluded from enjoying improvements nor can they avoid environmental degradation. Thus, these goods generally are not bought and sold and therefore are not listed on business balance sheets or included directly in the national income and product accounts.

Measuring the economic value of these consequences, or even identifying the beneficiary populations, is a difficult task requiring special concepts and tools. The earliest attempts were efforts to make economic sense of the nuisances caused by industrial smoke in London and Pittsburgh around the turn of the century (Russell 1899; O'Connor 1913). Most methodological progress has come only in the last 40 years, since Harold Hotelling (1949) suggested using travel costs to determine the demand for recreation. The bulk of the developments have occurred since roughly 1974. That year was marked by three major advances: the publication of Karl-Göran Mäler's treatise, *Environmental Economics: A Theoretical Inquiry;* the application of new survey techniques by Alan Randall, Berry Ives, and Clyde Eastman; and the publication of Sherwin Rosen's paper on hedonic price techniques. These works revolutionized the field. They inspired dramatic advances in sophistication and insight; for example, they resulted in the infusion of discrete choice models and the growing use of econometric techniques for handling limited dependent variables. The research frontier now stretches over a wide territory encompassing many different strategies and tactics.

1.1 Objectives of this Book

While research on techniques for estimating the value of environmental goods and services has flourished, there have been relatively few attempts to assess the progress. A. Myrick Freeman III (1979a, 1982, 1985a) has been the leading commentator. Joining Freeman's works have been a theoretical survey by Per-Olav Johansson (1987), an applied comparative study by V. Kerry Smith and William Desvousges (1986b), and assessments of survey techniques by Ronald Cummings, David Brookshire, and William Schulze (1986) and Robert Mitchell and Richard Carson (1989). With the exception of the Cummings, Brookshire, and Schulze volume, which is limited to survey methods, nowhere in the contemporary literature is there a broad assessment of the state of the art by leaders in the field. This book is intended to fill that gap.

Another objective is to assess the relative strengths of the various methods in evaluating preferences for important classes of environmental goods and services. Other economists have studied the value of quality changes for specific media (for example, cleaner air, cleaner water, less pesticides), but analysts frequently need to evaluate the effects of government policies on specific environmental attributes, such as health impacts, recreation services, and aesthetics. As compared with the media studies, which usually encompass several types of commodities, this book focuses on the specific types of goods and services, which allows us to be more specific about analytical methods. Thus, a major thrust of this book is to assess the most fruitful analytical approaches for five of the most important types of environmental impacts: human health, aesthetics, recreation, degradation of materials, and impacts not associated with direct consumption.

This book is intended to be a definitive reference for graduate students and practitioners on the subject of estimating the value of changes in environmental quality. Each chapter is an advanced instructional resource that brings the reader from first principles to the state of the art. The emphasis is on principles that guide the use of analytical methods. Because no two studies of the value of environmental services are the same, there is no pretext of trying to come up with a specific measure of the value of, for example, health effects, that could be applied in a range of circumstances. Specific applications and numerical estimates are used only to illustrate issues, not to define or resolve them. Our aim is to equip present and future analysts and those interested in research opportunities in this field to study the difficult issues that obscure the economic value of environmental goods and services.

1.2 Overview of Methods for Measuring Environmental Demand

Our use of the terms "goods" and "services" in connection with the environment may at first seem strange. One of the major accomplishments of the environmental movement has been to develop a sense of specialness about clean air, clean water, and unique environments; a specialness intended to set these resources apart from the market place where conventional commodities are bought and sold. And below we review some of the economic reasons why the environment does not shoehorn neatly into markets. However, this specialness has occasionally brought about reinvention of parts of microeconomic theory. Many aspects of the environment have close analogs in conventional commodities. This book focuses on methods of placing money values on various aspects of the environment. When attaching price-like numbers to quantities, be they milligrams per liter of biological oxygen demand (BOD), a probability of contracting cancer from ingesting pesticide residues, or a number of visits to the Grand Canyon, we are engaged essentially in estimating the *demand* for these environmental attributes. The methods that are appropriate for assessing demand for market commodities also apply in assessing demand for things not traded in the market; the major differences are not in the methods but in the data and how they are obtained. Our terminology is meant to emphasize that the principles of economic valuation used for dog biscuits, mink stoles, and tickets to soccer matches also have much to offer for evaluating the quality and quantity of environmental resources. Our aim is to more closely integrate the general and rich economic theory of demand with the particular circumstances of environmental goods and services.

If markets existed for environmental goods and services, their economic value would be the sum of actual payments for the commodities plus an appropriate measure of consumer surplus. Consumer surplus generally refers to the excess of individuals' willingness to pay for a good, as reflected in a demand curve, over actual payments.[1] Markets divulge the price and quantity data from which demand curves and actual payments can be deduced.

Market transactions entail exchanging ownership of commodities. When consumers are not easily excluded, as is the case with many public or collective goods, then ownership is not a requirement of consumption and market transactions do not capture consumer preferences. Moreover, one consumer's "consumption" of a collective good does not reduce others' access to the commodity, so conventional market pricing is inefficient. The ownership condition also fails when commodities are valued just because they exist or when they are "bads" that are not really owned but are consumed because they are costly to avoid. In the former case, many people who do not intend to make direct use of endangered species or natural environments, for example, willingly contribute to preservation and protection programs. The latter case describes most types of pollution.

In the absence of ownership and efficient pricing, special techniques are needed to place consumer preferences for environmental goods and services on common ground with the demands for more conventional commodities. Three types of procedures have been employed to measure these demands: *household production function methods* based on the demand for complements and substitutes; *hedonic methods* of decomposing prices of market goods to extract embedded values for related environmental attributes; and *experimental methods for elicitation of preferences,* either by using hypothetical settings, called *contingent valuation,*[2] or by constructing a market where none existed. All three methods stem from common roots in applied welfare economics.

---

[1] Formal distinctions between consumer surplus, which is derived from ordinary uncompensated demand curves, and "exact" measures of welfare change are made in chapter 2. Here, the term is used loosely to refer to a difference between consumers' valuation of a commodity and the payments made for that commodity in the market.

[2] A fourth approach is to derive values implicitly from policy decisions. For example, the government might enact a law against the use of lead in auto fuel with the knowledge that this law will cost $500 million in automotive repairs and retooling while reducing health care costs

In this section, we briefly develop the common theoretical core of these methods, and then we show how the three approaches branch off. The discussion introduces chapters 2 to 5, which present in greater detail the basic theory that underlies demand estimation, the relationship of demand estimation to measuring changes in welfare, and the procedures used to implement demand estimation for environmental commodities. We also indicate the relationship of the approaches to the environmental goods and services discussed in chapters 6 to 10.

1.2.1 Common Premises

A basic premise of all approaches is that individuals make welfare-optimizing consumption decisions. These decisions are captured in the consumer demand functions with respect to available goods and services. Environmental attributes enter those demands. For some environmental goods and services, such as recreation or visits to unique natural wonders, the consumer exercises direct choice over the amount consumed (assuming the commodity is available). Amounts of other environmental attributes are not chosen by individual consumers; however, changes in the amounts experienced can still cause the consumer to change the other goods and services he or she consumes.

To illustrate this framework, we let q be the quantity of an environmental attribute, v be the quality of that attribute, Y be income, x be a composite consumption good whose price is normalized to one, and p be the normalized price associated with q.[3] This could easily be generalized to the case where q, v, and x are vectors. We assume in the general case that all commodities, including the environmental attribute, have prices. We also assume initially that the consumer exercises choice over q but not v, and p is fixed. Different assumptions about consumer choices and fixed prices are the basis of differences between household production and hedonic approaches to valuation.

The consumer seeks to maximize utility, which is a function of q, x, and v:

$$\max_{q,x} \; u(q,x,v) \tag{1.1a}$$

$$\text{s.t.} \quad p \cdot q + x \le Y \tag{1.1b}$$
$$q,x \ge 0$$

The theoretically correct economic measure of welfare change is the payment that will make a consumer indifferent between having and not having a

by $300 million. The analyst might infer that the collective demand for the other environmental commodities affected by lead emissions is worth at least $200 million, the difference between known benefits and costs. There is no microeconomic basis to such estimates and they are crude approximations at best. They are not pursued in this book.

[3] The following discussion draws on the theoretical perspective provided by Schulze, d'Arge, and Brookshire (1981).

particular change in the quality or quantity of the environmental attribute. Assume that the rational consumer utilizes the available budget fully. For a particular level of Y and v, the consumer solves (1.1), yielding some level of utility, u^* and an optimal consumption bundle, (q^*x^*), both of which are functions of p, Y, and v. To investigate a change in v, holding utility constant, we can totally differentiate $u(q^*x^*v) = u^*$ and $p \cdot q^* + x^* = Y$ to obtain:

$$du = \partial u/\partial q \cdot dq + \partial u/\partial v \cdot dv + \partial u/\partial x \cdot dx, \tag{1.2a}$$

and

$$dY = q \cdot dp + p \cdot dq + dx. \tag{1.2b}$$

We will focus on how changes in q and v can be compensated by changes in Y. Thus we let $du = 0$. The assumption of fixed prices means that $dp = 0$, so the first term in (1.2b) drops out. Rearranging (1.2), we get:

$$-dx = \frac{\partial u/\partial q}{\partial u/\partial x} \, dq + \frac{\partial u/\partial v}{\partial u/\partial x} \, dv, \tag{1.3a}$$

and

$$-dx = p \cdot dq - dY. \tag{1.3b}$$

Now let v be the attribute for which a change is contemplated. Setting the right-hand sides of the expressions in (1.3) equal to one another and rearranging gives:

$$\frac{\partial u/\partial q}{\partial u/\partial x} \, dq + \frac{\partial u/\partial v}{\partial u/\partial x} \, dv - p \cdot dp = dY. \tag{1.4}$$

Equation (1.4) establishes that the payment must equal the difference between the personal worth of the change in quantity and quality [the first two terms on the left-hand side of (1.4)] and the change in expenditure on q (the last term on the left-hand side).

A fundamental condition in consumer theory is that optimizing consumers equate marginal rates of substitution to the ratio of product prices.[4] In our case, recall that p is normalized with respect to the price of composite commodity x:

$$\frac{\partial u/\partial q}{\partial u/\partial x} = p. \tag{1.5}$$

Substituting (1.5) into (1.4) and canceling terms results in:

$$\frac{\partial u/\partial v}{\partial u/\partial x} = -\frac{dY}{dv}. \tag{1.6}$$

That is, the marginal rate of substitution between v and x must equal the

[4] This condition must hold only if a positive amount of the commodity is consumed.

change in income that will keep utility constant as v changes. That income change is the "price" that reflects the consumer's maximum willingness to pay for a desirable change in v or be compensated for an undesirable change.[5]

1.2.2 Household Production Functions

The three approaches to measuring the demand for environmental attributes exploit different elements of this common framework. The first approach involves investigating changes in the consumption of commodities that are substitutes or complements for the environmental attribute. This is called the *household production function approach.*[6] The travel cost method, used widely to measure the demand for recreation, is a prominent example.[7] The costs of traveling to a recreation site together with participation rates and visitor attributes are used to derive a measure of the willingness to pay for the site itself. The analysis can be expanded to specify attributes of the site and estimate the demand for attributes.

Travel can be used to infer the demand for a recreation site only if it is a necessary part of the visit, or in economic terms, is a "weak complement."[8] Otherwise, the demand for transportation does not capture all of the demand for recreation.

Another example of the household production function approach is the use of averting costs to infer values.[9] Averting inputs include air filters, water purifiers, noise insulation, and other means of mitigating personal impacts of pollution. (These are also known as defensive or self-protection inputs.) Such inputs substitute for changes in environmental attributes; in effect, the quality of a consumer's personal environment is a function of the quality of the collective environment and the use of averting inputs. We measure the value of changes in the collective environment by examining costs incurred in using averting inputs to make the personal environment different from the collective environment. A rational consumer will buy averting inputs to the point where the marginal rate of substitution between purchased inputs and the collective environment equals the price ratio. By characterizing the rate of substitution

[5] This interpretation holds if the changes are being compared with the status quo without the change in v. If, instead, the benchmark is the level of welfare with the change, then we are measuring willingness to pay to avoid an undesirable change or willingness to be compensated to live with it.

[6] The literature on general household production approaches to consumer theory began with Lancaster (1966). Extensions to collective goods were provided by Hori (1975) and Mäler (1981) and a direct application to recreation was reported by Bockstael and McConnell (1981).

[7] Important studies that develop or use the travel cost method include: Burt and Brewer (1971), Cesario (1976), Clawson (1959), Clawson and Knetsch (1966), McConnell (1975), and Wood and Trice (1958).

[8] On weak complementarity, see the original treatment by Mäler (1974) and the generalized treatment by Bockstael and Kling (1988).

[9] Averting cost studies include: Bartik (1988a), Courant and Porter (1981), Gerking and Stanley (1986), Harford (1984), Shibata and Winrich (1983), and Watson and Jaksch (1982).

and knowing the price paid for the substitute, we can infer the price that consumers would be willing to pay for a change in the environment.

The common element in household production function methods is the use of changes in the *quantities* of complements (q and x) to estimate the value of a change in quality v. Say again that v changes and that we are interested in a compensated measure of welfare change.

Solving (1.3b) for dY, substituting into (1.6), and rearranging terms gives:

$$\frac{\partial u/\partial v}{\partial u/\partial x} = -(p \cdot dq/dv + dx/dv). \tag{1.7}$$

The left-hand side of (1.7) replicates (1.6), but the right-hand side is the sum of changes in the consumption portfolio, each weighted by a price (recall that the price of x is normalized to one). Tallying the changes in expenditures on substitutes and complements will yield a "price" for the environmental attribute.

The household production function method is important and valuable because it brings preferences for nonmarket goods and services into the arena of observable market relationships. In this method actual behavior serves as the basis of valuation, thus familiar types of data and analysis can be employed. However, these virtues are not unencumbered. The household production function method is limited to use values. Values that do not entail direct consumption cannot be estimated by looking at complements or substitutes. Moreover, weak complementarity must hold for a complement (or set of complements) and perfect substitutability must hold for substitutes in order for welfare measures to be valid and complete.

Chapter 3 contains a thorough discussion of the household production function methods. Uses of this method are discussed in the chapters on health, recreation, and materials damage.

1.2.3 Hedonic Price Analysis

Hedonic price analysis refers to the estimation of implicit prices for individual attributes of a market commodity.[10] Some environmental goods and services can be viewed as attributes of market commodities, such as real property. For example, proximity to noisy streets or factories, exposure to polluted air, and access to parks or scenic vistas are purchased along with residential property. Part of the variation in property prices is due to differences in these amenities. Other applications have been to wages for jobs that entail different levels of physical risk.

Hedonic methods operate on *prices* rather than the quantities used for

---

[10] The literature on hedonic methods includes: Anderson and Crocker (1971), Freeman (1979b), Griliches (1971), Harrison and Rubinfeld (1978), Nelson (1978), Ridker and Henning (1967), and Rosen (1974).

household production approaches. The price of an attribute v is considered as depending on the level of the attribute, $\partial p/\partial v \neq 0$. Furthermore, all compensation takes place through prices; specifically, the implicit price for v as projected on the price of q. With no income change ($dY = 0$), then (1.3a) and (1.3b) combine to become:

$$\frac{\partial u/\partial q}{\partial u/\partial x}\,dq + \frac{\partial u/\partial v}{\partial u/\partial x}\,dv - p\cdot dq = dp\cdot q. \tag{1.4'}$$

Recalling that marginal rate of substitution between q and x must equal p if $q > 0$, (1.4') can be combined and rewritten:

$$\frac{\partial u/\partial v}{\partial u/\partial x} = \frac{dp}{dv}\,q. \tag{1.8}$$

The left-hand side is again the marginal rate of substitution between v and income (since x has unitary price) while the right-hand side is the marginal value of v times the number of units of the commodity containing v. *All* of the compensation is assumed to be captured in the price for q.

Implementing this approach requires estimating the demand for the commodity attributes and then observing the price variation associated with different attribute levels. The implicit prices of specific attributes reflect supply and demand forces just as the observed commodity prices do. Hence, we cannot simply interpret the attribute price relationship as we do the attribute demand function. Rather, those implicit prices must be further related to consumer tastes and preferences in a "second stage" of estimation. The result of the second stage is an attribute demand function that can be used for value measurement.

Hedonic approaches share the main advantage of the household production function methods; that is, the use of observed market behavior. This avoids any confusion between what consumers intend and what they really do because only actual transactions are investigated.

As with the household production function method, hedonic methods confront some important limitations. The weak complementarity relationship must hold between the attribute and the commodity with which it is associated. Furthermore, a change in the attribute is assumed to be fully absorbed in the price or quantity of the weak complement. This is a very strong assumption. Finally, nonuse values cannot be measured with this approach.

Chapter 4 presents the hedonic method as it applies to environmental goods and services. The approach is further assessed in the chapters on health, aesthetics, and materials damages — matters for which hedonics have proven very useful.[11]

---

[11] The application to materials damages is primarily through the use of dual models from production theory. See chapter 9.

1.2.4 Direct Elicitation of Preferences

The third approach to demand estimation relies on questionnaires or experiments to elicit preferences. In general, respondents are asked what they would be willing to pay to have more recreation sites, cleaner water, better scenery, or some other environmental attribute. With this approach, quantity and price dimensions can be investigated and compensated demands can be developed directly. There is no need to call upon consumption of complements or substitutes or indirect pricing as outlined in the preceding sections, nor is there the difficulty of deriving compensated measures of welfare change from the uncompensated demands observed in the marketplace. Furthermore, unlike the surrogate market approaches, these methods can be employed to evaluate preferences that do not entail use.

Although direct elicitation methods circumvent the need for back door approaches through formal consumer theory, they confront large difficulties in implementation. The root of the difficulties is that most of these methods rely on intentions, ideals, or behavior expressed in hypothetical circumstances. Intentions are typically costless to express, or nearly so, which means that they may not be considered as carefully as are real consumption choices. Many economists are loathe to base economic values — values that will be used to allocate real resources — on information that does not grow out of real economic commitments.

The implementation issues grow largely out of the desire to bring consumers' intentions closely in line with their probable actions. The valuation exercise must be designed to be as real as possible and the econometric structure must account for various sources of bias, data discontinuities, and other factors. One line of research has attempted to construct actual experimental markets, using special auction procedures to elicit the demand for commodities like hunting licenses. Although the design of constructed markets remains an issue for economists, these markets do combine the advantages of the market-based and direct elicitation approaches. The consumers' behavior is real, entailing a real commitment of funds. The disadvantage of constructed markets is that they rarely can be utilized. They require a commodity for which consumption can be restricted, such as hunting in enclosed wildlife areas. In addition, money must change hands, and few socioeconomic research programs are geared to paying out or receiving more than token sums.

Because conditions are rarely suitable to simulate a real money economy for environmental goods and services, economists have retreated to other nonmarket approaches. One option is laboratory studies. *Experimental economics* — the study of markets and economic behavior in a laboratory setting — has been used to study how information conditions and market structure affect values. However, few such studies have addressed specific environmental commodities. They are constrained by the number and mix of subjects who can be studied, and where money changes hands, by limits on the amounts. Another option is simply to conduct a survey that is divorced from real economic choices so that valuation is contingent on the scenario outlined by the researcher.

The art of economics is in full bloom in the application of contingent valuation methods (CVM). Chapter 5 contains a detailed account of the theoretical underpinnings of direct elicitation methods and the design issues that surround them. Additional perspective is provided in the chapters on recreation, health, aesthetics, and nonuse values — categories in which these methods have played a major role in demand measurement.

1.2.5 Aggregation

Environmental changes typically affect several types of demand categories and may involve nonuse as well as use values. The process of putting together categorical estimates to arrive at a "total economic value" is subject to considerable misunderstanding and debate. This issue cannot be divorced from the choice of evaluation method. If the aggregation is to be proper in theoretical terms, the categorical estimates must be developed in special ways.

The aggregation issue is raised in chapter 2 and in several subsequent chapters. Alan Randall argues in chapter 10 that either a single value statement encompassing all impact categories must be sought or categorical values must be sequenced very carefully. To do otherwise risks double counting. In general, the only practical approach to composite evaluation is contingent valuation. This is the easiest way to get information that cuts across markets and the only way to elicit nonuse values. Nevertheless, it has been common to evaluate specific classes of use or nonuse demands independently and sum across categories for a total without allowing for interactions among categories. The result is systemic overcounting of demands.

1.3 Organization of the Book

This book is divided into two parts. Part I, encompassing chapters 1 through 5, provides a theoretical and methodological overview of environmental demand estimation. Chapter 2, by Charles Kolstad and John Braden, treats the measurement of environmental benefits as a branch of welfare economics. The coverage includes basic issues of demand specification and deriving defensible measures of welfare change from demand estimates. Because these issues arise repeatedly in the remainder of the book, chapter 2 serves as the foundation for all that follows.

Chapter 3, by V. Kerry Smith, covers the theory behind and uses of the household production function framework, which lies behind the travel cost

TABLE 1.1
Measurement methods and applications.

	Methods		
Applications	Household production	Hedonics	Constructed markets
Health	√	√	√
Aesthetics		√	√
Recreation	√		√
Materials damage	√	√	
Nonuse			√

and averting expenditure models of demand estimation. Raymond Palmquist provides similar coverage for hedonic techniques in chapter 4, and Richard Carson covers direct elicitation methods, under the rubric of "constructed markets," in chapter 5. These chapters provide a thorough grounding in the theory and use of these techniques.

In Part II, we look at how estimation methods pertain to five specific classes of environmental goods and services. The purpose is essentially perpendicular to that in Part I. Table 1.1 illustrates this perpendicularity and indicates the intersections between the methods and the impact categories.

The five categories are not exhaustive, but we believe that they cover the most important types of environmental services. Chapter 6, by Maureen Cropper and A. Myrick Freeman III, addresses human health impacts. Human health impacts are, arguably, the most compelling to policy makers. They have been the focus of many studies. Despite the extensive attention, issues such as the widespread reluctance to view discomfort or accelerated death in economic terms and the difficulty of perceiving risks, make demand measurement especially challenging in this area.

The focus in chapter 7 turns to aesthetics; for example, visibility, quiet, and scenic beauty. Philip Graves assesses the approaches used in this area. Two things make this an especially interesting topic. First, in most cases, the impacts are collective in the truest sense; a vista or noise can be shared by many consumers. Second, it turns out that some of the most interesting developmental and comparative work on valuation methods has been done in this area.

Chapter 8 is devoted to recreation impacts. Historically, recreation impacts are the bastion of the travel cost method, and consequently, the development of that approach is most advanced in this application. However, recreation impacts are increasingly approached in other ways, including ones that attempt to combine hedonic and travel cost methods. Authors Nancy Bockstael, Kenneth McConnell, and Ivar Strand provide a thorough assessment of work in this area.

Another important class of environmental damages includes impacts on materials, ranging from additional fabric cleaning costs and accelerated corrosion to reduced yields from agricultural crops. In chapter 9, Richard Adams and Thomas Crocker discuss methods of determining such damages. Since many of these damages affect regular market goods, there is considerable opportunity to draw on conventional economic valuation methods.

In chapter 10, Alan Randall discusses the evaluation of nonuse demand, an issue which cuts across other commodity categories. Randall also provides a careful treatment of the issues associated with breaking down total value into categories such as use and nonuse value.

Part II closes with a summary chapter by Braden and Kolstad. The summary highlights the major conclusions of the respective chapters and places special emphasis on the needs and opportunities for additional research.

Measuring the Demand for Environmental Quality
John B. Braden & Charles D. Kolstad (Editors)
© Elsevier Science Publishers B.V. (North-Holland), 1991

Chapter II

ENVIRONMENTAL DEMAND THEORY

CHARLES D. KOLSTAD and JOHN B. BRADEN

University of Illinois at Urbana-Champaign

2.1 Introduction[1]

Measuring the demand for goods and services is at the core of applied economics. Whether the commodity is labor or watermelons, economists have long been concerned with measuring how a consumer or industry values a commodity by examining trade-offs between product price and quantity consumed. Despite the large body of theory, estimating the demand for conventional goods is rarely easy; the problems are compounded in the case of environmental commodities.

One reason for the difficulties is that the notion of environmental goods and services is vague. For the purposes of this chapter, an environmental commodity will be considered as different from either a private good or a conventional public good in two ways. An environmental commodity will be defined as having at least one of the following two characteristics: either it is a negative good — a "bad" — which carries no price and thus is inefficiently allocated by the market;[2] or it is a public good endowed upon society (rather than purchased), such as a national park. In these cases, although the aggregate quantity of the good or bad supplied may be observed, the individual or aggregate expenditures or valuations of the good will not. This latter point distinguishes environmental goods from public goods in general in that researchers know the cost of supply of public goods like national defense or fire protection, and must trade off monetary costs with benefits; however, they do not know the costs of environmental goods. The valuation of assets such as national parks is, at best, imperfectly known.

---

[1] Comments from Richard Carson, Frank Wolak, and an anonymous referee have been appreciated.

[2] For private (rival and excludable) bads, perfect property rights (a rarity) can result in efficient allocation by the market. Public bads generally cannot be efficiently allocated by the market.

Measuring demand for conventional public goods, whether they be local or global, is not easy but is helped by the existence of a well-defined relationship between expenditures and quantity provided. In the case of environmental goods and services, usually all that can be observed is how consumption of private goods changes with the level of the environmental good. The challenge, then, is to recover the underlying demand for the environmental commodity. Alternatively, artificial or hypothetical markets may be constructed to elicit implicit prices or valuations for environmental goods and services. In the latter, demand estimation is easier although eliciting preferences becomes harder.

This chapter presents a unified treatment of demand theory, first as it is applied to standard private goods and then in the context of two variants that are particularly applicable to environmental goods and services: restricted expenditure/demand functions and demand under uncertainty. Also considered are demand theory as it applies to producers and the relationship between measures of environmental benefits and demand for environmental goods and services.

In section 2.2, conventional demand theory is developed from a dual perspective, focusing on a variety of issues including separability, integrability, and other theoretical restrictions on the nature of demand. Section 2.3 considers restricted demand functions for market commodities, involving levels of the environmental good as a parameter. The demand for the environmental commodity is obtained by integrating to the underlying expenditure function and then differentiating to obtain demand for the environmental commodity. Section 2.4 considers the connection between demand and measures of environmental benefits (for example, the value of clean air), followed by extensions to the theory of the firm. Section 2.6 considers econometric implementation issues; specifically, the random utility model. And last, section 2.7 introduces uncertainty into the analysis.

2.2 Theory of Consumer Demand

The goal of demand theory is to deduce as many properties of a demand function as possible, based on economic theory. This permits the extraction of the maximum amount of information from a set of data. The basic restrictions on a demand function arise from the fact that individual demand results from utility maximization. This fact alone allows researchers to deduce a rich set of properties of demand functions. Aside from restrictions on demand functions, another set of questions involves aggregation of individual demand functions. Because only individuals perform utility maximization, yet it is groups of individuals that are usually observed, this issue of aggregation is of some importance. However, aggregation will not be covered here; the interested reader is referred to Jorgenson, Lau, and Stoker (1982). Thus, with the exception of aggregation, these questions about restrictions on demand functions will be reviewed in this section. The treatment is intentionally brief; more detailed but still general treatments of demand theory can be found in Varian (1984), Deaton and Muellbauer (1980) and Philips (1974), among others.

2.2.1 *Indirect Utility, Expenditure, and Demand Functions*

Consider an individual facing a set of n goods, indexed by $i = 1, \ldots, n$. Let q be a vector of goods consumed by the individual and let Y be his or her income. Assume the individual possesses a quasi-concave utility function, $u(q)$.[3] Quasi concavity is equivalent to indifference curves being convex to the origin. Consumers are faced with a set of prices for goods, p, and choose a bundle of goods by maximizing utility:

$$\max_{q} u(q)$$
$$\text{s.t.} \quad p'q \leq Y \tag{2.1}$$
$$q \geq 0,$$

where the prime indicates transposition. The solution to problem (2.1), $q^* = x(p, Y)$, is a set of *ordinary,* or *Marshallian, demand functions,* giving quantity demanded as a function of price and income. Demand is uncompensated because as prices change, income is not adjusted to compensate for the resulting change in utility. Substituting q^* back into u yields a quasi-convex optimal value function $v(p, Y)$. This optimal value function, termed the *indirect utility function,* defines the highest level of utility attainable, given prices p and income Y. As can be seen from (2.1), v and x must be homogeneous of degree zero in prices and income. Roy's identity relates x and v:

$$x(p, Y) = - \frac{\nabla_p v(p, Y)}{\partial v(p, Y)/\partial Y}; \tag{2.2}$$

that is, the derivative of indirect utility with respect to the ith price yields the ith demand function, after normalizing by the marginal utility of income.

Dual to problem (2.1), which is utility maximization subject to a maximum expenditure, is expenditure minimization subject to attainment of minimum utility:

$$\min_{q} p'q$$
$$\text{s.t.} \quad u(q) \geq U \tag{2.3}$$
$$q \geq 0.$$

---

[3] u is quasi concave if for all constants, U, the level set $\{q \mid u(q) > U\}$ is convex. Thus a convex combination of bundles that each yield some level of utility also yields at least that level of utility.

The solution to this minimization problem, $q^* = h(p,U)$, is a set of *compensated, or Hicksian, demand functions,* which give the quantity demanded as a function of price and utility. Income is of no consequence; as prices change, expenditures are adjusted to maintain constant utility. Substituting h into the objective function of equation (2.3) yields an optimal value function, $e(p,U) = p'h(p,u)$, the *expenditure function,* which is concave in p. The function $e(p,U)$ defines the minimum expenditure necessary to achieve utility U at prices p. This is conceptually identical to the cost function in production theory and thus shares its properties. Clearly, from equation (2.3), e is homogeneous of degree 1 in prices and h is homogeneous of degree 0 in prices. The expenditure function, e, and compensated demand, h, are related according to Shephard's lemma,

$$h(p,U) = \nabla_p \, e(p,U), \tag{2.4}$$

which states that demand for the ith commodity is simply the derivative of the expenditure function with respect to the ith price.

Because of the concavity of the expenditure function, it is known that the Hessian of e (the matrix of second derivatives) is symmetric and negative semidefinite. This Hessian is the matrix whose (i,j)th element is $\partial h_i/\partial p_j$. Furthermore, concavity of e implies that $\partial^2 e/\partial p_i^2 \leq 0$ for all i; this yields, from equation (2.4), the law of demand, $\partial h_i/\partial p_i \leq 0$, that all compensated demand functions are downward sloping, or at least do not slope upward.

Two types of demand functions have been derived: ordinary and compensated. Ordinary demand functions have the disadvantage that price and income effects are bundled together; therefore, the effect of a price change as reflected by an ordinary demand function will involve price and income effects. Compensated demand functions do not have this problem because they focus on price effects alone. Unfortunately, economists typically estimate ordinary demand functions because they do not observe utility. Note that the curvature, slope, and other results derived above apply to compensated demand only, not necessarily to ordinary demand functions (indirect utility is only quasi-convex, thus does not have the same curvature properties as the expenditure function).

Properties of compensated and uncompensated demand functions are related through the Slutsky equations. Let $q^* = x(p^*,Y^*) = h(p^*,U^*)$; that is, suppose q^* maximizes utility at prices p^* and income Y^*, and yields utility U^*. For all prices, p, it must be true that

$$h_i(p,U^*) \equiv x_i[p,e(p,U^*)]. \tag{2.5}$$

Because both sides of equation (2.5) are identical, the partial derivative of each side with respect to p_j can be equated, yielding terms of the form

$$\frac{\partial h_i(p,U^*)}{\partial p_j} = \frac{\partial x_i(p,Y^*)}{\partial p_j} + \frac{\partial x_i(p,Y^*)}{\partial Y} \frac{\partial e(p,U^*)}{\partial p_j}. \tag{2.6}$$

Evaluating equation (2.6) at p^* and recalling that $x_j^* = \partial e(p^*, U^*)/\partial p_j$ yields

$$\frac{\partial h_i(p^*, U^*)}{\partial p_j} = \frac{\partial x_i(p^*, Y^*)}{\partial p_j} + x_j^* \frac{\partial x_i(p^*, Y^*)}{\partial Y}, \qquad (2.7)$$

which is the familiar Slutsky condition.

2.2.2 Integrability and Slutsky Conditions

When measuring demand for environmental commodities it is frequently desirable to recover the expenditure function from an estimated ordinary demand function. In other words, if economists use market data to estimate a demand function, they often wish to compute the underlying expenditure function. First of all, it is not always clear whether there exists an expenditure function that yields that demand function. The logic of the development in the previous section is unidirectional in that properties of demand functions follow from utility maximization rather than vice versa. However, this does not imply that any demand function obeying these properties can be obtained from maximizing some utility function. Given a set of compensated demand functions, $h(p,u)$, the important question is whether there exists a function $e(p,u)$ such that

$$\nabla_p e(p,u) = h(p,u). \qquad (2.8)$$

The Frobenius theorem of differential equations tells us that such a function e exists provided $\nabla_p h$ (which happens to be the Slutsky matrix) is symmetric; that is, $\partial h_i/\partial p_j = \partial h_j/\partial p_i$.

Knowing that an e function exists is not quite enough to know that a u function exists. Hurwicz and Uzawa (1971) proved that a sufficient condition for h to result from utility maximization is that the Slutsky matrix, $\nabla_p h$, be negative semidefinite and symmetric, in conjunction with certain regularity conditions on h. In essence, economists need to be sure that when they integrate h to find e that the integration is path independent. Symmetry of the Jacobian of h assures this. Starting from ordinary demand functions, if the matrix consisting of right-hand sides of equation (2.7) is symmetric negative-semidefinite, then demand results from utility maximization.

2.2.3 Compensated Versus Uncompensated Demand

There has been a great deal of debate in the literature about the differences between Hicksian (compensated) and Marshallian (uncompensated or ordinary) demand functions. It is Marshallian functions that are estimated, yet it is Hicksian functions that have well-defined slope and curvature properties and it is Hicksian functions that need to be used to compute measures of welfare (this is discussed later in this chapter). One approach to rationalizing these differences was taken by Willig (1976). He derived conditions on income elasticities and expenditure shares that imply a close agreement between Hicksian and Marshallian measures. (See also Randall and Stoll 1980; Weitzman 1988.)

Hausman (1981) took a more direct approach by using duality theory, and showed how to compute the compensated demand function directly from the ordinary demand function. Thus no approximation was necessary. He started from the left-hand side of equation (2.2), which he assumed had been estimated from data. The indirect utility function, $v(p,y)$, was then recovered by solving equation (2.2), a differential equation (this is the difficult part). The resulting indirect utility function, $U = v(p,y)$ was then "rewritten" as $Y = e(p,U)$, which is simply an expenditure function. Hicksian demand was then derived by differentiating the expenditure function, using Shephard's lemma, equation (2.4). Hausman gave as an example of the constant elasticity (ordinary) demand function:

$$\log x = \gamma'z + \alpha \log p + \delta \log y, \tag{2.9}$$

where z is a vector of "socioeconomic" variables, p is the price of the good consumed, the quantity consumed is x, y is income, and γ, α, and δ are coefficients that have been estimated. Writing equation (2.2) using equation (2.9), yields:

$$\exp(\gamma'z)p^\alpha y^\delta = \frac{\partial v(p,y)/\partial p}{\partial v(p,y)/\partial y}. \tag{2.10}$$

Hausman (1981) integrated this partial differential equation over a path of constant utility obtaining

$$v(p,y) = \bar{u} = -\exp(\gamma'z)\frac{p + \alpha}{1 + \alpha} + \frac{y^{1-\delta}}{1 - \delta}, \tag{2.11}$$

where $\bar{u}$ is a constant. Inverting this indirect utility function, one obtains

$$e(p,\bar{u}) = \left\{(1 - \delta)\left[\bar{u} + \exp(\gamma'z)\frac{p + \alpha}{1 + \alpha}\right]\right\}^{1/(1-\delta)}. \tag{2.12}$$

Applying equation (2.4) to this, one obtains:

$$\frac{\partial e}{\partial p} = \left\{(1 - \delta)\left[\bar{u} + \exp(\gamma'z)\frac{p + \alpha}{1 + \alpha}\right]\right\}^{\delta/(1-\delta)}\frac{\exp(\gamma'z)}{1 + \alpha} = h(p,\bar{u}) \tag{2.13}$$

which is the Hicksian demand function. The key step is integrating equation (2.10) into equation (2.11). Not only must the partial differential equation be solved successfully, but v must be quasi-convex; that is, demand must satisfy the Slutsky conditions.

The significance of this result is that it allows analysts to move "freely" between compensated and uncompensated demand functions. As a result, the old debate that the ordinary demand function must be used as an approximation to the "real thing" loses theoretical significance. Unfortunately, solving equation (2.10) is far from easy, at least in the closed form. Vartia (1983) discussed some numerical approaches to this issue.

2.3 Restricted Expenditure and Demand Functions for the Consumer

One of the most common characteristics of the demand for environmental commodities is that it is only indirectly observed. Typically, what is observed are changes in the consumption of a market good as the levels of the environmental commodity change. For instance, to measure the demand for air quality, researchers will measure a demand function for housing involving different levels of air quality and infer the demand for cleaner air. In contrast, the typical situation with conventional goods is that the researcher observes a time series or cross section of consumption and prices of the good, and from those data, estimates demand. The indirect approach that is necessary for measuring the demand for environmental goods and services presents unique challenges to consumer theory.

To outline the approach, start with an n dimensional vector of commodities, q, with all goods but q_n conventional market goods. The nth commodity is a nonmarket environmental commodity. Utility is given by $u(q)$, which induces a restricted expenditure function $e(p_1, \ldots, p_{n-1}, q_n, U)$, and a set of $n - 1$ restricted compensated demand functions $h_i(p_1, \ldots, p_{n-1}, q_n, U)$, $i = 1, \ldots, n - 1$. These functions are called "restricted" because not all quantities are "converted" to prices in the dual. Now if the set of $n - 1$ restricted compensated demand functions are estimated (or derived from estimated ordinary demand functions), under certain circumstances the restricted expenditure function can be recovered and then differentiated with respect to q_n to obtain the compensated demand function for the environmental commodity. Unfortunately when integrating back to the expenditure function, the constant of integration will remain unknown. If that constant involves q_n then it will not be possible to fully recover the demand function for q_n. These concepts are developed in more detail below (see also Freeman 1979a; Mäler 1974).

2.3.1 Restricted Utility, Expenditure, and Demand Functions

Start with n commodities, all conventional except for the nth. The consumer has a conventional quasi-concave utility function, $u(q)$. The consumer chooses levels of $q_1, \ldots, q_{n-1}$ but takes the amount of q_n as given. Denote quantities of all commodities but q_n by $q_{\langle n \rangle} \equiv (q_1, \ldots, q_{n-1})$. The consumer has income

Y to spend on $q_{\langle n \rangle}$. Eventually the researcher will be interested in the implicit valuation the consumer places on the nth commodity revealed by his actions regarding other commodities. The consumer faces the following utility maximization problem:

$$\max_{q_{\langle n \rangle}} u(q_1, \ldots, q_n)$$

$$\text{s.t.} \quad \sum_{i=1}^{n-1} p_i q_i \leq Y \tag{2.14}$$

$$q_{\langle n \rangle} \geq 0$$

The solution to equation (2.14), $q^*_{\langle n \rangle} = x_{\langle n \rangle}(p_{\langle n \rangle}, q_n, Y)$ is a set of $n - 1$ restricted, ordinary, Marshallian demand functions. Substituting $q^*_{\langle n \rangle}$ back into u yields a quasi-convex (in $p_{\langle n \rangle}$) optimal value function, $v(p_{\langle n \rangle}, q_n, Y)$, the restricted indirect utility function. Roy's identity implies that

$$x_i(p_{\langle n \rangle}, q_n, Y) = \frac{\partial v(p_{\langle n \rangle}, q_n, Y)/\partial p_i}{\partial v(p_{\langle n \rangle}, q_n, Y)/\partial Y}, \quad i = 1, \ldots, n - 1 \tag{2.15}$$

and

$$p_n(p_{\langle n \rangle}, q_n, Y) = \frac{\partial v(p_{\langle n \rangle}, q_n, Y)/\partial q_n}{\partial v(p_{\langle n \rangle}, q_n, Y)/\partial Y}. \tag{2.16}$$

Equation (2.16) results from duality between prices and quantities (Samuelson 1953) and implicitly defines an ordinary Marshallian demand curve for the environmental commodity.

Typically Y is the consumer's entire budget. Suppose the consumer is consuming $\hat{q}_n$ of the nth commodity. Hanemann (1989) has suggested that it is more appropriate to view income as Y plus sufficient income to purchase $\hat{q}_n$ at the marginal valuation of (2.16): $Y + \hat{p}_n \hat{q}_n$. Thus equation (2.16) can be rewritten as

$$p_n(p_{\langle n \rangle}, q_n, Y + \hat{p}_n \hat{q}_n) = \frac{\partial v(p_{\langle n \rangle}, q_n, Y + \hat{p}_n \hat{q}_n)/\partial q_n}{\partial v(p_{\langle n \rangle}, q_n, Y + \hat{p}_n \hat{q}_n)/\partial Y} \tag{2.16'}$$

Equation (2.16') implicitly defines an ordinary inverse demand function for the environmental commodity as if it were a private good:

$$p_n = f(p_{\langle n \rangle}, q_n, Y + \hat{p}_n \hat{q}_n). \tag{2.17}$$

Note that $\hat{p}_n$ and $\hat{q}_n$ are fixed and do not vary over the demand function. In essence, (2.17) is anchored at $(Y, \hat{p}_n, \hat{q}_n)$.

Dual to equation (2.14) is the expenditure minimization problem:

$$\min_{q_{\langle n \rangle}} p'_{\langle n \rangle} q_{\langle n \rangle}$$

$$\text{s.t.} \quad u(q) \geq U \tag{2.18}$$

$$q_{\langle n \rangle} \geq 0.$$

The solution to this minimization problem, $q^*_{\langle n \rangle} = h(p_{\langle n \rangle}, q_n, U)$ is a set of compensated or Hicksian restricted demand functions. Substituting h into the objective function of equation (2.18) yields an optimal value function, $e(p_{\langle n \rangle}, q_n, U)$, a restricted expenditure function, concave in $p_{\langle n \rangle}$.

If the consumer now were to face a price for q_n, say p_n, and had to choose q_n, the expenditure minimization problems would become

$$\min_{q_n} p_n q_n + e(p_{\langle n \rangle}, q_n, U)$$

$$\text{s.t.} \quad q_n \geq 0 \tag{2.19}$$

for which the first-order condition is

$$p_n = -\frac{\partial e(p_{\langle n \rangle}, q_n, U)}{\partial q_n}. \tag{2.20}$$

Equation (2.20) is the counterpart to equation (2.16) and implicitly defines a Hicksian compensated demand curve for the nth commodity. This can be coupled with results from Shepard's lemma applied to the restricted expenditure function to obtain

$$h_{\langle n \rangle}(p_{\langle n \rangle}, q_n, U) = \nabla_{p_{\langle n \rangle}} e(p_{\langle n \rangle}, q_n, U). \tag{2.21}$$

The optimal value function for equation (2.19) is, of course, concave. This implies that the matrix of second derivatives with respect to p ($\nabla^2_{p_{\langle n \rangle}} e$) that is bordered by $\nabla h_n(p)$ is negative semidefinite; consequently $\partial h_i / \partial p_i \leq 0$, for $i = 1, \ldots, n$.

2.3.2 Integrability and Weak Complementarity

As before, integrability conditions on e are that the $(n-1) \times (n-1)$ Jacobian of $h_{\langle n \rangle}$ be symmetric and negative semidefinite. If this is satisfied, there is a set of underlying utility functions, which if maximized, result in demands $h_{\langle n \rangle}$. In essence, the class of utility functions consists of members offset by constants of integration, which may be functions of q_n and U.

Suppose integrability conditions are satisfied. Then it is known from the fundamental theorem of calculus that for any fixed vector $(p_{\langle n \rangle}, q_n, U)$ and constant a,

$$\int_{p_1}^{a} \frac{\partial e}{\partial p_1}(x, p_2, \ldots, p_{n-1}, q_n, U)\mathrm{d}x = e(a, p_2, \ldots, p_{n-1}, q_n, U)$$

$$- e(p_{\langle n \rangle}, q_n, U). \tag{2.22}$$

(Without loss of generality, the focus is on good 1.) Equivalently, working in terms of derivatives of both sides with respect to q_n, equation (2.22) can be written as

$$\int_{p_1}^{a} \frac{\partial^2 e}{\partial p_1 \partial q_n} (x, p_2, \ldots, p_{n-1}, q_n, U) \mathrm{d}x = \frac{\partial e}{\partial q_n} (a, p_2, \ldots, p_{n-1}, q_n, U)$$

$$- \frac{\partial e}{\partial q_n} (\boldsymbol{p}_{\langle n \rangle}, q_n, U) \tag{2.23}$$

or, substituting equations (2.20) and (2.21) into (2.23),

$$\int_{p_1}^{a} \frac{\partial}{\partial q_n} h_1(x, p_2, \ldots, p_{n-1}, q_n, U) \mathrm{d}x =$$

$$\frac{\partial}{\partial q_n} e(a, p_2, \ldots, p_{n-1}, q_n, U) + p_n. \tag{2.24}$$

Because $h_{\langle n \rangle}$ is known, the left-hand side of equation (2.24) is also known. Thus it is clear from equation (2.24) that if it can be determined how expenditures change with q_n for *some* price a of good 1, then equation (2.24) will implicitly define a demand function for the nth good, the environmental commodity.

To accomplish this, Mäler (1974) developed the intuitively appealing notion of "weak complementarity" (see also Bradford and Hildenbrandt 1977). A good i is a weak complement with good n if there is some price $p_i = a$, such that

$$h_i(p_1, \ldots, p_{i-1}, a, p_{i+1}, \ldots, p_{n-1}, q_n, U) = 0 \tag{2.25a}$$

and

$$\frac{\partial}{\partial q_n} e(p_1, \ldots, p_{i-1}, a, p_{i+1}, \ldots, p_{n-1}, q_n, U) = 0 \tag{2.25b}$$

Obviously, equation (2.25b) is sufficient for these purposes, but equation (2.25a) makes the definition more economically meaningful. In essence, goods i and n go together and hence are "complements." When demand for good i drops to zero, then not only is good n not demanded but marginal changes in q_n have no effect on the expenditure function. A good example is the relationship between clean lake water (good n) and swimming (good i). If swimming becomes so expensive that it ceases to be consumed, then the cleanliness of the lake water becomes irrelevant (assuming lake water is only valued for swimming).

The reason equation (2.25b) works for inferring demand for an environmental good can be appreciated from figure 2.1. Of interest is the demand

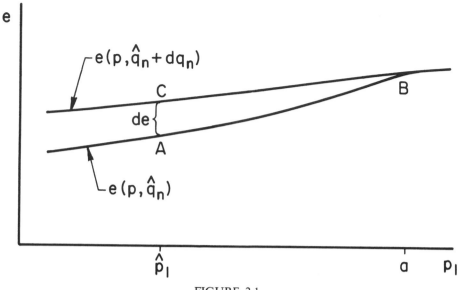

FIGURE 2.1
Demand with weak complements.

for the environmental, or nth, good at some price, $\hat{p}$, for other goods. To determine demand, consider good 1 only, keeping the prices of other goods fixed. What is of interest, therefore, is the demand for q_n at $p_1 = \hat{p}_1$. To find this demand function, the researcher has to know how much the expenditure function changes for a small change in q_n. Starting at the value of the expenditure function for $(p_1, q_n) = (\hat{p}_1, \hat{q}_n)$, he or she follows the changes in the expenditure functions along a line of constant q_n, from A to B. Then at B he or she perturbs $\hat{q}_n$ by dq_n. By assumption, expenditures do not change. Next, the researcher holds q_n constant at the new level of $\hat{q}_n = q_n + dq_n$ and follows a path from B to C, noting changes in expenditure. The net change in expenditures from A to C constitutes de, which is the information needed to compute de/dq_n. Note that the only information that was needed for this computation was the change in expenditures with respect to price changes for the first good; q_n and U, as well as other prices, were held constant. These changes in the expenditure function normalized by the change in price (de/dp_1) constitute, of course, a Hicksian demand function, which is what the problem began with. The approach is such that it is not necessary to know the expenditure function, just the integral of the Hicksian demand function from A to B to C:

$$de = \int_{p_1}^{a} h_1(x,p_2, \ldots, p_{n-1},q_n,U)dx$$

$$+ \int_{a}^{p_1} h_1(x,p_2, \ldots, p_{n-1},q_n + dq_n,U)dx \qquad (2.26a)$$

$$= - \int_{p_1}^{a} [h_1(x,p_2, \ldots, p_{n-1},q_n + dq_n,U)$$

$$- h_1(x,p_2, \ldots, p_{n-1},q_n,U)]dx. \qquad (2.26b)$$

Dividing both sides by dq_n and invoking equation (2.20) yields:

$$-p_n = \frac{de}{dq_n} = - \int_{p_1}^{a} \frac{\partial}{\partial q_n} h_1(x,p_2, \ldots, p_{n-1},q_n,U)dx, \qquad (2.27)$$

which is equivalent to equation (2.24), assuming equation (2.25b). Equation (2.27) gives the compensated demand function for the environmental commodity.

The significance of this result is in equation (2.27). Ultimately, the researcher is interested in a conventional demand function for the environmental commodity. This function can be directly computed if weak complementarity applies. And it weak complementarity is assumed, then finding this demand function only requires differentiating h_1 and then integrating the result. No expenditure function need be computed.

2.4 Demand Functions and Welfare Measures

Much of empirical environmental economics is concerned with the economic benefits of changing the level of environmental quality. In fact, it is not uncommon for an empirical paper to deal with the benefits without explicitly estimating a demand function. These two concepts are intimately related. The most common measure of welfare — consumer surplus — is the integral under the uncompensated demand curve. Compensating and equivalent variation are the integrals (for two different utility levels) under the Hicksian, or compensated, demand functions. Similarly, the derivatives of these benefit measures are the corresponding demand functions. Consistent with the rest of this chapter, the benefits of a price change will be considered separately from the benefits of a quantity change.

2.4.1 Welfare Implications of Price Changes

Consider a consumer with a specific income facing a set of commodity prices. Now suppose the price of one of the commodities changes. How can the well-

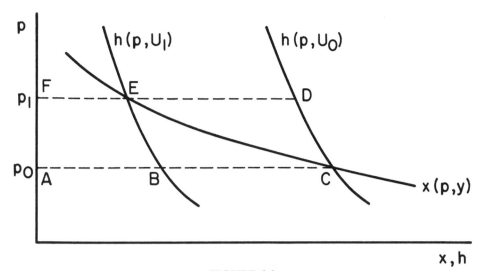

FIGURE 2.2
Ordinary and compensated demand.

being of the consumer be compared before and after the price change without dealing directly with utility, which is unobservable? Two concepts of welfare change first proposed by Hicks (1939) help address this question: compensating variation (CV) and equivalent variation (EV). *Compensating variation* is defined as the quantity of income that compensates consumers for a price change; in other words, returns them to the original level of utility. Similarly, *equivalent variation* is an income change that could be used in lieu of the price change to yield the same utility as the price change. Mathematically, let income be Y. Suppose prices change from p_0 to p_1, and as a result, utility from U_0 to U_1. Then compensating variation and equivalent variation are

$$CV(p_0, p_1) = e(p_1, U_0) - e(p_0, U_0) \tag{2.28a}$$

$$EV(p_0, p_1) = e(p_1, U_1) - e(p_0, U_1) \tag{2.28b}$$

Compensating variation provides income to maintain initial utility (U_0). Equivalent variation removes that amount of income necessary to yield the same utility as the price change (U_1). Because the derivative of the expenditure function with respect to price is the Hicksian, or compensated, demand function, equations (2.28) can easily be seen to be integrals under Hicksian demand curves from p_0 to p_1. This is illustrated in figure 2.2. The area to the left of $h(p, U_1)$ between the prices p_0 and p_1 is the equivalent variation (ABEF), and similarly for $h(p, U_0)$, the area is the compensating variation (ACDF). Also shown in the figure is the ordinary demand curve, $x(p, y)$. It intersects the Hicksian curve for U_0 at p_0 and the curve for U_1 at p_1. The area to the left of the ordinary demand curve, ACEF, is termed *consumer surplus* and is between EV and CV. Because only ordinary demand is usually estimated, consumer surplus is frequently used as a benefit measure. However, this usage is inappropriate because price effects are compounded by income effects. As was noted earlier, although Willig (1976) and Randall and Stoll (1980) showed that this may not be a severe problem, Hanemann (1989) showed that in the context of an environmental commodity for which there is no close substitute, differences between EV and CV may be large due to income effects. The income effects are magnified because the individuals have access to a highly valued environmental resource. This access dramatically increases their effective income (see discussion circa equation 2.16′). Also, Hausman (1981) showed how Hicksian demand functions can be recovered from ordinary demand functions.

2.4.2 Welfare Implications of Quantity Changes

As was discussed in section 2.3, in the context of demand theory, the researcher begins with observations on a set of restricted demand functions involving prices of some goods and quantities of others. The goal is to determine the welfare effect of changes in the quantity of the nonpriced commodity. Compensating and equivalent variations in the quantity case analogously to the price case are defined:

$$CV(q_0,q_1) = e(\boldsymbol{p},q_1,U_0) - e(\boldsymbol{p},q_0,U_0) \tag{2.29a}$$

$$EV(q_0,q_1) = e(\boldsymbol{p},q_1,U_1) - e(\boldsymbol{p},q_0,U_1), \tag{2.29b}$$

where U_0 is the utility resulting from (p,q_0,y) and U_1 results from $(\boldsymbol{p},q_1,y)$. Figure 2.3 illustrates the effect of a change in the quantity of q, keeping the price of the composite priced commodity fixed at $\overline{p}$. Shown in the figure are indifference curves between q and the composite commodity, x. The budget line is indicated by the horizontal line. After the quantity change, the compensating variation will reduce income sufficiently to reduce consumption of the private good by $CV/\overline{p}$, pushing the consumer back to utility level U_0. A similar discussion applies to equivalent variation.

An alternate way to view compensating variations is to rewrite equation (2.29a) as

$$CV(q_0,q_1) = e(\boldsymbol{p},q_1,U_0) - e(\hat{\boldsymbol{p}},q_1,U_0) \tag{2.30a}$$
$$- [e(\boldsymbol{p},q_0,U_0) - e(\hat{\boldsymbol{p}},q_0,U_0)] + E$$

where

$$E = e(\hat{\boldsymbol{p}},q_1,U_0) - e(\hat{\boldsymbol{p}},q_0,U_0). \tag{2.30b}$$

Clearly equation (2.30) holds for all $\hat{\boldsymbol{p}}$; however is there a $\hat{\boldsymbol{p}}$ for which $E = 0$? Weak complementarity between q and x yields such a $\hat{\boldsymbol{p}}$. Let $\hat{\boldsymbol{p}}$ be equal to $\boldsymbol{p}$

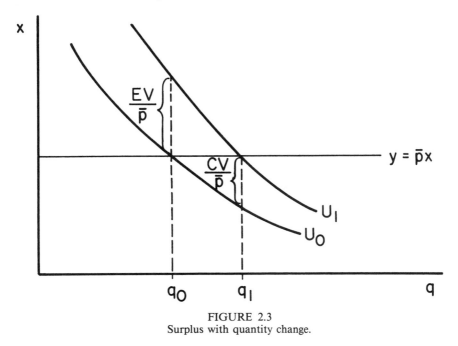

FIGURE 2.3
Surplus with quantity change.

except for the ith commodity where $\hat{p}_i$ will be a "choke price"; that is, no consumption occurs, either at $q = q_0$ or $q = q_1$. Further assume x_i and q are weak complements, as defined earlier. The definition of weak complementarity implies that equation (2.30b) is zero; that is, expenditures necessary to achieve a given utility level are independent of the level of q provided there is no consumption of the ith commodity.

Figure 2.4 shows two Hicksian demand curves for x_i, which is the weak complement with q. The two curves are for different q's but with utility at U_0. With weak complementarity $(E = 0)$, the compensating variation of equation (2.30a) is only the shaded area between the two demand curves. If the demand curves were drawn with utility at U_1, the area would be the equivalent variation.

2.5 Producer Theory

All of the theory presented thus far concerns consumer choice. It is a simple matter to extend the results to the firm (Epstein 1981). A firm possesses a convex production technology, $f(\mathbf{x},\mathbf{y}) \leq 0$, where $\mathbf{x}$ is a vector of inputs and

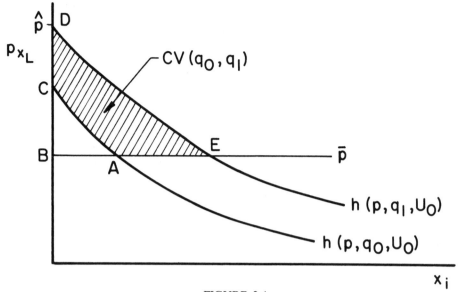

FIGURE 2.4
Weak complementarity.

y is a vector of outputs. The firm chooses inputs, faced with a vector of input prices, p, according to

$$\min_{x} \; p'x$$
$$\text{s.t.} \quad f(x,y) \leq 0 \qquad\qquad (2.31)$$
$$x \geq 0,$$

which defines a set of input demand equations $h(p,y)$ and an optimal value (cost) function, $c(p,y)$. This cost function is obtained in a manner virtually identical to the expenditure function equation (2.3). The exception is that problem (2.3) is constrained by $u(q) \geq U$ and problem (2.31) is constrained by $f(x,y) \leq 0$. However, both of these constraint sets are convex by assumption, so all of the results carry over. Because of the parallels, the results will not be developed for the case of the producer.

2.6 Random Utility Models

Contrary to what their name implies, random utility models (RUM) are not distinct microtheoretic models of consumer choice. Rather, random utility models are econometric models that allow estimation of preferences for commodities that are considered by the consumer but are not chosen. For example, because a consumer who purchases a house considers other houses with a variety of characteristics (including environmental quality), information is gained about the consumer's preferences regarding house A from the fact that the consumer turned down house A in favor of house B.

Random utility models are considered in a number of chapters in this book including the chapters on hedonic theory, household production theory, and recreation demand. These chapters go into detail on random utility models. The goal in this chapter is to present an overview of the technique.

Random utility models date to Thurstone (1927) but have their modern genesis in the work of McFadden (1974). There is a vast amount of literature on RUM applications and theory. Maddala (1983) provided an accessible review of theory, as did McFadden (1984). Hanemann (1984a) provided a development of the random utility model that is particularly germane to environmental commodities.

To outline the random utility method, suppose a consumer faces a set of commodities from which he or she will choose one. The commodities could be houses, parks, vacation sites, or a host of nonenvironmental goods such as cars. Index this set of commodities by $i = 1, \ldots, I$ and let X_i be the vector of characteristics associated with choice i. Let α be a vector describing the consumer (including income) and β a set of unknown parameters characterizing the utility function. (The econometric problem will be to estimate β.) The choice process the consumer goes through is to determine the utility that each choice yields, $U(X_i, \alpha)$, and to pick the choice that yields the highest utility. At this point, *randomness* enters the model. Randomness takes the form of errors of observation regarding consumer characteristics. Or, consumers may make errors in evaluating the utility arising from a particular choice. In any event, the perceived utility from choice i can be viewed as stochastic and of the form

$$\hat{U}_i = U(X_i, \alpha, \beta) + \epsilon_i \tag{2.32}$$

where ϵ_i is an i.i.d. residual for each i.[4] The utility realized by the consumer is given by

$$U^* = \max(\hat{U}_i | i = 1, \ldots, I). \tag{2.33}$$

Despite the many choices, only two things are observed: the consumer's choice, i^*, and the characteristics of the chosen and unchosen commodities. The problem is solved by considering the probability of choosing i^*. For a given set of goods characteristics $\{X_i | i = 1, \ldots, I\}$ and consumer characteristics, α, the probability of making a particular choice can be calculated. That probability will depend on β. The value of β sought is that which maximizes the probability of making the choice that was actually made, i^*.

---

[4] Brown and Walker (1989) discuss the significance of how the error term enters the utility function.

The probability of choosing i^* is the probability that $\hat{U}_i^* > \hat{U}_j$ for all $j \neq i^*$ (the possibility of ties is precluded):

$$
\begin{aligned}
\text{prob}(i^*) &= \text{prob}[\hat{U}_{i^*} > \hat{U}_j | j \neq i^*] \\
&= \text{prob}[U(X_{i^*},\alpha,\beta) + \epsilon_{i^*} > U(X_j,\alpha,\beta) + \epsilon_i | j \neq i^*] \qquad (2.34) \\
&= \text{prob}[\epsilon_j + U(X_j,\alpha,\beta) - U(X_{i^*},\alpha,\beta) > \epsilon_{i^*} | j \neq i^*] \\
&= \prod_{j \neq i^*} \int_{-\infty}^{\infty} F[\epsilon + U(X_j,\alpha,\beta) - U(X_{i^*},\alpha,\beta)] \, f(\epsilon) d\epsilon,
\end{aligned}
$$

where ϵ_j is distributed as f with cumulative density function F. If ϵ has a Weibull distribution, then

$$ f(\epsilon) = \exp(\epsilon - e^{-\epsilon}) \qquad (2.35) $$

and

$$ F(x < \epsilon) = \exp(-e^{-\epsilon}). \qquad (2.36) $$

This allows (2.34) to be rewritten as

$$ \text{prob}(i^*) = \frac{\exp[U(x_{i^*},\alpha,\beta)]}{\Sigma_j \exp[U(x_j,\alpha,\beta)]} \qquad (2.37) $$

Equation (2.37) gives the probability that an individual with characteristics α will choose i^*. To estimate β, the likelihood function is defined as the probability of observing the choices that were in fact observed in the sample. Index individuals in the sample by $s = 1, \ldots, S$ and possible choices by $j = 1, \ldots, J$. Let $I_{sj} = 1$ if individual s chose j, otherwise let $I_{sj} = 0$. The likelihood of observing the choices actually made is the product of the probabilities that each individual made the choice he or she actually made.

$$
\begin{aligned}
\log L(\beta) &= \sum_{s=1}^{S} \sum_{j=1}^{J} I_{sj} \log \left\{ \frac{\exp[U(x_j,\alpha_s,\beta)]}{\Sigma_k \exp[U(x_k,\alpha_s,\beta)]} \right\} \\
&= \sum_{s=1}^{S} \sum_{j=1}^{J} I_{sj} U(X_j,\alpha_s,\beta) \qquad (2.38) \\
&\quad - \sum_{s=1}^{S} \log \left\{ \sum_{j=1}^{J} \exp[U(X_j,\alpha_s,\beta)] \right\}
\end{aligned}
$$

There are a number of variants of the multinomial logit random utility model described above. Two primary ones are the *Probit RUM* and the *hierarchical nested RUM*. The probit RUM assumes ϵ is multivariate normal instead of each ϵ_i being Weibull and i.i.d. Computationally, the probit RUM is more difficult to apply and thus restricts the researcher to a much smaller set of choices (Hausman and Wise 1978).

The nested multinomial logit random utility model (McFadden 1981) involves a sequence of choices. Rather than choose all of the characteristics of a house in one step, the consumer first chooses whether to rent or buy; second, the neighborhood; third, a multi- or single-family dwelling; and so on. At each step of the hierarchy, the consumer's decisions are represented by a multinomial logit random utility model as described earlier.

2.7 Demand Under Uncertainty

A key environmental issue is the control of people's exposure to environmental risk. These risks may range from brief exposure to hazardous wastes to a lifetime of living near a nuclear reactor. To be sure, environmental externalities have always involved risk; when sulfur is emitted into the atmosphere, the health effects to a particular person are uncertain. This situation should be distinguished, however, from one in which the generator of the externality can control two variables: the level of the externality and the risk associated with it. For example, it is possible to control the number as well as the safety of nuclear reactors. People can control the amount of hazardous wastes deposited in landfills as well as the risk presented by the wastes, by taking care in constructing, maintaining, and monitoring the landfill. In the latter situations, risk has moved from the background to the foreground; from a passive characteristic of an externality to a consciously chosen variable on a par with the physical level of the externality. Because people have the ability to adjust both physical levels of an externality and the risk, it is important to explicitly introduce risk into the demand framework presented in this chapter; it is important even though there is no consensus on how best to treat the issue of risk associated with environmental commodities.

A common thread that runs through this area is the distinction between *ex ante* and *ex post*. If something is uncertain — be it prices, utility, income or a host of other things — two decision points are possible: one prior to uncertainty being resolved (*ex ante*) and one after uncertainty is resolved (*ex post*). It is possible, in fact likely, that decisions that are optimal *ex ante* turn out to provide a lower level of utility than decisions made *ex post*. In other words, hindsight is better than foresight. But policies generally aim to prevent or reduce risks; thus the appropriate evaluation perspective is *ex ante*.

There are two basic approaches to demand theory under uncertainty. The first involves the consumer. Uncertainty is embodied in state-dependent commodity consumption in which the state-of-nature is both uncertain and beyond the control of the consumer. Because single-period decision models require that consumers balance their budget, it is necessary to introduce dynamics to represent price uncertainty vis-a-vis consumer choice (see Chavas, Bishop, and Segerson 1986; Epstein 1975; Pollak 1969). Firms operating under price uncertainty are not restricted to a balanced budget, at least with functioning capital markets. Consequently, the second model examined is that of factor demand under price uncertainty.

For either the firm or the consumer, a basic tenet of the theory is that of expected utility (Hirshleifer and Riley 1979).[5] For a consumer, utility is defined with regard to a deterministic consumption bundle. For the producer, utility is defined with regard to profits (for the producer, the utility function may be the identity where the utility of profits is simply profits). Uncertainty in profits or the consumption bundle induces uncertainty in utility. The firm or consumer makes choices to maximize expected utility.

From an empirical perspective, Gaussian uncertainty, or quadratic utility, allows one to express expected utility only in terms of means and variances (Meyer 1987).

2.7.1 Consumer Choice Under Uncertainty

The basic model of consumer choice under uncertainty articulated in this section is based on Graham (1981) and Cook and Graham (1977), extended by Smith and Desvousges (1988).[6] Consider a two-period model. All consumer choice takes place in the first period. In the second period, "nature" shows its hand, determining the specific utility level the consumer will enjoy. Goods are partitioned into two classes: one class is *ex post,* or contingent on the state of the world;[7] in other words, the consumption of *ex post* goods depends on which state of the world is eventually revealed. Goods in the other class are *ex ante* in the sense that their consumption level cannot be made contingent on the state of the world that is revealed. Let the utility function for a representative consumer be

$$U(x,q,s) \tag{2.39}$$

where x is a vector of *ex post* goods, the consumption of which may depend on the state of nature; q is a vector of *ex ante* goods; and s is consumption of goods dependent on the state of nature but beyond the control of the consumer.

An example may be appropriate here. Drawing on Smith and Desvousges (1988), let there be a composite contingent claim x_i, (*ex post* good with states of nature i), such as health care, and a single *ex ante* good, such as drinking water. The drinking water may be contaminated and make an individual sick or it may be clean and safe to drink. Therefore, the two states to consider are sick (S) and well (W). To be consistent with expected utility theory, it is important that utility be a function only of commodities consumed, not of state. Thus it is necessary to introduce the commodity health, $h,$ less of which

[5] Expected utility theory is not without its detractors; see Kahneman and Tversky (1979).
[6] See also Gallagher and Smith (1985) and Freeman (1988).
[7] These are commonly referred to as contingent claims because the consumer is purchasing in the first period (with certain money) the right to consume the good in state $s,$ should state s develop in the second period.

will be consumed in the sick state than in the well state. The amount of health *care* consumed (x) is a choice variable and will depend on the state of nature. Let the probability of being sick be π_s. Expected utility is thus

$$EU(x_s,x_w,q,\pi_s) = \pi_s U(x_s,q,h_s) + (1 - \pi_s)U(x_w,q,h_w) \tag{2.40}$$

To add a little more realism and illustrate the distinction between contingent and noncontingent goods, let π_s be a function of q: the more water consumed, the higher the probability of being sick.

Returning to the theoretical model, assume a finite set of states. The consumer's utility maximization problem is

$$\max_{\{x_j\},q} EU(x_j,q,h_j) \tag{2.41}$$
$$\text{s.t.} \sum_j p'_j x_j + r'q \leq y,$$

where y is available income, p_j is the state-dependent vector of prices for the contingent goods, x_j, h_j is health in the two states, and r is the vector of prices of the *ex ante* goods, q. Now define a state-dependent expenditure for the contingent goods. Smith and Desvousges (1988) termed this an *ex post* expenditure function:

$$e_j(p_j,q,\overline{U}) = \min p'_j x_j \tag{2.42}$$
$$\text{s.t.} \quad U(x_j,q,h_j) \geq \overline{U}.$$

Equation (2.42) defines the minimum expenditure on contingent goods for consumption in state j to assure utility level $\overline{U}$. Now an *ex ante* expenditure function can be defined as

$$e(p_1, \ldots, p_s,r,\pi_1, \ldots, \pi_s,\overline{U}) - \min_{q,\{U_s\}} \sum_s e_s(p_s,q,U_s) + r'q \tag{2.43}$$
$$\text{s.t.} \sum \pi_s U_s \geq \overline{U}$$

where π_s is the probability of state s being realized.

Equation (2.43) defines a "conventional" expenditure function in the sense of minimum expenditures necessary to achieve expected utility, U. Thus, applying Shephard's lemma:

$$\nabla_{p_ie} = x_i(p,r,\pi,\overline{U}) \tag{2.44}$$
$$\nabla_r = q(p,r,\pi,\overline{U}).$$

Also, the additional expenditures (positive or negative) associated with a shift in the probability of realizing a particular state can be computed. This

essentially defines the demand curve for risk reduction. This can be obtained
by totally differentiating the Lagrangian of equation (2.43),

$$L = \sum_s e_s(\boldsymbol{p}_s, \boldsymbol{q}, U_s) + \boldsymbol{r}'\boldsymbol{q} + \lambda(\sum_s \pi_s U_s - \overline{U}),$$ (2.45)

using the identity $\Sigma_s \pi_s = 1$. Substituting $\pi_1 = 1 - \Sigma_{s>1} \pi_s$ into equation (2.45),
differentiating with respect to π_j and applying the envelope theorem results in

$$\frac{\partial e}{\partial \pi_j} = \sum_s \frac{\partial e_s}{\partial \pi_j} + \frac{\partial \boldsymbol{r}'}{\partial \pi_j}\boldsymbol{q} + \lambda(\overline{U}_j - \overline{U}_1).$$ (2.46)

Equation (2.46) implicitly defines the demand for risk reduction. On the right-
hand side of equation (2.46), the partial derivatives with respect to prices and
expenditures reflect the fact that prices, $\boldsymbol{p}_s$ and $\boldsymbol{r}$, and thus e_s, may depend on
the probabilities of realizing a state. The λ is the inverse of the marginal
expected utility of income. This is obvious from equation (2.43). The U_j and
U_1 are the state-specific utility levels that solve equation (2.43).

Return to the example regarding drinking water to evaluate the significance
of equation (2.46), which becomes

$$\frac{\partial e}{\partial \pi_s} = \frac{\partial e_s}{\partial \pi_s} + \frac{\partial e_w}{\partial \pi_s} + \frac{\partial r}{\partial \pi_s}\boldsymbol{q} + \lambda(\overline{U}_w - \overline{U}_s).$$ (2.47)

The parameters p, π, and r are arguments of the expenditure function;
however, in equilibrium, some of these parameters may be endogenous. For
instance, state-dependent prices, p_s, may depend on state probabilities. This
would be the case with an insurance market in which the price of the
contingent good depends on the probability of realizing that state. For instance,
suppose the price of contingent goods is just π_s. Then

$$\frac{\partial e_s}{\partial \pi_s} = \frac{\partial e_s}{\partial p_s}\frac{\partial p_s}{\partial \pi_s} = + x_s$$
$$\frac{\partial e_w}{\partial \pi_s} = \frac{\partial e_w}{\partial p_w}\frac{\partial p_w}{\partial \pi_s} = - x_w,$$ (2.48)

which reduces equation (2.47) to

$$\frac{\partial e}{\partial \pi_s} = x_s - x_w + \lambda(\overline{U}_w - \overline{U}_s).$$ (2.49)

The interpretation of equation (2.49) is that the change in expenditures due
to an increase in the probability of being sick is due to increased health care
costs and lost utility.

This section has developed a framework for examining demand when there
is uncertainty in how a commodity will be valued. It has distinguished between
ex ante and *ex post* goods as well as *ex ante* and *ex post* expenditure and
demand functions. The key point is that the existing framework of consumer
choice can be used, if it is modified.

2.7.2 Producer Choice Under Uncertainty

A somewhat different form of uncertainty, is *ex ante* price uncertainty. The basic idea underlying this theory of demand is that a firm faces stochastic factor and output prices at the point at which purchases are made and production plans are formulated. After those decisions are made, the state of the world is revealed and uncertainty is eliminated. Many authors have contributed to this literature.[8] Pope (1980) offered one of the most complete developments of the theory of the firm under price uncertainty.

Suppose a firm produces outputs *y* using inputs *x* according to the production function $f(x,y) = 0$. Suppose the prices of *y*, *p*, and the prices of *x*, *r*, are stochastic in the sense that the firm must make decisions based on stochastic prices. After the production decisions have been made, specific prices will be realized. The firm's problem, then, is to find *x*, *y* such that

$$EU^*(p,r) = \max_{x,y} EU(p'y - r'x)$$
$$\text{s.t.} \quad f(x,y) = 0. \tag{2.50}$$

This expected utility of profit function is akin to the profit function except that a utility function is interposed. Nevertheless, Roy's identity yields factor demands

$$x_i^* = \frac{\partial EU^*(p,r)}{\partial r_i} \Big/ EU'(\pi). \tag{2.51}$$

Demand is equal to the change in expected utility associated with a price change, normalized by the expected marginal utility of profit.

Because the theory of producer demand under price uncertainty has not played a very significant role in environmental economics, it will not be considered further. More information may be found in the cited references.

2.8 Conclusions

The purpose of this chapter has been to present a skeleton review of demand theory with a particular emphasis on environmental commodities. The aim has been to provide a common foundation for the remaining chapters of the book. The two unique features of environmental commodities are: (1) that quantities rather than prices may be observed; and (2) that significant uncertainty exists in demand and preferences. Despite its centrality to the entire economics profession, demand theory continues to have a number of unanswered questions, particularly concerning uncertainty. It is hoped the reader will be motivated to extend this theory.

[8] For example, Leland (1972), Sandmo (1971), Blair (1974), Pope (1980), Batra and Ullah (1974) Ratti and Ullah (1976), Lippman and McCall (1981), Baron (1970), and Wolak and Kolstad (forthcoming).

Measuring the Demand for Environmental Quality
John B. Braden & Charles D. Kolstad (Editors)
© Elsevier Science Publishers B.V. (North-Holland), 1991

Chapter III

HOUSEHOLD PRODUCTION FUNCTIONS AND ENVIRONMENTAL BENEFIT ESTIMATION

V. KERRY SMITH

North Carolina State University and Resources for the Future

3.1 Introduction[1]

Observed behavior combined with maintained hypotheses about people's motivations has been the primary basis for economists' inferences about consumer preferences. Because these maintained hypotheses arise from a constrained maximization model, economists usually focus on how changes in the parameters of the constraint(s) influence observed choices. With private goods, these parameters are prices, the budget constraint is linear, and the powerful insights of duality theory assure that the analyst can generally recover sufficient information from households' decisions to characterize their preferences (up to a monotonic transformation).[2]

When the objects of choice are available to consumers under a variety of access conditions, many without directly observable prices, the task of describing consumer demand becomes more complex, especially with attempts to measure the values people place on environmental resources. In these circumstances, economists naturally tend to add to the maintained assumptions embedded in the modeling strategy used to estimate these values. This chapter is about one such strategy that adds information by specifying structural connections between the environmental services of interest and other private goods.

These additions will be described in this chapter within the general setting of household production functions (HPF). In some respects, the discussions

[1] Partial support for this research was provided by the Institute for Environmental Studies, University of Illinois, and by the National Science Foundation under Grant No. SES8911-372. Thanks are due John Braden and Charles Kolstad for helpful comments on an earlier draft of this paper.
[2] See Varian (1984) for a review of general principles of duality theory as they relate to describing consumer preferences.

of some theoretical issues will duplicate the definitions offered in the previous chapter. This is unavoidable because the HPF method provides two types of insights into measuring the values people place on environmental resources.

The first insight is specific. The structural restrictions of weak complementarity, weak substitutability (whether for one private good or several), perfect substitutability, and commodity separability conditions specify the features of a person's preference structure. They can be argued to arise because people use marketed commodities, nonmarketed environmental resources, and their time to produce services. In some situations, it will be useful from both conceptual and analytical perspectives to use the HPF framework as a basis for organizing the goods and services a person consumes according to the activities they support. Because each of these restrictions is a maintained hypothesis, it cannot be tested. The HPF orientation helps to identify where potential links might occur between marketed goods and nonmarketed environmental services. Moreover, as will be developed in the following paragraph, when people's responses to changes in environmental resources can be observed, HPF can be helpful in developing approximate gauges of whether economists have the "story" right.

Second, the HPF framework offers qualitative insights into empirical methods that could be easily attempted without "telling the HPF story." Consider the following example. Suppose a set of geographically disparate households' purchases of goods and services are observed; for example, both the prices and quantities are known).[3] Moreover, assume that measures of the air and water quality conditions for each of these locations are also known. Economists are routinely willing to use this type of information as potential determinants of housing prices or wage rates (as the chapter by Palmquist describes), but they have been reluctant, until recently, to specify a reduced-form behavioral function that includes these environmental variables.

Why hasn't an indirect utility function in terms of the household's prices, income, and the environmental quality variables been specified? And, why hasn't Roy's identity and estimated the features of preferences been used to recover equivalent information? With specifications that identify the relevant parameters in the indirect utility function, economists could use these estimates to compute how each person would value a change in a nonmarketed environmental resource. Shapiro and Smith (1981) demonstrated how such a strategy would work. Yet, few applications have been attempted. The three known applications will be discussed in section 3.2.

The answer to these questions lies in economists' inability to describe how environmental quality measures should enter these functions together with the recognition that such modeling decisions will often have a substantial

---

[3] Even if the prices they pay for the commodities involved are the same, the price indices for the aggregates they purchase will not be. This follows because households will purchase different amounts of the commodities involved, so with the expenditure shares weighting prices in composing these aggregates, prices will be different.

influence on the results. The household production theory provides one basis for making these choices. By describing how goods and services are used, it is easier to conjecture how environmental services affect the activities involved. While admittedly informal, some of the most enduring benefit estimation methods have resulted from this type of information. The travel cost model, for example, arose from Hotelling's (1947) insight that many types of outdoor recreation activities take place *in situ* at recreation sites, so the time and travel costs required to get to a site function as implicit prices for the site's services. His characterization of the relevant costs of this behavior, together with the trip-taking decisions of households at different locations, can be used to estimate the demand for a recreation site's services. In the parlance of the HPF framework, this is a *derived demand.*

Because the HPF approach is both a language for explaining modeling strategies and a potential source for explicit restrictions, its use varies with the application. The focus of this chapter will be on both types of insights provided by the HPF framework. This requires explaining how prior restrictions on preferences used to estimate the demand for a nonmarketed good can be reinterpreted and extended in the HPF setting. It also requires summarizing a few applications to illustrate how the approach has influenced existing empirical models.

3.2 The Household Production Function and Valuation Methods

3.2.1 Background

Becker's (1965) insightful description of individuals' time allocation decisions in terms of a household production process stimulated the use of this framework for describing consumers' choices in the intervening 25 years.[4] The household production framework argues that marketed (and nonmarketed) goods and services are demanded as intermediaries in a household's consumption process. They are inputs that, together with the time of household members, are used to produce service flows. To understand the nature of the demands for these goods, income and time constraints must be considered together with the features of these household technologies. Of course, production functions involving purely private goods could also be treated as separability restrictions

[4] Muth (1966) and Lancaster (1966) are often identified as other contributors to the household production framework. However, Lancaster's approach was different in that it focused on characteristics as the fundamental sources of utility. These were inherent in commodities. While his framework did allow combinations of commodities to yield more characteristics than the simple sum of each one's specified quantities, this was not the same as the process described by Becker. Muth's analysis was more consistent with a set of household production activities.

on preferences. When Muth (1966) first described his version of the framework, he clearly noted this interpretation.

Most economists would likely argue that the HPF framework has not enhanced the predictive power of the conventional model for consumer behavior. That is, they would characterize the net result of models describing "consumers as producers" as providing a good vehicle for the "story-telling" component of model development, *but* offering a paucity of new testable hypotheses.[5] However, as the introduction to this chapter suggests, the usefulness of the HPF approach may not be so limited. This claim is substantiated with some familiar restrictions between commodities that illustrate how HPF-based stories can be especially helpful in implementing empirical models for measuring the benefits from improvements in environmental quality.

This overview begins with perfect substitution and complementarity relations between a marketed good and a nonmarketed environmental service. Following this discussion, the more commonly applied case of weak complementarity is described for the same pairing of goods. As Feenberg and Mills (1980) suggested [and Bockstael and Kling (1988) demonstrated], extending the logic behind these linkages to relationships between groups of private commodities and a nonmarketed public good provide a comparable means of inferring the value of the nonmarketed good.

Because these groups are perhaps most easily justified as inputs to a specific household production activity, the HPF framework offers a means of identifying the groups. Although not generally recognized, these linkages to different groups of private commodities offer opportunities to bound the values individuals place on nonmarketed services. Indeed, weak complementarity or substitutability restrictions are special cases where the bounds reduce to the value of the environmental resource.

3.2.2 Perfect Substitutes and Complements

To develop these cases in simple terms, assume that an individual engages in one household production activity by combining a private good, *X,* and the nonmarketed good (or environmental service), *q.* The output of this activity will be described as a final service flow, *Z.* This person's utility function includes *Z* and a composite good, *y,* as in equation (3.1):

$$U = U\,(y,Z). \tag{3.1}$$

The production technology is described by equation (3.2):

$$Z = \phi(X,q). \tag{3.2}$$

The case of perfect substitutes implies that $\phi(\cdot)$ is linear in *X* and *q,* with

---

[5] This conclusion is the basic verdict most analysts have taken from the Pollak and Wachter (1975) and Barnett (1977) papers on the household production description of consumer behavior.

a constant b describing the rate at which X substitutes for q, as in equation (3.3):

$$Z = X + b \cdot q. \qquad (3.3)$$

Equation (3.3) could be readily generalized to a situation involving multiple private goods, whose aggregate was a perfect substitute for q, by replacing X with a function of private goods.

As Freeman (1985a) and Mäler (1985) developed in more specific terms, this case underlies the use of defensive expenditures to measure the value of environmental quality. To see this point, begin by recognizing that the value W a person places on an additional unit of q is given (regardless of the form of the household production technology) by the amount he or she will save in expenditures to realize the same level of utility at given prices, as in equation (3.4). The expenditure function, $e(\cdot)$, is derived by minimizing the expenditures required to realize a given utility level subject to the constraints of the utility function, the household production function, and any other restrictions associated with the conditions of access to q:

$$W = -\frac{\partial e}{\partial q} = P_y \frac{U_z \phi_q}{U_y} = P_x \frac{\phi_q}{\phi_x}, \qquad (3.4)$$

where P_y and P_x are prices for the composite good y and the market good X used in the production of Z; and U_i and ϕ_j are the partial derivatives of the utility and household production functions with respect to i and j. In (3.4), the second and third equalities result from equating marginal rates of substitution to price ratios. For the use of perfect substitutes, equation (3.4) becomes (3.5); for example, by using the expression in (3.3) for $\phi(\cdot)$:

$$W = P_x \cdot b \qquad (3.5)$$

This means the marginal value of an increase in q is the savings in expenditures on X. Because they exchange in production at a fixed rate of b, a one-unit change in q implies b less of X will be required to produce Z.

In many applications, the interpretation of (3.5) simply reverses. That is, if environmental quality deteriorates, then by spending additional resources on the b more units of X to adjust for the reduction in q, a change in the service flow is avoided. Of course, to apply this basic model, the central issue is whether the defensive action offers a perfect substitute for the nonmarketed resource.

Consider the following example of how this might arise. Suppose a household's ordinary supply of water is from a well. A public water main is available to the family, but they choose to use the well because the implicit price of the well water to them is cheaper. If the actions of a local firm cause the well to become contaminated, the household is offered two alternatives: either (1) the firm will pay to provide bottled water for drinking and cooking, although the well water can continue to be used for cleaning and bathing; or (2) the firm will pay for the required connections to the public water supply. To implement the defensive expenditure approach, the analyst must judge which, if any, of the alternatives is a perfect substitute for the situation prior to contamination.[6]

This example also illustrates a second issue in implementing the HPF model. When economists observe the household's responses, any effort to consider the value people place on clean drinking water must distinguish three possible measures: the change in expenditures on X, given a fixed income; the change in expenditures required to hold the final service flow, Z, constant; and the expenditure change associated with holding utility constant at the precontamination level. The expression given here for the marginal value of q is the last of these three measures. The nature of the technology assures it will also be the second measure. Once the assumption of perfect substitutes is relaxed, this correspondence will not hold.

To derive the relationship between the first measure and all others, the comparative statics of household behavior in this simple model need to be examined. The budget constraint defined by equation (3.6), with I the income, yields the first-order conditions for constrained maximum utility (equation 3.1 and 3.2) given in (3.7). Totally differentiating (3.7) yields (3.8). The responses to changes in q or I are described by solving (3.8) for the relevant choice variable in terms of changes in q or I:

$$I = P_X \cdot X + P_y \cdot y \tag{3.6}$$

$$U_y - \lambda P_y = 0$$
$$U_z - \lambda P_X = 0 \tag{3.7}$$
$$-P_y y - P_X X = -I$$

$$\begin{bmatrix} U_{yy} & U_{yz} & -P_y \\ U_{zy} & U_{zz} & -P_X \\ -P_y & -P_X & 0 \end{bmatrix} \begin{bmatrix} \mathrm{d}y \\ \mathrm{d}X \\ \mathrm{d}\lambda \end{bmatrix} = \begin{bmatrix} -bU_{yz}\mathrm{d}q \\ -bU_{zz}\mathrm{d}q \\ -\mathrm{d}I \end{bmatrix}. \tag{3.8}$$

Equations (3.8) allow the change in expenditures on X to be related to the marginal value of q defined by (3.5). Assuming $\mathrm{d}I = 0$ and solving produces an expression for $\partial X \partial / q$. Multiplying that relationship by P_x, yields equation (3.9):

$$P_X \frac{\partial X}{\partial q} = - \frac{bP_y P_X (P_X U_{yz} - P_y U_{zz})}{|A|}, \tag{3.9}$$

where U_{ij} is the second partial derivative of $U(\cdot)$ with respect to i and j, and

[6] This example is not as contrived a situation as it may seem. It is comparable to the calculations that have been developed as part of several recent natural resource damage cases under the Superfund legislation. See Kopp and Smith (1989).

$|A|$ is the determinant of matrix given on the left side of (3.8). Thus, the *observed* changes in expenditures on the input used to produce Z will not measure the monetary value that would hold utility constant with changes in the amount of the environmental resource, even for technologies in which X and q are perfect substitutes. The analysis must also account for the amount of income the individual has to reallocate to spend more on X. This point is made by Courant and Porter (1981) and Harrington and Portney (1987) in their discussions on interpretations of observed averting expenditures.[7] Scaling (3.10) to reflect the income effects of the reallocation induced by the change in q, yields (3.5). More specifically, with a constant income, the increased expenditures on X to adjust to reductions in q must come from y, as in (3.10):

$$P_x \frac{\partial X}{\partial q} = - P_y \frac{\partial y}{\partial q}. \tag{3.10}$$

To avoid this, either compensate to hold utility constant or focus on:

$$\frac{P_y \, (\partial y / \partial q)}{P_y \, (\partial y / \partial I)} = \frac{-P_x \, (\partial X / \partial q)}{P_y \, (\partial y / \partial I)}. \tag{3.11}$$

Solving for $\partial y / \partial I$ from (3.8), produces (3.12):

$$\frac{\partial y}{\partial I} = \frac{P_x U_{yz} - P_y U_{zz}}{|A|}. \tag{3.12}$$

To derive the form for the utility constant, marginal value of a change in q, substitute $P_y(\partial y / \partial I)$ into the right-hand term in (3.11) and use (3.9) for $P_x(\partial X / \partial q)$. This process takes account of the difference between observed expenditures on X and the real value of q measured in terms of X; for example, equation (3.4). This difference is the income adjustment required to hold utility constant:

$$W = - \frac{P_x(\partial X / \partial q)}{P_y(\partial y / \partial I)} = bP_x. \tag{3.13}$$

Once the assumption of perfect substitutes is relaxed or q is allowed to contribute to utility in addition to the contribution made through the production of Z, these influences must be accounted for in relating observed changes in private goods expenditures to a person's value for an environmental resource. All of the applications discussed later in this chapter return to this point. The restrictions imposed on the relationship between one (or more)

---

[7] Of course, the precise nature of the adjustments depend on whether the quality terms enter only the household production function or if they also enter the utility function separately. While this case was considered by Courant and Porter, it is treated in detail by Harrington and Portney.

private good(s) and a public good determine what can be learned from a household's observed expenditures.

The case of perfect complements is harder to motivate and more difficult to characterize in general terms. The demand for a public good can be recovered from observing a private good involved with a public in a household production technology that exhibits this type of input association, but this relies on the existence of a range of prices for the private good where the value of the public good is zero (Mäler 1974, 180-182). Because this added restriction is also a key feature of weak complementarity, economists have been less interested in imposing the additional restrictions associated with perfect complementarity. More generally, the value placed on a public good in these types of production technologies, where $\phi(X,q) = \min (X/\beta,q)$ with $\beta > 0$, will depend on the segments of (P_x,P_y,q,I) space they examine (see Hanemann 1989).

3.2.3 Weak Complements[8]

Weak complementarity was introduced by Mäler (1974) as a restriction to preferences. (See chapter 2 for a discussion of the concept.) It restricts the marginal utility of the public, or nonmarketed, good to zero when the consumption of the private good is zero. It has been expressed in a wide variety of ways. To describe these alternative forms for the definition of weak complementarity, assume the preference function $U(\cdot)$ is redefined eliminating Z; that is, $u(y,X,q)$ instead of $U[y,\phi(X,q)]$. Also assume X and q are the weak complements, and define P_X^C as the choke price for X; that is, price at which demand for X is 0. Equation (3.14) defines weak complementarity in four equivalent expressions. The first two use the direct and indirect utility functions with $X = 0$ and $P_X = P_X^C$ describing the same situation where the individual does not demand the private good. The last term simply expresses the relationship involving the direct utility function in terms of the marginal rate of substitution (MRS_{qy}). The third term expresses the condition using the expenditure function:

---

[8] Feenberg and Mills (1980) also describe the case of weak substitution relationships. However, their discussion has been misconstrued. To use their approach for benefit measurement in this case, we must assume both weak substitution relationships and that a form of benefit transfer is acceptable. Feenberg and Mills use the case of public and private education. If the price of a private education is zero, presumably the individual would consume no public education, so its quality would not be valued. As the price of private education rises, the consumption of public education increases, so there is a value for quality improvements in public. However, in this example we have used what might be termed weak complementarity in terms of relative prices of substitute commodities to justify the benefit measurement process (e.g., when price of substitute private good is zero, relative price of public is infinite; while as price of private increases, the relative price of public declines).

$$\frac{\partial u(y,0,q)}{\partial q} = \frac{\partial V(P_y,P_X^C,I,q)}{\partial q} = \frac{\partial e(P_y,P_X^C,q,u)}{\partial q} = \frac{\partial MRS_{qy}}{\partial q} = 0, \qquad (3.14)$$

where $V(\cdot)$ is the indirect utility function:

$$MRS_{qy} = \frac{\partial u(y,0,q)/\partial q}{\partial u(y,0,q)/\partial y}$$

However, in terms of a HPF framework, weak complementarity corresponds to treating the private input to the production of Z as essential. In this case, the technology would imply that without X, production of Z would not be possible, as in (3.15a):

$$Z = 0 = \phi\,(0,q) \qquad (3.15a)$$

and

$$\partial Z/\partial q = 0 = \phi_q\,(0,q). \qquad (3.15b)$$

Because these requirements correspond to equation (3.16), now expressed in terms of the original preference function, this formulation reduces to the conventional definition,

$$\frac{\partial U[y,\phi(0,q)]}{\partial q} = 0. \qquad (3.16)$$

Thus, this equation illustrates how the first definition in equation (3.14) can be readily expressed in terms of the household production technology because $u(y,x,q) = U[y,\phi(x,q)]$.

Weak complementarity has been used in at least three ways. Two of them are most easily understood in the context of a household production framework. Consider the first — estimating the consumer's value for a change in q, knowing the demand for X. Weak complementarity implies that by knowing the compensated demand for X and integrating over a defined region — from the current prices to the choke price specific to each level of q — yields the Hicksian measure of the value of that change. More formally, if $X(P_x,P_y,q,u)$ is the Hicksian derived demand (for X as an input), the compensating variation (CV), measure for an individual's value of an increase in q from q_0 to q_1 is given in (3.17):

$$CV = \int_{P_X^0}^{P_X^C(q_1)} X(P_X,\overline{P}_y,q_1,u_0)\mathrm{d}P_X - \int_{P_X^0}^{P_X^C(q_0)} X(P_X,\overline{P}_y,q_0,u_0)\mathrm{d}P_X \qquad (3.17)$$

where $P_X^C(q_i)$ is the choke price for X at the level of environmental good corresponding to q_i, and u_0 is the utility realized at the initial prices for X

and y $(P^0_X, \overline{P}_y)$ and q_0. Equation (3.17) can be reinterpreted using the "partial" or restricted expenditure function, $\tilde{e}(\cdot)$, corresponding to each integral. The complete function is not obtained because variations in $\overline{P}_y$ cannot be observed:

$$
\begin{aligned}
CV = \ & \tilde{e}(P^C_X(q_1), \overline{P}_y, q_1, u_0) - \tilde{e}(P^0_X, \overline{P}_y, q_1, u_0) \\
& - [\tilde{e}(P^C_X(q_0), \overline{P}_y, q_0, u_0) - \tilde{e}(P^0_X, \overline{P}_y, q_0, u_0)] \\
& + k(\overline{P}_y, u_0, q_1) - k(\overline{P}_y, u_0, q_0),
\end{aligned} \tag{3.18}
$$

where: $k(\cdot)$ is the constant of integration. Weak complementarity assures that the expenditures will *not* be affected by the level of the environmental service if Z is not produced. X is not used at P^C_X so $Z = 0$. Thus,

$$
\tilde{e}[P^C_X(q_1), \overline{P}_y, q_1, u_0] = \tilde{e}[P^C_X(q_0), \overline{P}_y, q_0, u_0] \tag{3.19a}
$$

and

$$
k(\overline{P}_y, u_0, q_1) = k(\overline{P}_y, u_0, q_0) \tag{3.19b}
$$

Thus (3.18) reduces to (3.20):

$$
CV = \tilde{e}(P^0_X, \overline{P}_y, q_0, u_0) - \tilde{e}(P^0_X, \overline{P}_y, q_1, u_0). \tag{3.20}
$$

A subtle but important difference exists between this interpretation of the simple case of weak complementarity and the Bockstael-McConnell (1983) evaluation of welfare measurement in a household production setting. They assumed the public good enters the utility function directly and is weakly complementary to all the final service flows, Z. The first case in this chapter describes weak complementarity as a characteristic of one household production technology. The analog to that case is to have one Z entering preferences and to assume q is a weak complement with it. If this is done, more final service flows can be assumed to be affected by q (as weak complements in preferences) without changing the conclusions. One household production activity is assumed to remain consistent with the basic framework for developing HPF restrictions.

In Bockstael and McConnell's description of the household technology, one input must be essential to all activities; the same one if more than one activity is assumed. Of course, if the technology involves only one private input, then the essentiality requirement is trivial. If the input is not essential, it is only a restriction on activities. Moreover, in some cases essentiality of a factor may not hold. One example is marine fishing and boats: people can enjoy fishing without having a boat because they can fish from a pier or the bank; nonetheless, the quality of their experience will no doubt be enhanced if they use a boat because they will be able to reach more areas and increase their likelihood of catching fish. Therefore, the availability of the boat can be hypothesized to increase the productivity of fishing experiences without it being an essential input.

Bockstael and McConnell's analysis demonstrates that even though demand functions for final service flows may not exist, benefit measures for the non-marketed good can be constructed from the derived demand for the essential input.[9] Because the input is essential to the service flows using it and these service flows are weak complements to q, this is equivalent to weak complementarity between the input and the environmental service (q).

The role of the HPF framework differs significantly in these two uses of weak complementarity. In the first, the household production function simply motivates the restriction. Weak complementarity is equivalent to essentiality of a factor input to a household production activity, provided both the input and the environmental service only entered that activity. In contrast, for the Bockstael and McConnell analysis, weak complementarity is a feature of preferences linking one *or more* final services to q, and the essentiality of the input to all these activities focuses attention on just one factor demand — the essential factor.

The last way in which weak complementarity can be used is, as Mäler (1974) originally suggested, to recover the full expenditure function from the full set of Marshallian demand functions, provided they satisfy the integrability conditions. There are two important differences in this use of weak complementarity versus the previous uses. First, the analysis now begins with Marshallian and not Hicksian demands. Second, the result assures recovery of the full expenditure function, including the role of the nonmarketed good. As Mäler (1974) acknowledged, the results obtained with his method will be sensitive to the specification of the demand functions. His approach also requires that prices exist when the compensated demand for the private good (the weak complement) is zero. In this last use of weak complementarity, the demand specification becomes an issue because the behavior of the compensated demand around the choke price is important to the recovery of the expenditure function. No direct empirical applications of this version of the Mäler proposal have been attempted.

Furthermore, applications of the first two issues using Mäler's method are not straightforward because the primary results are for the Hicksian demands and *not* the Marshallian demands. In the case of price changes, whether single (Hanemann 1980a; Hausman 1981) or multiple (La France and Hanemann 1989), it is possible to recover sufficient information from Marshallian demands to measure Hicksian surpluses if integrability conditions are satisfied.[10] The same process is not necessarily possible when the change is in a nonmarketed good like environmental quality. The distinction in this case rests with

---

[9] For further discussion, see Pollak and Wachter (1975) and Barnett (1977).

[10] Hausman (1981) also sketched a similar but less complete version of the LaFrance-Hanemann argument. However, in all cases the analysis relates to valuing price, not quality changes.

economists inability to observe enough information on the role of the environmental resource in these demand functions.

Willig (1978) demonstrated that it is possible to bound Marshallian measures of the value of changes in the nonmarketed good given the following three conditions: weak complementarity; Z (as noted in this chapter) is *not* an essential commodity in consumption; and the Marshallian surplus is unaffected by income.[11] These three conditions assure that the Hicksian measures of compensating and equivalent variation bound the Marshallian measure. However, as Bockstael and McConnell (1987) and Hanemann (1989) demonstrated, satisfying these conditions does not assure that only small errors will result from using Marshallian measures of the value of a change in q.[12] Moreover, if Willig's third condition is not met, this does not imply that large errors will result from using a Marshallian measure of the value of the nonmarketed good. Only small income effects insure small errors with the Marshallian measure.

3.2.4 Using Groups of Commodities

Bockstael and McConnell's (1983) analysis of final service flows illustrated how the effects of an environmental resource on groups of household final services could be evaluated by considering the demand for a single common *and essential* commodity that served as a factor input to all the production processes. From this, it is possible to see directly how the initial discussion of weak complementarity, in terms of the role of X and q in $\phi(\cdot)$, can be extended to groups of private goods. This is readily accomplished by replacing X with a separate subfunction in these private inputs (designated as $r_1, \ldots, r_k$). In formal terms, equation (3.21) describes the implied household production technology,

$$\phi(X,q) = \phi[h(r_1, \ldots, r_k),q]. \tag{3.21}$$

In this case, it is assumed that the separable inputs r_1 to r_k are each essential to the aggregate X and that the structures are the same as in (3.15a) and (3.15b), now generalized to allow for k private goods. This is a household production interpretation of the situation recently described by Bockstael and Kling (1988).

---

[11] This feature is the same as the index Cook and Graham (1977) used to characterize the irreplaceability of commodities in their analysis of values for risk changes with state-dependent preferences. This is also the index I found to affect the relative size of option value to expected consumer surplus in a simple case defined to be consistent with the Schmalensee (1972) treatment of option value. See Smith (1984).

[12] While the focus of Bockstael and McConnell's (1987) discussion was directed to quality changes, the conceptual issues are the same.

To illustrate the implications of the requirement to consistently sequence the quantity changes in the nonmarketed goods, consider the three-goods case, with $r_i(P_{r1}, P_{r2}, P_{r3}, q, u)$ defining the Hicksian derived demands. The value of a change in q from q_0 to q_1 becomes:

$$\int_{P_{r1}^0}^{P_{r1}^C(q_1)} r_1(P_{r1}, P_{r2}^0, P_{r3}^0, \overline{P}_y, q_1, u_0) dP_{r1} - \int_{P_{r1}^0}^{P_{r1}^C(q_0)} r_1(P_{r1}, P_{r2}^0, P_{r3}^0, \overline{P}_y, q_0, u_0) dP_{r1}$$

$$+ \int_{P_{r2}^0}^{P_{r2}^C(q_1)} r_2[P_{r1}^C(q_1), P_{r2}, P_{r3}^0, \overline{P}_y, q_1, u_0] dP_{r2}$$

$$- \int_{P_{r2}^0}^{P_{r2}^C(q_0)} r_2[P_{r1}^C(q_0), P_{r2}, P_{r3}^0, \overline{P}_y, q_0, u_0] dP_{r2} \tag{3.22}$$

$$+ \int_{P_{r3}^0}^{P_{r3}^C(q_1)} r_3[P_{r1}^C(q_1), P_{r2}^C(q_1), P_{r3}, q_1, u_0] dP_{r3} - \int_{P_{r3}^0}^{P_{r3}^C(q_0)} r_3[P_{r1}^C(q_0), P_{r2}^C(q_0), P_{r3}, q_0, u_0] dP_{r3}$$

Unfortunately, all of this detail still relates to Hicksian demand functions. The issues identified by Willig arise with the separable composite of private goods, $h(\cdot)$, as was discussed for X in the private-commodity case. Thus, the same qualifications raised by Bockstael and McConnell (1987) now apply to the group of goods. The net implication of their extension is that the added complexity of more goods linked to the nonmarketed service makes judgments on benefit measures based on Marshallian demands more difficult. Moreover, it seems less likely that the income effects will be small as the group of affected commodities increases.

It is possible, however, to bound the value of a quality improvement by recognizing that the HPF framework provides a logic for grouping private goods into categories. Economists can explain and interpret these groupings as collections of inputs to different production activities undertaken by the individual. If it is plausible to assume that changes in q will affect activities differently, then the analyst can adapt recent results of Neil (1988) to bound Hicksian measures of the marginal value of a quality change.

Perhaps the best way to illustrate the logic of Neil's argument is to use an example. Consider the case of an air quality improvement. Suppose an individual engages in two different types of recreation, and improvements in air quality can be assumed to have different effects on the activities. The first type of recreation occurs outdoors; for example, bike-riding, walking, or having a picnic at the local park.[13] If all else is equal, improvements in air quality can be assumed to increase the Hicksian demands for the inputs associated with this activity. The second type of recreation occurs indoors; for example, working on crafts, watching television, or reading. If economists assume in their analysis that the Hicksian demands for inputs to these activities declines with improvements in air quality, then they have the ingredients for Neil's bounds.

Neil's bounds are straightforward. They reiterate a point central to the logic underlying the HPF approach: that is, economists can exploit relationships between commodities and the nonmarketed environmental service provided they know enough to be able to identify them in the first place. Although they do not need to assume the existence of household production functions, this orientation provides a convenient starting point for "telling the story."

More specifically, his argument has three key requirements:

1. It must be possible to identify two subsets of the set of all private goods that do not, when combined, form the full set of private commodities. Furthermore, each set must exhibit a different relationship to the non-marketed good: one must have a net substitution relationship and the other net complementarity.
2. Both sets *must be* gross substitutes for the environmental resource.
3. Both sets must exhibit positive income effects.

Using the link between Hicksian and Marshallian demands and a three-private-good generalization to our utility function, where X_1 and X_2 are assumed to enter different production activities with q a "public input" across them, $Z_1 = \phi^1(X_1,q)$ and $Z_2 = \phi^2(X_2,q)$, the logic can be outlined in simple terms:

$$X_1^H = X_1[P_1,P_2,P_y,q,e(P_1,P_2,P_y,q,u_o)]$$ (3.23)

where $e(\cdot)$ is the Hicksian expenditure function, and X_i^H and X_i are the symbols to distinguish Hicksian (utility constant) and Marshallian (income constant) demands for X_i,

$$\frac{\partial X_1^H}{\partial q} = \frac{\partial X_1}{\partial q} + \frac{\partial X_1}{\partial I} \cdot \frac{\partial e}{\partial q}.$$ (3.24)

Recall from equation (3.4) that $\partial e/\partial q = -W$. Thus, equation (3.25) describes the connection between changes in observed uses of X_1 and W:

$$\frac{\partial X_1}{\partial q} = \frac{\partial X_1^H}{\partial q} + W \frac{\partial X_1}{\partial I}.$$ (3.25)

[13] Of course, this example assumes the improvements in q would take place where the activities are undertaken. One form of adjustment to low levels of quality is to travel to another location, presumably less accessible, to realize a level of quality more compatible with a person's demands.

A comparable relationship can be derived for X_2. Moreover, net substitution and complementarity relationships imply $\partial X_1^H/\partial q < 0$ and $\partial X_2^H/\partial q > 0$, respectively.[14] Thus, if each private good exhibits a different linkage (as in this example), the observed changes in expenditures relative to the effects of income on the expenditures for each good bound the marginal value as in equation (3.26).

$$\frac{\partial(P_1 X_1)/\partial q}{\partial(P_1 X_1)/\partial I} < W < \frac{\partial(P_2 X_2)/\partial q}{\partial(P_2 X_2)/\partial I} \tag{3.26}$$

where X_1 is the net substitute $\partial X_1^H/\partial q < 0$ and X_2 is the net complement $\partial X_2^H/\partial q > 0$.

A number of interesting parallels can be drawn using this relationship. First, when $\partial X_1^H/\partial q = 0$, it produces the same result as did the case of perfect substitutes, $W = (\partial X_i/\partial q)/(\partial X_i/\partial I)$.[15] As the next subsection describes, this result implies that while the assumption of a perfect substitutes relationship between the private and nonmarketed goods was sufficient to provide a linkage between marginal value of the environmental resource and observed choices of the private good, it is not necessary. Other linkages between one or more private goods and the nonmarketed service have potential as well. Moreover, this is when the HPF structure can be helpful in understanding real-world counterparts to these restrictions.

Second, the framework may offer an alternative to the Willig (1978) approach for bounding Marshallian measures of the value of a change in a nonmarketed resource. Recall that the Willig criteria requires the consumer surplus for an improvement in environment quality per unit of the weakly complementary private good to be insensitive to income. The Neil bounds suggest that by examining private goods for differences in their linkages to the nonmarketed good, bounding for Marshallian measures will be improved.

Finally, these bounds are different from Bartik's (1988a) use of defensive expenditures to provide bounds for the Hicksian consumer surplus measures of a change in an environmental resource. To bound the value his analysis relied on evaluating the expenditures for a set of net substitutes for environmental quality at the pre- and post-adjustment levels of that quality.[16]

3.2.5 Separability Restrictions

When private commodities are unrelated to the nonmarketed good, then the Neil bounds for the marginal value of a nonmarketed good imply that the

---

[14] The specific derivation of this result is outlined in Neil (1988). Briefly, it simply used the change in the shadow price for the nonmarketed good in relation to a change in the level, q, to associate conventional price-based definitions with those based on quality changes.

[15] It is important to note that measure is derived from the other private good, *not* the private good in a quasi-linear household production subfunction.

[16] Bartik develops his analysis using pollution rather than quality. If the model proceeded with this definition, then the private good would be a net complement with pollution.

marginal value of the nonmarketed good can be measured directly from changes in the Marshallian demands. Returning to the simple formulation of the household production model with the private goods, its implications can be established by assuming $\phi(X,q)$ is written in general terms rather than as $x + b \cdot q$ as in the earlier analysis. This generalization implies that equation (3.8) describing the comparative static adjustment becomes equation (3.27):

$$\begin{bmatrix} U_{yy} & U_{yz}\phi_x & -P_y \\ U_{yz}\phi_x & U_{zz}\phi_x^2 + U_z\phi_{xx} & -P_x \\ -P_y & -P_x & 0 \end{bmatrix} \begin{bmatrix} dy \\ dX \\ d\lambda \end{bmatrix} = \begin{bmatrix} -U_{yz}\phi_q dq \\ (-U_{zz}\phi_x\phi_q - U_z\phi_{xq})dq \\ -dI \end{bmatrix}$$

(3.27)

Solving for $\partial y/\partial q$ and $\partial y/\partial I$ produces:

$$\frac{\partial y}{\partial q} = \frac{P_x\phi_q \, [U_{yz}P_x - P_y U_{zz}\phi_x - U_{zz}\phi_x - P_y U_z(\phi_{xq}/\phi_q)]}{|\bar{A}|}$$

(3.28a)

$$\frac{\partial y}{\partial I} = \frac{\phi_x \, [U_{yz}P_x - P_y U_{zz}\phi_x - P_y U_z(\phi_{xx}/\phi_x)]}{|\bar{A}|},$$

(3.28b)

where $|\bar{A}|$ is the determinant of the matrix on the left side of equation (3.27). The ratio of $(\partial y/\partial q)/(\partial y/\partial I)$ reduces the expression derived earlier for marginal value of q $[P_x(\phi_q/\phi_x)$; see equation (3.4)] provided equation (3.29) holds:

$$\phi_{xx}/\phi_x = \phi_{xq}/\phi_q$$

(3.29)

where X and q must form a particular type of separable production function so that $\phi_x - \alpha(q) \cdot \phi_q = 0$. Mäler (1985) has shown that $\phi(X,q)$ must be $\phi[X + f(q) \cdot q]$. This is exactly the function Neil suggested will be comparable with his definition of independence of the private commodity and the nonmarketed good q. This restriction on the technology involving X and q allows the change in consumption of the "unrelated" private good y to be used to recover the marginal value of q, and corresponds to quasi linearity in the household production technology. When imposed as a restriction on preferences, compensated and uncompensated demands are equivalent because the marginal utility of income is constant (see Varian 1984). In this case, the restriction is not imposed on the full preference function but only on the separable subfunction, or the household production technology, that involves X and q.

Thus far, this discussion has not considered Hori's (1975) early analysis of the role of household production technologies in recovering the demand for nonmarketed goods. The reason is because Hori's approach is the most restrictive. It requires complete knowledge of all household production functions. His paper considers four cases. The first had one private good and one nonmarketed good entering each household production function. Neither good affected other activities. His second case allowed the private good to be allocated among activities, but each nonmarketed good remained uniquely associated with each production function. Solving the first case was straight-forward given knowledge of the production function and the household's observed choices. Solving the second case may have actually involved multiple solutions; therefore, at issue was the uniqueness of the results derived.

Hori's latter two cases involved joint production, and under reasonable conditions for the technologies involved, again admitted solutions. The central question is whether it was reasonable for Hori to assume that this much information was known *a priori*. Indeed, resolving this question is what motivated economists to use the HPF approach in the first place. However, it is not necessary to add further complexity by assuming a specific set of household production functions as restrictions to the subfunctions present in an indirect utility function, including the nonmarketed goods. The critical issue that the analysts must address when they propose a model to describe a household's uses of the nonmarketed resources is whether the model can begin with a specific form for the indirect utility function. If this specification cannot be offered because more information on how the specified environ-mental resource affects the household is needed, then attention should shift to imposing the minimum set of prior restrictions that are necessary to recover the required estimates of people's preferences.[17] Hori's strategy adds unnec-essary complexity and diverts attention from this question. It assumes that the analyst knows the household's production technologies and that the problem is simply one of specifying the conditions under which a set of simultaneous, nonlinear equations will yield a unique solution.

3.2.6 Summary of Theoretical Developments

This section began by arguing that the household production framework serves two roles. First, it organizes the "story" an analyst uses to handle existing empirical information about how people use nonmarketed environ-mental resources. Although not essential to the empirical model, the framework is often quite helpful in explaining how specific empirical insights can be used.

Second, the HPF framework provides a rationale for grouping private commodities with the nonmarketed goods and for imposing restrictions on their interrelationships. These restrictions can be explained in terms of preferences; however, both the criteria for organizing the commodities and the restrictions can be more easily understood within an HPF framework.

This section has focused primarily on these types of restrictions. Three have emerged as potentially useful for empirical analysis of the values people place on environmental resources. Indeed, in some cases the restrictions were uncovered "after the fact" when plausible empirical strategies were proposed.

---

[17] The character of the required external information is quite different in the two strategies. This is part of the point of my earlier critique (see Smith 1979) of the Hori proposal.

Two of the three are widely recognized, although perhaps not exactly in the forms developed here.

Substitution Links. Perfect substitution between a private good and an environmental resource has been widely used as a rationale for using defensive expenditures and averting behavior models to measure people's values for environmental resources. This model was presented here in an HPF framework because this framework best illustrates the concerns of Courant and Porter (1981), Harford (1984), and Harrington and Portney (1987) for carefully distinguishing the observed Marshallian responses, which reflect income and substitution effects, from the Hicksian measures, which are required to measure the value of a change in the nonmarketed good based on averting behavior.

With adjustment for income effects, economists can obtain such measures from averting responses, provided the private good and environmental resource are either perfect substitutes or form a separable additive subfunction (that is, a restriction on the form of the household production function) where the marginal rate of substitution is a function of only the level of the nonmarketed good.

Weak Complementarity. Weak complementarity is the functional restriction linking private and nonmarketed goods that has received the most attention in the literature. In this chapter and in Mäler's (1985) work, it is described as the equivalent of a household production technology where the private good is an essential input.

Economists have found weak complementarity to be a potentially useful restriction. However, it does not assure that the value of a nonmarketed good can be derived. Because the restriction relates to Hicksian demands and because there is no clear connection (as in the case of price changes) between Marshallian and Hicksian demands with respect to the role played by the nonmarketed good, the process of developing theoretical properties for Marshallian measures, even with weak complementarity, remains incomplete. For the present, the available theoretical evidence suggests that characterization of the size of these errors arising from the use of a Marshallian measure must rely on simulation analyses.[18]

Recognizing Different Relationships Between Private and Nonmarketed Goods. The newest area to emerge from this review of theory is using the relationships between private goods and the nonmarketed resource. For example, this strategy can bound the marginal value of the nonmarketed good if economists can identify two groups of private goods with different relationships to the nonmarketed good. It may also offer additional insight if analysts are willing to impose more structure on the relationships between different subsets of private goods and the nonmarketed commodity. In some respects, this strategy amounts to an intermediate between weak complementarity and complete knowledge of household production functions. The first example of

---

[18] For examples, see Kling (1988a, 1988b).

such an intermediate case involves assuming quasi linearity of the household technology.

3.3 Implementation and Applications

Because the chapters in Part II provide detailed reviews of actual applications by the types of affected commodities or services, this discussion focuses on how the HPF framework influences the interpretation of the estimating models. The applications are classed into two categories — (1) models that try to describe completely a set of interrelated activities, where the motivation for the point estimation of these activities' responses to change in environmental quality either was or could be explained by a household production model of consumer behavior; and (2) models that involve a single activity or partial descriptions of how a household responds to deteriorations in one or more dimensions of environmental quality.

Both groups of models will be considered in the subsections that follow. The subsections will describe some aspects of the data used and the results obtained but will not report specific findings, which are covered elsewhere in this book.

3.3.1 HPF Models for Interrelated Activities

Four applications of the HPF model have used an interrelated set of equations to describe how nonmarketed resources affect the composition of a household's expenditures. All four studies attempt to describe the role that measures of environmental quality have in influencing households' expenditures on marketed goods; one of them also tried to describe the roles of other measures of nonmarketed resources. The first study to propose using HPF models for interrelated activities was Shapiro and Smith (1981). In their study, the analysts classified commodity aggregates as taxable, nontaxable (for example, food, medicine, and services), and housing expenditures. Using county level data for California in 1970, they specified a quadratic indirect utility function in prices, total expenditures, and four nonmarketed resources — air pollution (a linear index of carbon monoxide, sulphur dioxide, nitrogen oxide, nitrogen dioxide, and hydrocarbons), rainfall, temperature, and the level of local public expenditures in each county.

Equation (3.30) provides the basic structure of the specified indirect utility function.

$$-V = \sum_{i=1}^{3} \alpha_i \left(\frac{P_i}{M}\right) + \sum_{i=1}^{3} \sum_{j=1}^{3} \beta_{ij} \left(\frac{P_i}{M}\right)\left(\frac{P_j}{M}\right)$$
$$+ \sum_{k=1}^{4} \sum_{i=1}^{3} \gamma_{ik} \left(\frac{P_i}{M}\right) q_k \tag{3.30}$$

where M is total expenditures, q_k is the kth nonmarketed commodity, P_i is the price of the ith private good, and V is the value of the maximized utility function subject to the budget constraint.

Although their results did indicate that the proxy measures for some of the nonmarketed goods were significant influences on the expenditure shares, their study is best interpreted as a methodological proposal rather than a model for benefit estimation. One reason is that the analysts' commodity aggregates were too broad to reveal the likely role of these measures of the nonmarketed resources. Indeed, by introducing them into the indirect utility function, the analysts missed an opportunity to investigate the plausibility of the relationship between changes in the nonmarketed resource and expenditures.

These are the types of opportunities identified by Neil's bounds. The composition of a commodity aggregate should tell an analyst whether the private goods should be interpreted as net complements or as substitutes to the measure for the nonmarketed good. Based on these insights, the analyst could then hypothesize that the environmental quality measures affect different commodity groups differently. For example, if the available measure of q, the nonmarketed resource, was described by air pollution concentrations (for example, higher pollution concentrations implies lower q and vice versa), then the analyst could expect to observe different effects on the price indices for the different commodity aggregates when air pollution increases. If one of the aggregates was health and personal care activities, then the analyst could expect air pollution to have a positive influence because, for example, more pollution implies it takes more resources to produce a given health service flow. The rationale for this hypothesis is that increases in pollution make it more difficult to produce the *same* amount of health.

Shechter (1989) recently followed this strategy, but limited his analysis to the housing and medical services subsets of total expenditures. Moreover, he followed Shapiro and Smith's format by introducing air quality in the first-order and second-order terms for both private commodities. According to his model, increases in air pollution should reduce housing expenditures (for example, rents will decline because of an increase in a site disamenity). In contrast, medical expenditures should increase with increases in air pollution. Because Shechter's analysis treated these categories as the only two components of expenditures, he could not apply Neil's arguments, which state that an increase in one class of expenditures must lead to a decrease in the other, if the individual faces a constant budget. He measured air quality using perceptions rather than technical measures of pollution concentrations. As in the Shapiro-Smith analysis, Shechter found that the data did support the general structure. In spite of the limited nature of his commodity definitions, his analysis was based on a more plausible data set that used information at the individual level on the expenditure categories. It also adjusted for health-related consumption patterns and self-reported symptoms.

The next two studies can be interpreted as substantial refinements in this general strategy, although they were both undertaken before Shechter's work was completed and before Neil's theoretical bounds were proposed. The studies, Math-Tech (1982) and Gilbert (1985), sought to use the HPF approach more explicitly for developing their respective models. They chose the HPF structure because it motivates the role assigned to both air pollution and other nonmarketed services in specificying how each resource influenced expenditures. The Gilbert study extended the Math-Tech application in several ways: it refined the expenditure categories; improved the price measures used for each household-produced final service flow; investigated the effects of alternative specifications for the air pollution variables; and included durables as quasi-fixed commodities in the household's expenditure function.

Each analysis used average household data for selected cities from the 1972 to 1973 Consumer Expenditure Survey for the United States. The researchers could not use individual data because its analysis required merging the air quality readings with each expenditure record. Due to confidentiality limitations, a microlevel version of the Consumer Expenditure Survey with locations was not available to the researchers and they could not identify the locations of each respondent for the merge. Math-Tech used 24 Standard Metropolitan Statistical Areas (SMSA) in each of two different years to obtain an average of household expenditures. Gilbert used the average expenditures over the two years for 27 SMSAs. Table 3.1 compares each study's specifications for its commodity aggregates as well as the role of environmental variables. Both studies used linear expenditure models, based on Stone-Geary utility functions, to characterize household production activities. These functions described both the "cost" subfunctions for each activity as well as the aggregated expenditure system. The environmental variables were introduced as translating variables (see Pollak and Wales 1981). For example, when the Stone-Geary function is hypothesized to describe a weakly separable collection of private goods, this is equivalent to assuming the household production technology can be written as

$$Z_i = \prod_{j \epsilon S_i} (X_{ij} - \gamma_{ij})^{\beta_{ij}} \tag{3.31}$$

where Z_i is the ith service flow produced by combining X_{ij}, $j = 1$ to K; X_{ij} is the amount of each private good j used in the production of Z_i; β_{ij} is the

TABLE 3.1
A comparison of the Math-Tech and Gilbert household production models.

Gilbert	Math-Tech
Commodity description	
7 aggregated commodities	7 aggregated commodities
4 prices constructed as division aggregates: food, household durables, clothing, and transportation	All prices estimated from expenditure subfunctions: food (F), shelter (S), home operations (HO), furnishings and equipment (FE), clothing (C), transportation (T), and personal care (PC)
3 prices estimated from expenditure subfunctions: shelter (S), household operations (HO), health and personal care (HC)	
Environmental variables[b]	
Maximum of second high readings, SO$_2$ (S); Average of second high readings, SO$_2$ (HC*); Maximum of second high readings TSP (HO*)	Maximum of second high, SO$_2$ (S*, FE*, T*); Maximum of second high TSP (S*, HO*) Rainfall Temperature

Note: There were differences in the composition of the aggregates and in the sources used for the price indexes in each study.

[b] The letters are used to indicate the variables included in each subfunction.

* Indicates whether they were found to be significant plausible influences on the expenditure shares.

parameter usually interpreted as the marginal budget share when used in the context of explaining expenditures (the subscripts describe the ith service flow and jth private good input); γ_{ij} is the parameter often interpreted as describing the threshold or subsistence requirements for X_j in the "production" of Z_i; and S_i is the set of subscripts for the private goods used in the production of Z_i.

Translating refers to the fact that the γ_{ij} are specified to be functions of exogenous variables. It was originally introduced by Pollak and Wales (1981) as a strategy for including demographic variables into demand analyses, but it is also well-suited to specifying a role for expenditures related to avoiding behavior. By specifying the threshold requirements for X_{ij} as a function of pollution, the analyst implies that increases in pollution will require individuals to purchase more X for the activities affected by that pollution before there will be further positive contributions to the relevant final service flows.

Although appearing to incorporate the effects of pollution measures in a plausible way, this specification for the production functions can have an unanticipated effect on the pattern of marginal costs of activities. *Ceteris paribus,* it is desirable to have the "prices," or marginal costs, of the activities affected by the environmental resource change when the level of the resource changes. This result does *not* follow with the Stone-Geary specification. The total cost function corresponding to (3.31) is:

$$C_i = Z_i \prod_{j \in S_i} \left(\frac{P_j}{\beta_{ij}} \right)^{\beta_{ij}} + \sum_{j \in S_i} \gamma_{ij} P_j \tag{3.32}$$

This means the price of Z_i (the marginal cost) will be independent of the nonmarketed resource when these variables are entered in a translating form; for example, $\gamma_{ij} = f_{ij}(q)$, as in equation (3.33).

$$\frac{\partial C_i}{\partial Z_i} = \prod_{j \in S_i} \left(\frac{P_j}{\beta_{ij}} \right)^{\beta_{ij}} \tag{3.33}$$

Changes in q would induce changes in the composition of the private goods used to produce Z_i, but *not* the marginal cost of an additional unit of Z_i. This composition effect arises because the cost shares for each X_j in the production of Z_i will change as given by equation (3.34):

$$s_{ij} = \left(\frac{P_j}{C_i} \right) \gamma_{ij} + \beta_{ij} \left[1 - \sum_{j \in S_i} \left(\frac{P_j}{C_i} \right) \gamma_{ij} \right] \tag{3.34}$$

where s_{ij} is the cost share for private good j in production activity i.

The two empirical studies undertaken using this overall framework hypothesized that environmental variables affected a person's ability to realize both specified levels of well-being and the composition of activities undertaken. Therefore, each application had to develop alternatives to equation (3.33) to estimate the price indices for each Z_i so that both effects are reflected in the overall expenditure function that describes the role of each of the final service flows in the overall expenditure system.[19]

Gilbert used the estimated parameters in each subfunction to predict the cost shares and formed prices using Divisia price indices based on these predicted shares for each private good in a household production activity. This was argued to be an approximation to the appropriate "prices" and to be consistent with the treatment of the price indices used for aggregates of purchased private goods that were not treated as part of a household's production process.

Math-Tech (1982) followed a modified version of Green's (1980) suggestion to define the aggregative price index in terms of the net consumption of each commodity and net income. Because the Stone-Geary function is homogeneous of degree one in terms of the net measures of private goods consumed — for example the terms $(X_{ij} - \gamma_{ij})$ — this practice resolved the issue raised earlier with the use of marginal cost derived from the Stone-Geary cost function [see equation (3.32)] for each activity. However, their approach did not follow Green's suggestion exactly. Green developed his index by drawing an analogy

[19] Another important issue arises because the Stone-Geary specification is not a constant returns-to-scale production technology. Consequently, even though the marginal costs are constant, use of them for the prices of final services will not assure that total costs calculated with these prices equal total expenditures on private goods and the time costs of the production activities.

between the price index for a Cobb-Douglas specification and the consistency requirement for price and quantity indices.

For a Cobb-Douglas specification of the household production technology as in equation (3.35), a final service flow is given in equation (3.36):

$$Z_i = \prod_{j \in S_i} X_{ij}^{b_{ij}} \tag{3.35}$$

$$P_{Zi} = \prod_{j \in S_i} \left[\frac{P_j}{b_{ij}} \right]^{b_{ij}}. \tag{3.36}$$

In this equation, b_{ij} are the cost shares for each private good, X_j, used in the production of household activity i; for example, $b_{ij} = P_j X_{ij} / \Sigma_{k \in S_i} P_k X_{ik}$. Substituting into (3.36) yields:

$$P_{Zi} = \prod_{j \in S_i} \left[\frac{\Sigma_{k \in S_i} P_k X_{ik}}{X_{ij}} \right]^{b_{ij}}. \tag{3.37}$$

Thus, the production technology given by (3.35) assures that $P_{Zi} Z_i = \Sigma_{k \in S_i} P_k X_{ik}$.

This result would not follow using marginal costs from the Stone-Geary function given in (3.33). However, it is possible to have an aggregate price index for Z_i equivalent to (3.37) in terms of the *net usage* of each X_{ik} if each X_{ik} by $X_{ik} - \gamma_{ik}$ and $\Sigma_{k \in S_i} P_k X_{ik}$ is replaced by $\Sigma_{k \in S_i} P_k (X_{ik} - \gamma_{ik})$. Math-Tech's proposed index stops short of this simple reinterpretation and instead uses the form given in (3.38) for $\tilde{P}_{Zi}$.

$$\tilde{P}_{Zi} = \prod_{j \in S_i} \left[\frac{\Sigma_{k \in S_i} P_k S_{ik}}{(X_{ij} - \gamma_{ij})} \right]^{b_{ij}}. \tag{3.38}$$

This will satisfy the adding up requirement $\tilde{P}_{Z_i} Z_i = \Sigma_{k \in S_i} P_k X_{ik}$ and increases the effective cost attributed to each X_j based on the effects of the quality variables on the γ_{ij}'s. However, it does not logically follow from the technology.

Neither strategy is correct, and unfortunately, no further attempts have been made to pursue this line of research. Both of these studies found plausible effects of air pollution measures, but as their respective authors acknowledged, they relied heavily on the authors' judgments of the relevance of each environmental variable included in each household cost function. The authors based their judgments on the variables' effects on the price indices derived from them. The authors also rejected estimates with implausible effects. Because of this practice, anyone evaluating this strategy for explicitly using the HPF framework in estimating the role for nonmarketed goods must consider the procedures used to construct the approximate price indices.

Two alternative strategies have been used to develop these cost functions; however, they cannot be fully evaluated because there has not been an application with microdata, detailed treatment of household production activities, and theoretically correct price indices for the HPF services. The first, represented by Shapiro and Smith and Shechter, uses a consistent price index, but it cannot parameterize a role for the environmental variables that links them to expenditures-specific commodities. The second, by Math-Tech and Gilbert, uses a specific link such as translating to provide an intuitively plausible link between the measures of environmental services and private goods, but uses an approximate price index.

3.3.2 Partial Equilibrium Descriptions of Household Production Activity

The partial equilibrium applications are divided into three categories. Each can be interpreted as using some variant of a household production framework either to estimate the value of a nonmarketed resource or to confirm that behavior is consistent with positive values for the resource (and therefore consistent with averting behavior to avoid quality deteriorations). The first group attempts to estimate functions that approximate household production functions or reduced-form descriptions of averting actions. The key feature of these models is that they have a fairly "loose" connection to a behavioral model. In general, they involve situations where insufficient information on people's decision processes prevents analysts from formulating conventional demand functions.

The second group is the newest set of research. It uses some form of discrete choice model to explain the averting behavior taken by households. The two studies that have been done on this subject involved indoor radon — an air pollutant that poses a potentially serious risk of lung cancer. One of the studies was based on actual mitigation decisions and required that the researchers estimate the costs of mitigation in order to implement the model. The other was a contingent behavior study and examined households' demand for radon tests, which is a form of risk avoidance, before and after a public information campaign designed to promote testing. In the second study, the objective was to determine if the program influenced households' demands for radon testing kits, even if the residents chose not to test their homes.

The third category of partial equilibrium applications is the most extensive. It involves travel cost demand models for recreation sites. Because these analyses are interpreted as derived demand relationships, which arise when people demand that a recreation site's services "produce" specific recreational activities such as boating, fishing, or swimming, this model is probably the most widespread application of the HPF framework. All three categories are described in the subsections that follow.

Production or Reduced-Form Models. Table 3.2 summarizes the key features of four studies in this category. All of these studies focused on a subset of people's decisions and attempted to evaluate whether the respondents were motivated by knowledge of some change in a nonmarketed environmental resource. One of the most direct examples of this strategy is the Smith and

TABLE 3.2

A summary of the production/reduced form models of averting behavior.

Authors	Model activity	Environmental variable	Data/empirical model	Results
Gerking and Stanley (1986)	Household Production Function Model for Health Stock; single equation framework recognizing role of medical care; health treated as a choice variable because of prospects for avoiding responses and health expenditures	Air pollution — ozone, sulfur dioxide, total suspended particulates, and oxides of nitrogen	Individual sample for St. Louis; probit model to describe discrete outcome — visit doctor in past year at least once	Ozone a significant positive influence; chronic health conditions also a positive determinant of likelihood of visiting doctor
Smith and Desvousges (1986b)	Reduced form model — undertake activity — purchase water filters, boiled water, attend public meetings about contamination of water with hazardous substances	Attitude toward hazardous waste	Individual sample of households for suburban Boston; probit model for discrete outcome of each action	Attitudes important to purchase of filters and bottled water; news accounts and location in town with contamination incidents affect participation in public meeting
Dickie and Gerking (1989)	Set of health symptoms hypothesized to be associated with air pollution; extension to Gerking-Stanley with behavior reflected in estimation of functions hypothesized to be associated with air pollution	Air pollution carbon monoxide, nitrogen dioxide, ozone, sulfur dioxide	Individual sample with each respondent having history of respiratory disease; Panel study over 17-month period; 3 to 6 contacts with each respondent; reported experience with subsets of 26 symptoms grouped into chest and throat and all other; Tobit estimator	Air pollution measures not significant
Desvousges, Smith, and Rink (1989)	Reduced for model describing discrete choice — monitor home for radon; no observed variation in price of monitoring	Knowledge of radon; information treatment received; attitudes toward environmental risk	Controlled experiment involving public risk communication program; panel study in each of 3 towns in Maryland; 1,500 households, 2 interviews of each respondent; probit model for testing after information program	Prior knowledge of radon, discussion with friends and neighbors, and intensive information program positive determinants of likelihood of testing

Desvousges (1986a) study of people's decisions to purchase either water filters or bottled water in response to episodes of contaminated groundwater in suburban Boston wells. It is important to point out that some of the respondents' decisions may have resulted from them anticipating that this action would yield several outcomes. Therefore, not all of the expenditure on the filters or bottled water can be attributed to a willingness to pay to avoid risks from contaminated water if these actions also improve the taste, appearance, or other valued qualities of water, regardless of whether it is contaminated. These other outputs may have a relatively small role in the respondents' decisions; however, the extent of this role is likely to vary with each application. For example, Cropper and Freeman's discussion (in chapter 6) of the problems associated with valuing health effects that arise from deteriorations in environmental quality focuses primarily on the effects of averting behavior on the risk of health outcomes associated with either sickness or death. Their discussion distinguishes between those decisions that are intended to reduce risk but require only marginal changes in activities and those that demand discrete all-or-nothing choices. The specific interpretations of each type of behavior depend on how the models involved represent a person's decisions.

The Gerking and Stanley (1986) and Dickie and Gerking (1989) studies are part of a larger set of economic analyses of the role of environmental quality variables for measures of a person's health stock. In the first study, the health stock is altered by medical services measured by self-reported doctor visits. The second study uses various aspects of a person's health status to measure health stock directly. The explicit recognition that observed health outcomes are not purely technical relationships distinguishes these studies from economic models of epidemiological relationships. Observed health outcomes must explicitly include the results of behavioral decisions.

For example, the Gerking and Stanley analysis uses two measures of health capital as determinants of the likelihood of at least one health visit per year. The first determinant is self-reported existence of a chronic illness; the second is the number of years the illness has been present. Both variables result, in part, from behavioral choices. Because of this assumption, Gerking and Stanley had to develop instruments for both determinants. They did this by hypothesizing that each variable is a function of the price of medical care, the person's hourly wage, and a set of demographic variables, as well as the pollution measures. A similar strategy was used in Dickie and Gerking's later analysis of health symptoms.

The last study in the table reports the effects of a controlled information program to promote averting behavior. The source of the health effect in this case is radon. By using a panel design, Desvousges, Smith, and Rink (1989) were able to investigate how individuals responded to different types of information campaigns. They investigated two types of informational campaigns directed at households in two towns in Maryland. A third community was used as a "control." Because they could not control what other sources of information were available to households during the course of the study, it is reasonable to expect a weak effect for these types of programs. Nonetheless, a probit analysis of the new testing decisions after completion of the study-designed programs did detect a significant positive effect from the most intensive program, which included large amounts of citizen involvement, public meetings, and the participation of local government officials. Although this type of study cannot describe the demand for risk-averting behavior, it does indicate that these responses are likely to be influenced by public information campaigns.

None of these applications establishes a direct link to the structural relationships of a behavioral model. Instead, the estimates of the reduced-form relationships are used along with other information to estimate a person's marginal willingness to pay for an improvement in the nonmarketed resource. For example, the Gerking and Stanley analysis can be interpreted as estimating $P_x(\phi_q/\phi_x)$ of equation (3.4). The P_x in their case is the price of medical care and the (ϕ_q/ϕ_x) estimated from their household production function for health.

Discrete Choice Models for Averting Behavior. One difficulty with implementing the HPF framework arises from incomplete data. For the most part, economists do not know how people allocate their time among the activities they produce at home. While it might be possible to conjecture which private goods contribute to which activities (indeed this was the basic premise of the Math-Tech and Gilbert analyses), in many cases the goods will contribute jointly to multiple household production processes.

Often the best data available consists of records stating that households undertook certain specific actions in response to a change in one or more environmental services. Implementing the HPF framework as conventionally developed would require describing these choices as marginal adjustments in some overall pattern of activities. Thus, knowledge of the other activities becomes essential to the model's characterization of people's responses.

An alternative strategy is to assume the decisions are separable — not marginal adjustments but discrete choices. A person undertakes an averting action or attempts to mitigate a perceived deterioration in environmental quality based on his or her comparison of the *total* utility realized with and without the action. As McFadden (1974) suggested in introducing a random utility model (RUM) for these types of decisions, the decisions relate to the extensive margin of choice. Provided economists are willing to accept the comparison of conditional indirect utility functions as separable from other decisions associated with a household's expenditures, this type of RUM is consistent with the HPF framework.

The first example of this model strategy was by Äkerman, Bergman, and Johnson (1989). They considered a situation in which households reduce risk by undertaking mitigating activities for the radon concentrations in their homes. Their analysis considered individuals living in private homes in a

Stockholm suburb who monitored their homes for radon and experienced readings above the Swedish national guidelines for action. Using a sample of 317 households, 45 percent of whom had mitigated, the researchers estimated a RUM to explain these decisions based on engineering estimates of the cost, the level of radon, household income, and age. Their model also included a dummy variable for the type of heating system — that is, forced air or exhaust ventilation — because it is likely to affect radon exposure. While the researchers were not explicit, it appears they assumed that nonmitigators faced the average cost experienced by those mitigating.[20]

Their results indicated that the radon concentration was a positive and significant influence on the likelihood of mitigating. The effects of their cost estimates were sensitive to the functional form assumed to characterize the indirect utility fuctions. For example, using a log-linear specification, the cost measure was a significant negative factor in explaining these decisions; with a linear specification, it was not.

The second study that adopted a similar framework was based on another aspect of the Maryland experiment that was described earlier. Recall that the Maryland experiment evaluated how the informational programs influenced the homeowners' actual testing decisions. However, the research also considered how the informational programs influenced homeowners who did not test. The model of particular interest in that study was a contingent behavior analysis that used a RUM to estimate a household's willingness to purchase a radon monitoring kit before and after the informational programs designed to promote interest in testing. Each respondent was given a hypothetical monitoring choice with a kit at a specific price that varied across respondents.

Because of the "panel nature" of the sample (the same sample described earlier), the research could focus on households that did not test after the program and estimate their demands and values for radon tests before and after the informational programs. By asking the same contingent behavior question before and after the program, the researchers could use their responses to estimate and compare the resulting RUMs.[21] Each household was randomly assigned a different price for a monitor. Linear, conditional, indirect utility functions were assumed to describe the structure.

This approach was initially proposed in Smith et al. (1987) for an evaluation of a different type of informational program. The same basic strategy used

[20] This assumption is important because the costs of mitigation are likely to be associated with the radon levels in the structure. The use of the average ignores the effects of the radon level, as well as any characteristics of these households' homes that might influence the cost they would experience to mitigate.

[21] The specific text of the question included in telephone interviews was: Suppose your local health department was offering a radon test for a one-time cost of $10, $25, $50, $100. The cost would cover two radon detectors, the results, and a booklet about radon. Would you take part in such a radon testing program? (CIRCLE ONE NUMBER.)

    a. YES 01
    b. NO 02
    c. DON'T KNOW (DON'T READ) 94

TABLE 3.3a
Probit models for decisions to purchase radon monitors: Maryland panel.

	Before information program	After information program
Intercept	0.2108 (0.611)	1.7196 (4.838)
Price of monitor	-0.0113 (-8.871)	-0.0146 (-10.854)
Income	0.451×10^{-5} (1.610)	0.148×10^{-5} (0.519)
Health attitude (1 = concerned about health)	0.1927 (2.046)	-0.0230 (-0.243)
Age	-0.0048 (-1.526)	-0.0154 (-4.775)
Education	0.0210 (1.045)	-0.0192 (-0.946)
Number of correct answers in multiple choice quiz on radon (total = 12)	0.0628 (2.716)	
Smoker (=1)	0.1894 (1.877)	-0.0059 (-0.058)
Frederick[b] (=1)	-0.2754 (-2.436)	-0.225 (-1.922)
Hagerstown[b] (=1)	-0.980 (-0.867)	0.030 (0.258)
Number of correct answers on same multiple choice quiz on radon after program		0.0084 (0.355)
n	909	909

Source: Smith and Desvousges (1989).

[a] The numbers in parentheses below the estimated coefficients are ratios of the coefficients to the estimated asymptotic standard errors.

[b] These qualitative variables are intended to reflect the effects of the different information programs in comparison to the "control" community of Randallstown.

by Äkerman, Bergman, and Johnson was repeated in this Maryland study. Table 3.3 reproduces some of the results from Smith and Desvousges (1989), illustrating the estimated probit models and the calculated willingness to pay from each model. These results provide direct evidence that the programs were not effective in changing the nontesters' demands for information about the radon concentrations in their homes.

These types of models have advantages over the production and reduced-form models described earlier in that a consistent behavioral model is imposed on the analysis of household responses. The model requires that researchers

TABLE 3.3b.
Willingness to pay (WTP) of nontesters for radon monitors based on purchase intentions.

Sample	n	Before information program		After information program	
		Mean	Range	Mean	Range
1. Nontesters with (WTP > 0)	886	$54.54	$1.53-$122	$53.38	$0.11-$90.65
2. Nontesters who changed decisions from no test to test for radon	167	$51.18	$5.04-$106	$55.72	$3.91-$85.34

Source: Smith and Desvousges (1989).

either observe or are able to control the "prices" households experience in making decisions to undertake specific activities. As in the case of the reduced-form models, the HPF structure is not essential to the analysis. It simply facilitates the description of the model and interpretation of the results.

Travel Cost Recreation Demand Models. The travel cost recreation demand model is probably the oldest indirect method for measuring the value of nonmarketed environmental resources. In 1947 Harold Hotelling suggested that the travel costs an individual incurs to visit a recreation site be used as an implicit price for that site's services. The first empirical applications of the method came ten years later by Wood and Trice (1958) and Clawson (1959). In the intervening period, especially in the past decade, an extensive literature has developed on the issues of modeling recreation demand.[22]

The household production framework has proved especially helpful in describing the basic structure of the model and in developing specifications for site demand models. As a rule, the travel cost demand function is interpreted as a derived demand for a site's services. These services, along with other inputs, are used by a recreationist to produce recreational activities (see Smith 1990). A complete review of the technical issues associated with estimating these models is beyond the scope of this summary; moreover, in Part II, Bockstael, McConnell, and Strand describe the conceptual and empirical issues associated with using these models for benefit analysis. Therefore, to illustrate the usefulness of the framework, this section will focus on two issues exposed by interpreting the travel cost demand model in an HPF setting.

First, an important use of estimates of recreation demand functions concerns the implied values on a per-trip or per-day basis either for activities, such as freshwater or saltwater fishing, or for different types of recreation sites. Implied values are used for different purposes in policy decisions, depending on the form in which the information for evaluating an investment or management decision is available. The reason for raising the distinction is to identify the need for consistent connections between measures using values for an activity

[22] See chapter 8 or Smith and Kaoru (1990) for an empirical review of these results.

versus values for a trip to a particular site. Because the site demand is interpreted as a derived demand, changes in the access or quality conditions for a site can be measured by using the site or the activity demand functions with equivalent results. A review of Bockstael and McConnell's (1983) analysis of the HPF framework's usefulness in the absence of constant returns to scale and nonjoint production suggests that focusing on input demands would be a more appropriate starting point if there are concerns about assumptions.

However, suppose an analyst wanted to estimate the values people place on activities. He or she can easily interpret either trips or site usage as measures of the amount of the recreational activity produced. Price is the only element in the model that distinguishes the two interpretations. Activity prices must be the marginal costs of producing the trip. To the extent that travel costs are the primary elements, these demands may provide a good approximation for both. Of course, this assumes that recreationists using the sites all engage in only that one activity. To the extent that they undertake different activities, they would be expected to have different demands for site services.

David Gallagher first identified this issue in unpublished notes; Smith and Desvousges (1986b) discussed it more specifically. Conventional demand specifications may need to consider explicitly the different mixes of activities undertaken by different recreationists as part of the model in order to reflect the influence of any differences in how the recreation site's services contribute to each type of activity. Such differences would affect the structure of each person's derived demand for the site's services. Randall, Hoehn, and Swanson (1990) made a similar observation recently when they discussed the use of these methods for a benefit analysis of the Tongass National Forest in Alaska. An example of the potential importance of these issues can be taken from current research. Using a factor analysis of the recreational activities people say they undertake at coastal beaches, Robert Leeworthy, Norman Meade, and I found that "natural" groupings of these activities are derived from the factor analysis and that certain sites are more likely to be used for some groupings than others. These groupings are given in Table 3.4. These preliminary results have two implications. First, in future travel cost demand analyses (at the micro level), we may want to consider separating the samples into subsets based on activities users undertake. That would assure that each respondent had a comparable derived demand for site services. Second, the subsets would provide a basis for pooling responses across sites because each grouping represents demands for recreation site services supporting the same recreational activities.

This strategy reinforces the importance of considering the connections between activities undertaken and the sites used, as implied in the HPF of recreation. Moreover, it calls for the use of participation records to identify the activities usually undertaken together.

The connection between the value of a site's services and the value of recreational activities may not be as straightforward as the simple case discussed

TABLE 3.4
Grouping of recreation activities undertaken at coastal beaches.

Factor	Activities with large positive weights
1	Wildlife observation and photography; other nature study; photography; sightseeing; walking for pleasure, running or jogging; bicycling; driving vehicles or motorcycles off road; driving for pleasure
2	Attending outdoor sporting events, outdoor concerts, plays, fairs, festivals, an evening campfire program or ranger guided walk; visiting prehistoric structures, historic areas or museums; day hiking; horseback riding; using a self-guided trail; backpacking
3	Outdoor swimming; picnicking; family gatherings; sunbathing; collecting seashells
4	Camping in developed or primitive campgrounds; cold freshwater fishing; warm freshwater fishing; saltwater fishing; anadromous fishing
5	Outdoor pool swimming; golf; tennis outdoors
6	Boating, canoeing or kayaking, sailing; motor boating; water skiing

Source: This is based on unpublished research developed by V.R. Leeworthy, N. Meade, and V.K. Smith using the 1987 Public Area Recreation Visitors Survey for Coastal Recreation Areas.

earlier. To this point, most applications have involved public recreation sites where the most important cost to the household has probably been the travel and time costs to the site. Thus, the implicit cost of the site's services usually accounted for the majority of the costs of producing the recreational service flow. Moreover, when the household production technology is assumed to have the form given in equation (3.39), then a measure of site usage can also function as an index of output for technically efficient households producing recreation activities:

$$Z_R = \min[F(X_1, \ldots, X_k), \beta V] \qquad (3.39)$$

where $X_1, \ldots, X_k$ is private goods used in production of the recreation service flow, Z_R; V is visits to the site as measures of site usage; and β is the input-output coefficient. Technical efficiency would imply:

$$F(X_1, \ldots, X_k) = \beta V \qquad (3.40)$$

Based on these assumptions, the analyst should be able to describe the relationship between the private goods using production or cost functions that treat V as a measure of output, $V = 1/\beta \, F(X_1, \ldots, X_k)$.

Of course, if he or she believes that prospects for substitution exist between one or more of the X's and site usage, this simple connection will not be possible. When other private goods such as boats or gear for sport-fishing account for a substantial fraction of the costs of the activity, this seems to be a reasonable working hypothesis. To develop further insight into this question, measures of the activity outputs will be required along with the input measures. Both of these conclusions follow from interpreting the travel cost within the HPF framework. While they don't necessarily lead to testable hypotheses, they clearly do affect the questions raised in formulating demand specifications and in using the results derived from them.

The second set of issues raised by using the HPF framework to describe recreation demand arises in modeling the role for site quality in a travel cost setting. Here, the formulation of the production function hypothesized for the pure defensive expenditure model is contrasted with the one required to implement the varying parameter model (Smith and Desvousges 1986b). These two production functions are given in equations (3.41) and (3.42):

$$Z_R = X + b \cdot q \qquad\qquad\qquad (3.41)$$

$$Z_R = f[X \cdot h(q)] \qquad\qquad\qquad (3.42)$$

where Z_R is a recreational activity, X is a measure of a site's services, and q is site quality. In equation (3.41), quality changes can be compensated for by using more for the site's services and the value is measured by these expenditures [see equation (3.4)]. In contrast, with equation (3.42) describing the production process, each recreationist acts *as if* a site has an effective price of $P_X/h(q)$. That is, the price is measured in terms of effective quality units. To the extent that analysts can observe differences in the demand functions for sites when these sites vary in quality, they can recover the influence of the conversion function, $h(q)$. However, estimates of the Hicksian value of a quality change face the very issues discussed earlier in considering whether the assumption of weak complementarity would allow the use of Marshallian demand functions (with quality variables) to measure the monetary value of quality changes.

3.4 Future Research

To this point, the household production framework has provided a conceptual basis for several of the modeling strategies used to value nonmarketed environmental resources. The models usually associated with this formulation for describing consumer decisions are the averting behavior and travel cost recreation demand models. Moreover, the majority of the available empirical results are for travel cost models. Data limitations have hampered development of empirical models for averting behavior.

Until detailed household-level data are available on the patterns of expenditures, time allocations, commodity prices and wage rates, along with measures of levels of environmental quality experienced by the same households reporting the data, few empirical advances will be made in applying the HPF framework more generally. This explains why so little empirical research has attempted to use the HPF model to extend the Shapiro and Smith approach along the lines proposed by Math-Tech and Gilbert. The efforts need not be exhaustive, although these are certainly data intensive methods. Nonetheless, partial demand systems along the lines proposed by LaFrance and Hanemann (1989) could be adapted as alternatives to the full expenditure system approach.

A second future line of research involves exploring the performance of these system approaches in the presence of specification errors. This type of analysis could follow the general outline used by Kling (1988a, 1988b) for evaluating methods to estimate recreation demand. Because it is a controlled experiment, such an evaluation could consider the merits of a general specification of expenditure systems without attempting to group private goods (as in Shapiro and Smith) versus the more selective modeling and pretesting efforts in the work of Math-Tech and Gilbert. The first strategy imposes few prior restrictions on the role of the nonmarketed good and relies on flexible specification of the indirect utility function to capture the effects of the environmental quality measures. Of course, the first-order effects of these variables cannot be measured in such models, so they must be assumed away. Therefore, this assumption becomes an example of a restriction linking the nonmarketed service to private goods.

This interpretation warrants further research on the nature of these connections. Such research could serve to evaluate the feasibility of using expenditure bounds, such as those proposed by Neil (1986, 1988) as either a basis for approximating the value of environmental resources or as a source of checks on the estimates derived from systems analyses of consumer expenditure models incorporating environmental services.

Finally, another potential use of the system approach follows from the HPF grouping of commodities. By explicitly incorporating nonmarketed goods in preference functions, the responses of expenditures on private goods could be estimated to actual changes in the nonmarketed services across households. In addition, these same models could be used to describe how households *should* respond to contingent behavior or valuation questions. First applied by Cameron (1988) in a partial equilibrium framework involving recreation demand models, this strategy requires that contingent valuation surveys collect information on households' expenditures on private goods hypothesized to be related to the resource being valued. Following conventional practice, the value that households should bid for a quality change could then be related to the same behavioral model used to explain the expenditure patterns. Roy's identity provides the commodity demands — $X_i = -(V_{P_i}/V_I)$, with $V(\cdot)$ the indirect utility function, P_i = price of X_i, and I = income and the marginal value of the resource derived from a similar condition — $-e_q = (V_q/V_I)$, where $e(\cdot)$ is the Hicksian expenditure function and q the nonmarketed level on environmental quality. Because both relationships can be derived from the same hypothesized indirect utility function, it should be possible to jointly estimate both models restricting the parameters to be equal across estimating equations.

Clearly this is not a complete list of future research possibilities based on the household production concept. Rather it is a selection of the more specific ways that insights from the household production framework can be incorporated into the practice of benefit measurement for environmental resources.

Measuring the Demand for Environmental Quality
John B. Braden & Charles D. Kolstad (Editors)
© Elsevier Science Publishers B.V. (North-Holland), 1991

Chapter IV

HEDONIC METHODS

RAYMOND B. PALMQUIST

North Carolina State University

4.1 Introduction

The measurement of the benefits of environmental improvements is difficult because typically there are no markets for environmental quality. However, one can observe behavior in markets that are related to environmental quality, and it is sometimes possible to measure people's willingness to pay for the environmental goods by using data from those markets. Several techniques for doing this are discussed in this book. This chapter focuses on one of these — hedonic methods.

Hedonic methods are based on the realization that some goods or factors of production are not homogeneous and can differ in numerous characteristics. One of these characteristics is environmental quality. Economists use hedonic studies to analyze the effects these different characteristics have on the price of a good or factor and then to extend the analysis to the underlying demands for the characteristics. The benefits of the environmental change can be measured after either one of these stages, depending on the nature of the change. This chapter is concerned with hedonic theory, the econometric issues that arise in hedonics, and how hedonic analysis can be used to measure environmental benefits. Also presented are alternative techniques that are closely related to hedonics. (The results of hedonic analysis will be discussed in Part II of this book.[1])

The next section presents theoretical models that can form the basis for hedonic estimation. The most familiar of these deals with differentiated consumer products, such as housing. The section also discusses models of

---

[1] There are other surveys of hedonic methods that should be noted. Bartik and Smith (1987) is an excellent survey of the application of hedonic techniques to the study of urban amenities. Freeman (1979a or 1985a) contains sections on the use of hedonic methods in environmental economics. Finally, Follain and Jimenez (1985) consider the applications of hedonics to the demand for urban housing.

differentiated factors of production, such as land or labor, and considers the possibility of obtaining closed-form solutions to hedonic models. Sections 4.3 and 4.4 discuss econometric issues that arise in the estimation of the hedonic equation and the demands for characteristics, respectively. Section 4.5 describes the uses of hedonic results for measuring benefits under a variety of situations. Section 4.6 presents benefit measures in a number of closely related models of land values, whereas section 4.7 discusses ways of combining wage and property value hedonics to estimate willingness to pay. Section 4.8 discusses models that use repeat sales of properties. And section 4.9 describes discrete choice models of residential location as alternatives to hedonic studies.

4.2 Theory of Hedonic Models

4.2.1 Differentiated Consumer Products

Most of the recent work in environmental economics using hedonic models has been based on the seminal article by Sherwin Rosen (1974). In that article, Rosen provided a theoretical model underlying the hedonic regressions that were already common at that time.

There are consumer products that are distinct, yet closely related in consumers' minds. For example, automobiles received considerable attention when hedonic regressions were first used; however, a product with greater relevance to environmental economics is housing as houses differ in the environmental quality at their location. In this chapter, the differentiated products will be houses unless otherwise specified. Economists could consider houses with various amounts of living space to be different commodities, and estimate a separate demand function for each size of house. Individual consumption of almost any of these commodities, however, is zero. Rosen, on the other hand, modeled houses or other appropriate goods as single commodities that are differentiated by the amounts of various characteristics they contain. The consumers of the different varieties of the commodities derive utility from the characteristics of the commodity, whereas the producers or sellers of the commodities incur costs that are dependent on the varieties they provide. The interactions of the consumers and the producers in a competitive market for a differentiated product determine the equilibrium hedonic price schedule. A model of this interaction is the greatest contribution of Rosen's article.

The Rosen model assumes that a particular variety of the differentiated product can be represented by a vector $z = (z_1, z_2, \ldots, z_n)$ of the characteristics of the product. The price for which the product sells is a function of the characteristics of the product, which is the hedonic function $P = P(z)$. The range of product choices is assumed to be continuous, although some researchers have objected to this assumption (see section on discrete choice modeling). If a characteristic in a product can be separated from the product so that the amount of that characteristic can by varied independently, then the hedonic price function is linear. However, such repackaging often is not possible, so nonlinear as well as linear hedonic price functions are possible. Because there are large numbers of consumers and producers, an individual cannot affect the price schedule.

The determination of the hedonic price schedule in the market can be explained by considering the behavior of consumers and firms. Consumers differ according to a vector of socioeconomic characteristics, α. A typical consumer gets satisfaction from the consumption of the characteristics of the differentiated product and a composite good x by maximizing utility $U(x, z_1, \ldots, z_n; \alpha)$ subject to the budget constraint, $y = x + P(z_1, \ldots, z_n)$, where y represents the consumer's income. The first-order conditions for this optimization problem require that the marginal rate of substitution between one of the characteristics and the composite good be equal to the marginal price of the characteristic, $\partial P / \partial z_i$.

Rosen represented a consumer's actions with the bid function $\theta(z, u, y; \alpha)$, which represents the consumer's willingness to pay for a product with characteristics z, given a variable for income and a level of utility. The bid function is defined implicitly by $U(y - \theta, z_1, \ldots, z_n; \alpha) = u$. The bid function is increasing and concave in the characteristics, and decreasing in the given level of utility. If income changes, there is an equivalent change in the bid. The marginal bid for z_i, θ_{z_i}, equals the marginal rate of substitution, U_{zi}/U_x, discussed earlier. Thus, optimization requires that the marginal bid be equated to the marginal price in the market. The budget constraint shows that the total bid must also be equal to the commodity price. Different consumers will wind up choosing different products because of differences in the socioeconomic characteristics, α.

On the other side of the market are producers (or sellers for nonproduced goods). In Rosen's model for produced goods, the producers must select which version of a product they will produce as well as the number of units. Their costs will depend on both of these decisions and may vary among firms if they have different technologies or factor prices. The cost function can be represented by $C(M, z; \beta)$, where M is the number of units produced and β is a vector of firm-specific technologies and factor prices. Revenues depend on the price schedule and the number of units sold; profits are simply revenues minus costs, $\pi = M \cdot P(z) - C(M, z; \beta)$, where the producer selects the model manufactured, z, and the number of units, M. The firm takes the price schedule as given. The first-order conditions require that a version of the product be chosen such that the marginal price for each characteristic is equal to the marginal cost per unit of increasing the amount of that characteristic. The output level must equate product price and the marginal cost of output.

Because Rosen's purpose was to study market equilibrium, he translated the firms' behavior into offer functions $\phi(z,\pi;\beta)$, where ϕ represents the unit price a producer can accept for a product with characteristics z and make profits π given producer attributes β.

If some or all of the characteristics of the differentiated product are not produced, then the quantity of those characteristics is exogenous to the seller. The offer prices depend only on the level of profits, and the equilibrium price is determined completely by demand once the level of the characteristic is given.

The equilibrium price schedule is determined by the interaction of the consumers and the suppliers. Obviously, consumers would like the lowest possible bid to win, thus maximizing their utility. Firms would like the highest possible offer to be accepted, thus maximizing their profits. The market reconciles these conflicting goals by matching consumers to suppliers such that consumers cannot increase their satisfaction by choosing a different product, nor can firms increase their profits by varying the quantity or the version of the product they produce. Graphically, this relationship is represented by Rosen's well-known diagram in which the bid and the offer functions are just tangent. In figure 4.1, the quantity of one of the characteristics, z_i, is shown on the horizontal axis, while the other characteristics are at a fixed level. Each consumer's bid contour represents a particular level of utility. The level of utility increases as the bid price is reduced while holding constant the level of the characteristics. Similarly, the firm's offer contours represent a given level of profit that increases as the offer price increases, *ceteris paribus*. The equilibrium price schedule shown is determined by the market interaction of the two groups.

Although it is important to include both consumer and producer behavior when describing the market, for many environmental issues the critical information is contained on the consumer side of the market. Fortunately, it is often possible to focus on the equilibrium price schedule and on consumers' decisions. In these cases, ignoring the producer side does not create theoretical or econometric problems.

The model of equilibrium presented here has proven quite useful, although economists (e.g., Mäler 1977 and Freeman 1979b) have raised two questions as to its application. First, is it reasonable to assume that the market for a differentiated product, such as housing, achieves equilibrium? And second, do the versions of the differentiated product vary enough that the price function and the bid and offer functions can be assumed to be continuous? The latter question is closely related to discrete choice modeling (which is discussed later in this chapter), whereas the former can also be applied to markets for undifferentiated products. The presence of moving costs and other transactions costs in housing markets makes the former question especially relevant since the uses of hedonics in environmental economics usually involve changes in variables such as air quality, which are probably of less importance

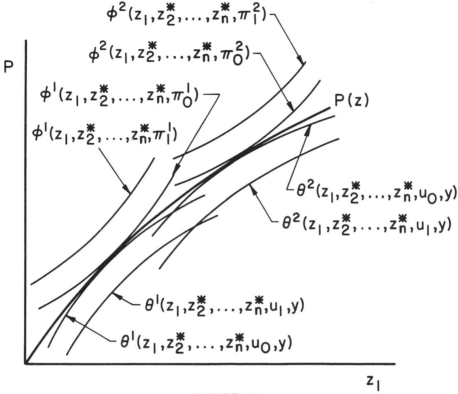

FIGURE 4.1
Hedonic equilibrium.

to the decision maker than are variables such as living space. However, there are households that move for reasons that may or may not include the environmental conditions. Yet, when they choose their residence they will certainly consider the environmental conditions. The price schedule will reflect these considerations.[2]

In addition to the model of the determination of the equilibrium price schedule, Rosen's article (1974) contains suggestions for estimating the underlying demands for the characteristics. These suggestions have proven troublesome and will be discussed in section 4.4; however, the theoretical issues concerning the demands of the characteristics can be discussed here. Most of these issues arise because the nonlinear hedonic price schedules cause nonlinear budget constraints for consumers. When budget constraints are nonlinear, the standard duality results no longer hold because marginal prices are not parametric for the individual consumers.

---

[2] It will be shown in the section on benefit measurement that it is actually easier to measure welfare change if a change in an environmental variable does not induce residents to move.

Rosen (1974) suggested that the coefficient estimates from the hedonic price schedule could be used to calculate the marginal prices for the characteristics. The partial derivatives of the bid functions with respect to the characteristics are marginal rates of substitution between the characteristic and the composite good, which is measured in dollars. In equilibrium, the marginal bid function is equated to the marginal price. Rosen stated that the marginal bid functions are "reservation demand prices for additional amounts of z_i at a constant utility index," and suggested that they were inverse compensated demands. However, Rosen next suggested that the marginal bid functions could be replaced by ordinary inverse demands with utility eliminated from the functions. If this were correct, it would be possible to estimate a system of demand and supply equations for the characteristics. This procedure is followed in most empirical applications (for example, Harrison and Rubinfeld 1978; Nelson 1978; Witte, Sumka, and Erekson 1979).

McConnell and Phipps (1987) pointed out that there is a difference between the marginal bid functions — which they call marginal rate of substitution function to emphasize the information it contains — and ordinary inverse demand functions, which can be derived from the Hotelling-Wold theorem. The marginal rate of substitution functions are derived only from $n - 1$ of the first-order equations, where n is the number of first-order conditions. They do not include the information in the budget constraint, nor equivalently, the equating of the total bid with the hedonic price.

It is not surprising that marginal bid functions, which depend on the level of utility but not income, are not the same as the uncompensated inverse demands, which depend on income but not on utility. Palmquist (1988) utilized the distance function to show that it is possible to normalize the marginal bid functions by income to obtain normalized compensated inverse demand functions. These functions depend on unobservable utility, but duality results can be used to transform them into uncompensated inverse demands for estimation.

4.2.2 *Differentiated Factors of Production: Land*

The consumer models described in the preceding paragraphs were appropriate in many cases because most of the hedonic studies involved residential real estate. However, other studies involved differentiated factors of production such as land. The quality of land varies considerably in activities such as farming. Some characteristics of farmland have substantial environmental effects. For example, the susceptibility of the land to erosion can affect not only the productivity of the farm but also the downstream environment. Similarly, the drainage of wetlands on farms has numerous environment impacts. Thus, studies similar to hedonics have frequently been applied to farmland (Ervin and Mill 1985; Gardner and Barrows 1985; Miranowski and

Hammes 1984), but only recently has a formal model of a market for a differentiated factor of production been available.

According to Palmquist (1989), the rental price of land R, depends on the characteristics of that land, $R = R(z_1, \ldots, z_n)$. A farmer who uses the land has a multiple-input/multiple-output production function that depends on, among other things, the characteristics of the land, $g(x,z,\alpha)$, where x is a vector of net outputs exclusive of land and α is a vector of farmer-specific skills. The farmer maximizes profits; however, to focus on the farmer's bid for a parcel of land, it is useful to focus on a type of variable profits, which is the value of outputs minus the cost of nonland inputs:

$$\max_{x} \pi^{DV} = \sum_{j=1}^{m} p_j x_j$$

$$\text{s.t.} \quad g(x,z,\alpha) = 0, \ \pi^{DV} \geq 0,$$

where π^{DV} is the farmer's variable profit and the p's are the prices of outputs and nonland inputs. Maximizing these "variable" profits subject to the production function yields output supply and nonland input demand functions, $x = x(p,z,\alpha)$. Substituting the supply and demand functions in the objective functions yields the variable profit function $\pi^{*DV} = \pi^{*DV}(p,z,\alpha) = \sum p_j x_j(p,x,\alpha)$. A farmer's bid for a parcel of land will be his or her variable profits minus the desired profit level $\Theta(z,p,\pi^D,\alpha) = \pi^{*DV}(p,z,\alpha) - \pi^D$. This bid will depend on the characteristics of the land, the prices of agricultural outputs and nonland inputs, the level of profit, and the farmer's production skills. The resulting bid function is increasing and concave in the characteristics; thus, marginal bid functions are downward sloping.

Differentiating the bid function with respect to the prices of output and nonland inputs is equal to the net output, a result which is analogous to Hotelling's lemma for profit functions.

In this example, the land is supplied to the rental market by landowners, although if the farmer owns his or her land, then he or she can be considered as renting from himself or herself. The supply side is more similar to the Rosen (1974) model, although the characteristics have to be separated according to whether or not the landowner is able to alter them. An offer function can be derived that depends on both types of characteristics, input prices, landowner profits, and a vector of technical parameters that may vary between landowners.

Both farmers and landowners take the market rental price schedule as given, although in actuality it is determined by the interaction of these two groups. The number of individuals in each of the groups is not fixed. Equilibrium is established when a combination of land price adjustments and entry or exit of farmers or land from farming results in an equilibrium of land to farmers similar to the previous example of consumers and producers. Such a model enables researchers to correctly specify the estimating equations and can be used for welfare measurement.

4.2.3 Differentiated Factors of Production: Labor

Another differentiated factor of production that has been quite important in measuring the benefits of environmental quality is labor. Though there are numerous similarities between markets for land and labor because both are factors of production, there are also key differences. On the demand side, the differences are only minor. Firms wish to hire differentiated labor services and are interested in the characteristics of those services that increase their productivity. The bid functions of the firms would be similar to those discussed in the previous section. The differences on the supply side of the market are more noteworthy. The suppliers of the labor services are the workers, and their levels of utility are directly influenced by their working conditions. For this reason, their supply decisions will be affected by their expectations as to the uses of their services. A more important factor for environmental economics, however, is that the locations of workers' jobs influences where they choose to live. The level of amenities in an area may influence the workers' offer functions and, therefore, the wages that firms in the area must pay to attract workers. Thus, the specification of the hedonic wage equation must include a wider variety of characteristics than did the hedonic equations discussed in the previous sections.

Smith (1983) classified the variables that are likely to influence wage rates as characteristics of the individual, of the job, and of the location. In a competitive labor market, the individual's characteristics that affect wage rates and may influence a worker's productivity are measures of training and ability. Job characteristics affect wage rates because firms are often forced to offer workers compensating differentials in wages to entice them to accept less desirable positions. The characteristics of the location of the firm, however, may or may not influence the productivity of the firm. Some amenities assist with production in a firm, whereas others interfere. In addition, the size of the population in an area may influence productivity because of agglomeration economies. However, even if the firm is not directly influenced by the nonpecuniary aspects of a location, wages may be because the residents may value the amenities in an urban area. They may be willing to accept lower wages if it allows them to live in a location with more amenities.

The size of the market area for labor services is an important issue in hedonic equations. If labor markets in different urban areas are separate, the locational attributes of the firm will have less influence on wages than if the labor market is nationwide. This is because the firm draws employees from throughout the urban area. Even if amenities differ within the city, the wages at the firm cannot differ according to the amenities at the residences. For this reason the environmental studies that have used wage rates have assumed that the labor market is national and generates a single hedonic schedule. These studies assume that when workers job hunt they consider employment opportunities throughout the country. However, since they must live in the same urban area in which they work, their choice of a job may be influenced by nonwork-related amenities associated with the area.

When deciding upon a job, the individual also has to consider the cost of living in that area. An important component of the cost of living is housing prices, thus a property value hedonic equation can be relevant to the decision. Recently, several theoretical and empirical studies have considered both wages and rents (property values) in valuing amenities. These studies assumed that both labor and housing are traded in national markets. However, an alternative way of modeling the location decisions is to divide the decisions into two steps. In the first step, the individual selects a job after considering the choices available to him or her throughout the country. When making this decision, the individual considers the distribution of housing prices in each urban area, as well as the wages and amenities. Moving costs may be incorporated at this stage so that moves take place only if some threshold for differences in utility is passed. Once the job location is chosen, the next step is for the individual to choose a residential location within the selected urban area.

4.2.4 Closed-Form Solutions to Hedonic Models

The hedonic equation is an equilibrium that results from the interactions of the suppliers and demanders of a differentiated product. Thus, the hedonic equation contains information on the underlying preferences and technologies of these two agents, if only ways can be found to extract this information. Rosen (1974) briefly discussed this possibility for a situation that has only one characteristic of the product. But the calculations required even for this simplified case are quite complex. The difficulty of deriving the information from the hedonic equation led Rosen to propose a methodology for estimating the demands and supplies of characteristics in a second stage rather than using the hedonic equation directly.

The *closed-form approach* of extracting information from the hedonic equation was first pursued extensively in a contemporary hedonic framework by Epple (1984, 1987), although he acknowledged that a series of articles by Tinbergen (e.g., Tinbergen 1956) on labor markets contributed to his work. Epple took a differentiated product with an arbitrary number of characteristics. Consumers were assumed to have a utility function that was quadratic in each of the characteristics but additively separable between characteristics. The composite good was entered linearly. All consumers had the same utility function except for translating factors (Pollak and Wales 1981), which differ among individuals. These differing taste parameters were assumed to be normally distributed with no covariance. In Epple's first model, the supply of the differentiated product is exogenous and is also distributed normally with no covariance. Given this relatively simple structure he showed that the hedonic equation will be quadratic in the characteristics. The parameters of the equation will be functions of (1) the common taste parameters, (2) the mean and the covariance matrix of the distribution of the taste parameters that differ between individuals, and (3) the mean and covariance matrix of the distribution of products. In his 1984 article, Epple also went through the same process when supply is endogenous — first when inputs are owned and then when they are purchased — with a geometric increase in the complexity of the coefficients of the hedonic equation.

The goal of closed-form solutions is to extract the parameters of the utility function from the parameters of the hedonic equation. However, even when this is possible, the assumptions on which the estimates are based may or not be realistic. The assumption that the utility function is additively separable in the characteristics seems undesirable in hedonic models where joint decisions on characteristics are emphasized. The assumption that the taste parameters and product characteristics are normally distributed with a diagonal covariance matrix might also be questioned, although Epple (1984) has shown that under some circumstances the zero covariance assumptions can be relaxed. Finally, the utility function is quasi-linear, so the marginal utility of the composite good is constant and the income elasticity of demand for the characteristics is zero. This assumption seems doubtful given the empirical evidence that is available on the demand for characteristics, but Epple (1984) indicated that this is necessary to obtain closed-form solutions.

Giannias (1988) sought to relax some of these assumptions and use the model to estimate people's willingness to pay for better air quality. He allowed nonzero off-diagonal terms in the distributions of taste parameters and characteristic supply parameters. However, he obtained this generality by having the utility function depend on an index of the characteristics instead of on the characteristics themselves. For this assumption to work, all consumers must have the same marginal rates of substitution between characteristics regardless of their socioeconomic characteristics. Giannias used a modification of the model to examine Houston residents' willingness to pay for better air quality.

At present, closed-form solutions to hedonic models may not be valid methods for measuring environmental benefits because they are based on questionable assumptions. Nonetheless, such analyses are instructive for understanding the workings of the hedonic model.

4.3 Estimating the Hedonic Equation

For certain types of benefit measures, it is only necessary to estimate the hedonic equations; in other cases, the hedonic equation is used to generate marginal prices that are then used to estimate the demands for environmental characteristics. In either case, the estimates for the hedonic equation must be reliable. The many different econometric issues that must be addressed to obtain reliable estimates are discussed in the following paragraphs.

4.3.1 Functional Form for the Hedonic Equation

Theoretical considerations generally do not dictate the functional form for the hedonic equation. The only exception is if the product characteristics can be costlessly repackaged. In this case, a nonlinear price schedule would offer profitable arbitrage opportunities for individuals who reallocated the characteristic from goods where it is relatively cheap to goods where it is more valuable. For a differentiated product such as housing, costless repackaging would only apply in the long run. Although the long-run equilibrium is relevant to some of the theoretical work that has been done on hedonic models (Scotchmer 1986), such complete adjustment generally has not been anticipated nor found in empirical hedonic models. Thus, the functional form for the hedonic equation must be determined empirically.

A variety of nonlinear functional forms can be made linear by transforming the variables. The most common transformations are the semilogarithmic, inverse semilogarithmic, and log-linear. These were the forms, together with linear forms, that were used widely in the years immediately following the publication of the Rosen (1974) article. It must be remembered that if the price of the differentiated product is transformed, the researcher cannot legitimately compare residual sums of squares in choosing among these forms since the variance in price is not invariant to changes in units (Rao and Miller 1971). However, if the price is multiplied by the inverse of the geometric mean of the price, and if this adjusted price is used in the various regressions, then the functional form can be chosen by minimizing the residual sum of squares.[3]

More recently, a wider range of functional forms have been considered. In one influential article, Halvorsen and Pollakowski (1981) proposed using a highly general flexible functional form called the quadratic Box-Cox:

$$P^{(\theta)} = \alpha_0 + \sum_{i=1}^{m} \alpha_i z_i^{(\lambda)} + 1/2 \sum_{i=1}^{m} \sum_{i=1}^{m} \gamma_{ij} z_i^{(\lambda)} z_j^{(\lambda)}$$

where $P^{(\theta)} = (P^\theta - 1)/\theta$ if $\theta \neq 0$ and $P^{(\theta)} = \ln P$ if $\theta = 0$, and $z_i^{(\lambda)} = (z_i^\lambda - 1)/\lambda$ if $\lambda \neq 0$ and $z_i^{(\theta)} = \ln z_i$ if $\lambda = 0$. Special cases of this general form include translog, log-linear, quadratic, linear, generalized square root quadratic, generalized Leontief, and semilog, among others. Halvorsen and Pollakowski (1981) utilized ordinary least squares (OLS) in a grid search to obtain an estimate of this equation. Spitzer (1982) provided an overview of several

---

[3] In Palmquist and Danielson (1989) it is shown that minimizing the residual sum of squares with the transformed price is equivalent to maximizing the Box-Cox log-likelihood function for the candidate functional forms.

estimation methods that can be used with the Box-Cox transformation. However, the transformation results in a truncation in the distribution of the dependent variable and the error term (Amemiya and Powell 1981), therefore the assumption that error terms are normally distributed cannot be used in the likelihood function.

Despite of the flexibility of the quadratic Box-Cox functional form, it is still somewhat restrictive. First, it provides only a local approximation to the true function. Globally flexible functional forms (e.g., Gallant 1981) may prove superior for some purposes, particularly when the test statistics for the parameter estimates are of central importance. Second, while the parameter λ is allowed to have a value other than θ, all of the independent variables have the same λ. There is no *a priori* reason to expect that any or all of the environmental variables enter in the same form as do the structural characteristics. However, allowing each independent variable to have a different λ would be computationally prohibitive.

Cassel and Mendelsohn (1985), in a comment on the Halvorsen and Pollakowski paper, suggested that the estimates of the coefficient of the environmental variable may be more reliable with simple functional forms than with flexible forms. This is because the environmental variable plays a minor role in determining the price, and thus it also plays a minor role in determining the parameter λ. If the environmental variable enters in a simple form, the quadratic Box-Cox may force it into a more complex form with less accurate parameter estimates. Perhaps a better alternative to Cassel and Mendelsohn's suggestion is to enter the environmental variable with a separate λ since that is the variable of interest, while continuing to use the quadratic Box-Cox. This would increase the flexibility of the functional form without making the computational costs prohibitive.

Graves et al. (1988) experimented with a dataset from Southern California to see how large were the differences in the estimated parameters of the environmental variables when different functional forms were used. They found that with one specification the differences were negligible. Their marginal price estimates for total suspended particulates and visibility were fairly stable between the various functional forms they tested. On the other hand, when they expanded the specification, the estimates varied dramatically with the different functional forms and from those obtained with the first specification. Of course, they could not determine which estimates were "true"; furthermore, the variation could be caused by various factors.

Probably the most ambitious experimental work on functional form to date is Cropper, Deck, and McConnell (1988). They conducted Monte Carlo experiments to determine the accuracy of the marginal prices that are estimated with various functional forms for the hedonic price equation. To make the simulations approximate actual conditions, they used houses in and around Baltimore. To obtain variation in true hedonic price function, they considered alternative functional forms and attributes for the utility functions of consumers as well as different distributions of utility function parameters, the buyer characteristics, and the housing characteristics. Equilibrium was established using an assignment model. They considered cases in which the estimated hedonic equation was both correctly and incorrectly specified. Their results were interesting. When the hedonic equation was specified correctly, the quadratic and linear Box-Cox forms yielded the closest estimates. However, when the hedonic equation was misspecified because of unobserved or proxied variables, the simpler forms and linear Box-Cox performed better. Since the correct specification probably is difficult to achieve, the Cropper, Deck, and McConnell results suggest that the linear Box-Cox functional form may be the most promising compromise.

Once the appropriate functional form is selected and estimated, it is usually necessary to transform the results to their original form so they can be used for interpretation and prediction. Simply applying the inverse transformation to the estimate yields the median response in the untransformed space rather than the mean. Goldberger (1968), Duan (1983), Miller (1984), and Vaughn (1988) described ways to avoid retransformation bias for continuous variables. If an environmental variable is dichotomous, the appropriate retransformation is discussed in Halvorsen and Palmquist (1980) and Kennedy (1981). Palmquist (1982b) discussed the appropriate retransformation to use when the variable is discrete (polychotomous).

4.3.2 Market Size

The hedonic price schedule represents a market's equilibrium prices. Thus, the way in which the constituents of a market for a differentiated product are defined is important. For example, some environmental hedonic studies for housing have assumed that the scope of the appropriate market is national, whereas other studies have assumed that the geographic extent of the market is as small as a census tract. Still other studies have suggested that housing markets are segmented according to income, accessibility, race, or other variables. If economists assume that there is a single market when it is actually segmented, their coefficients will be biased. On the other hand, if they assume that the markets are segmented when they are not, their estimates will be imprecise and they may have insufficient data in the segments. The prevalent view in housing studies is that the housing market within a city is a single entity, whereas the housing markets in cities that are separated by significant distances represent separate entities. In hedonic wage studies, however, economists usually assume that the housing market is a single national market.

Theoretical considerations may help the economist in defining the constituents of a market. Freeman (1979b) defined a market as segmented if the segments have different hedonic price functions. This situation is possible only if a barrier, such as geography, discrimination, or lack of information, prevents the purchasers in one segment from participating in the other segment. It also requires that there be differences in the structure of supply or demand in the various markets (Freeman 1979b). Market segmentation between cities may occur if information and moving costs between cities are high enough. Within cities, the presence of discrimination can lead to segmented markets; although discrimination seems to be declining, environmental hedonic studies have never segmented markets along racial or ethnic lines. Saying that there are segments with respect to other variables requires that the reasons for the segmentation be justified; an individual's refusal to locate in a particular neighborhood is not evidence that he or she cannot locate there.

The alternative to theoretical arguments about market segmentation is to address the question empirically. F-tests can be used to test whether the sets of coefficients from potential segments are equal, but for the tests to be valid, the true hedonic specification must be used. Furthermore, because of the many specification considerations discussed in this section, neither F-tests nor less stringent comparisons of the standard errors of the estimates in the constrained and unconstrained cases, will prove conclusively whether a market is segmented.

The possibility of a national housing market has been advocated by Linneman (1980) and implemented by Linneman (1981), Cobb (1984), Smith and Deyak (1975), and Deyak and Smith (1974). Butler (1980) attempted to test the hypothesis of a national real estate market by using census data for 36 cities. Using an F-test, he accepted the hypothesis of a single market for renters but rejected it for owners. The comparison of the standard errors suggested that in either case the assumption of a single market would not have serious impacts on the explanatory power of the regressions. However, in environmental economics the interest is in the environmental variables rather than the housing prices, so the latter may not be much consolation.

On the other side of this argument, Straszheim (1973) suggested that real estate markets within cities are segmented. This suggestion was followed by Goodman (1978). Several attempts to test this hypothesis have had very mixed results. Straszheim (1974) used an F-test on geographic segments in San Francisco and strongly rejected the hypothesis of a single market. Schnare and Struyk's (1976) studies in the Boston area yielded mixed results when they tested market segmentation by accessibility, income, and structure size. Their F-tests showed some segmentation; however, when they compared the standard errors, the results suggested that the differences were small relative to the housing price variations. Finally, F-tests by Ball and Kirwan (1977) found that clusters of housing types in the Bristol area did not result in separate submarkets with different hedonic prices.

A different approach to market segmentation has been followed recently by Michaels and Smith (1990). They asked realtors, who are the "experts" in the real estate field, which segments they should use. This approach is not without precedent. Buyers and sellers of houses use the services of these middlemen because otherwise it would be expensive and time-consuming for them to obtain information about the real estate market. For the Michaels and Smith study, the agents separated the towns in the Boston area into groups that they perceived to be homogeneous. The statistical tests that were then performed indicated that some segmentation was appropriate. For example, the results as to the effect of hazardous waste sites differed significantly when the market was considered as segmented versus unified. Even though Michaels and Smith's definition of segmentation would not satisfy Freeman's conditions because of the complexity of the specification issues in hedonics, it may prove to be a practical means of discovering empirically tractable "submarkets."

4.3.3 Data Issues

As with any field in economics, the data in environmental hedonic analysis are not as good as one might wish; yet, they are better than in many fields. Disaggregate data, which are rarely available in some economic disciplines, are frequently used in both housing and wage hedonics. The data are collected by private organizations such as multiple listing agencies, by local tax assessors, and by agencies of the federal government. Large, cross-sectional datasets are available for many areas and many years, although panel datasets are generally unavailable because of infrequent house sales.[4] On the other hand, compiling a dataset can be time-consuming because it usually requires obtaining data from a wide range of sources. In early studies, researchers usually used census data on the average owner-estimated house value within the tract because these data were easy to obtain; however, now most of them prefer disaggregate data even though it is less readily available.

Even with relatively good data available, there are still some important issues. In environmental economics a key question concerns the appropriate form for the environmental variables; that is, how the environment is perceived by individuals. Originally, economists debated whether or not the individuals were aware of the quality of their environment (Mäler 1977; Freeman 1979b). Today, evidence has proven they are. Now the question is more difficult; it concerns how they perceive environmental quality.

Environmental quality can be measured objectively. For example, air quality can be measured in "parts per million" or "micrograms per cubic meter." There is a fairly extensive monitoring network for most pollutants, although sometimes there are problems with the accuracy of the instruments, the accuracy of the recording of the data, and the number or the placement of the instruments. However, even as these problems are gradually eliminated,

---

[4] Data sets containing houses with repeat sales are sometimes available. See the section on resale indexes.

the question still remains as to whether objective measures are good proxies for individual perceptions.

Almost all studies of the effects of environmental disamenities, particularly air pollution and noise pollution, on property values have used a single objective measure for the disamenity. Some researchers have considered several pollution measures before they select one empirically (Harrison and Rubinfeld 1978; Nelson 1978). Yet, the various air pollutants have different effects on the population and there is not a high enough correlation in their levels that one pollutant can act as a proxy for all of the pollutants (Palmquist 1982a). Some researchers have begun to include more than one pollutant in their hedonic studies (Palmquist 1982a and 1983; Graves et al. 1988). Most have used the arithmetic or geometric mean pollution level over some time period, although some researchers have used measures such as the second highest reading.[5] Recently, Murdoch and Thayer (1988) have found that when the distribution of visibility (measured by the probability that visibility will be in a given range of distances on any given day) was included, it added to the explanatory power of the regression when compared with using only the mean visibility. Palmquist (1983) considered the possibility that individuals consider the "aggregate" quality of air rather than the individual levels of the pollutants. Pollutant levels were scaled by the primary National Ambient Air Quality Standards to standardize units according to the severity of their adverse effects. The mean and the highest of these scaled measures were then used in hedonic regressions. The technique of aggregation proved to be useful.

To date, there has not been a comparison of the usefulness of objective versus subjective measures for pollution in hedonic regressions. However, when Lang and Jones (1979) compared subjective and objective measures for nonpollution neighborhood amenities, they got some encouraging results. They used a mail survey to determine purchasers' perceptions of neighborhood quality with respect to prestige, attractiveness, schools, and so on. Hedonic results using these measures were compared with the results from objective measures, such as median neighborhood income and education. They found little improvement using the subjective measures. Similar studies should be done for the pollution variables. Contingent valuation studies have become quite sophisticated at soliciting subjective measures of pollution, and contingent valuation results have been compared with hedonic studies (Brookshire et al. 1982). However, perception measures only rarely have been used in environmental hedonics.[6]

Graves et al. (1988) considered the effect of measurement error in the neighborhood and environmental variables on the estimation of the coefficients of the pollution variables. Their results are not encouraging because they

---

[5] See Freeman (1979a) for summary of many studies.

[6] Recently Schulze, et al. (1986) elicited households' perceptions of the dangers of nearby hazardous waste sites and incorporated these in hedonic regressions. The distances from sites used in other studies may also be a proxy for perceptions rather than objective measures.

found that measurement errors in their sample could result in extreme changes in the coefficients of the environmental variables. Of course, this did not mean that the measurement errors were necessarily present or affecting the coefficients, but it was a possibility. Their work did indicate that improving the measurement of the pollution variables may have the greatest payoff.

A final data issue concerns whether it is more appropriate to use rental prices or asset prices for houses. The theoretical models generally use rental prices, yet data are usually more readily available on housing sales. For this reason, most hedonic studies are done using sales prices.[7] Sales prices represent the capitalized value of expected future rents. Those expected future rents are influenced by expected changes in the property, so the characteristics at the time of the sale may not be adequate to explain the selling price (Palmquist and Danielson 1989; Grieson and White 1989). A related issue arises when hedonic results based on sales prices are used to estimate the benefits of environmental improvements. Because property taxes affect the net return to owners, the sales prices will reflect future rental prices net of property tax. And as a result, the benefits of an improvement would be underestimated if attention is focused on hedonic results from sales prices (Niskanen and Hanke 1977). Freeman (1979a) has pointed out that this effect is not the complete story because property taxes are deductible and the income tax does not apply to imputed rents on owner-occupied housing. These issues should be considered in evaluating benefits, even though the net effect is probably small.

4.3.4 Other Econometric Issues

The specification of the hedonic equation can have significant effects on the estimates of the coefficients of the environmental variables, yet the theory does not generate a single specification that is unambiguously correct. The theory only suggests which types of variables should be included. Butler (1982) sought to discover if a change in the specification of the hedonic equation from a few key variables to a more extensive list had significant effects. He found that leaving out the less important variables had little effect on the coefficients of the key variables, but he suggested that if the interest was in a coefficient of a less important variable the more complete specification might be important. Graves et al. (1988) experimented with the effects of changes in specification on pollution coefficients. They found that as the specification changed the visibility coefficient ranged from positive and significant, as expected, to insignificant and of mixed sign. On the other hand, the particulate

---

[7] Linneman (1980) used both sales and rents from the Annual Housing Survey by converting both to "annualized housing expenditures," which includes property taxes and utility payments and used a capitalization rate to convert asset prices to a rental flow. Linneman ignored expected future changes in the houses and neighborhoods that would be incorporated in the sales prices. Sonstelie and Portney (1980) used similar techniques to measure "gross rent."

variable was reasonably stable, particularly when both pollution variables were in the equation. More research on specification, probably with Monte Carlo studies, is necessary before it will be possible to draw clear conclusions. Until then, care is warranted.

Just as the geographical extent of the market is an important issue, the temporal stability of the hedonic equation must also be considered. If there were active forward markets for houses, arbitrage would insure that the hedonic equation was relatively stable over time unless new information became available. However, such forward markets do not exist, so such stability is not guaranteed. On the other hand, aggregation of data from different time periods cannot be ruled out *a priori* either, so it is an empirical question. Edmonds (1985) found that hedonic regressions on two Japanese data sets from 1970 and 1975 could not be aggregated according to an F-test. Palmquist (1980a) found that aggregation over a 13-year period using time dummies was unacceptable, but aggregation over adjacent pairs of years could not be rejected. In general, it is probably wise to view aggregation over time with caution, but if the market has not received significant shocks during the time period, aggregation may be justified. Viewed from the other side, temporal segregation of markets may not produce significantly different hedonic price schedules for use in estimating the demand for characteristics.

A related issue is establishing the point at which an environmental change affects the real estate market. For example, are property values influenced when the location of a highway or hazardous waste site is announced, when construction begins, or when the use of the facility begins? Such questions can be addressed by using interaction terms between the proximity to the site and the time of the sale relative to the various information events (Michaels and Smith 1990).

Since hedonic regressions often involve large numbers of variables, a frequently expressed concern is that multicollinearity may degrade the estimates. If one of the coefficients is insignificant or has the wrong sign, multicollinearity provides a ready justification, but there have been few attempts to analyze whether near multicollinearity may actually have degraded the estimates. Palmquist (1983) applied the techniques refined by Belsley, Kuh, and Welsch (1980) to hedonic regressions in 14 cities with an extensive specification that included measures for four different air pollutants. The condition numbers and coefficient variance decompositions indicated that for only two of the pollutants in one of the cities was multicollinearity between pollutants potentially degrading. In another city there was evidence of degrading collinearity between a pollutant and a structural variable. Since more than 50 pollution coefficients were estimated, these results are quite encouraging. Among the nonenvironmental variables, the only cases of potentially degrading collinearity were among neighborhood variables or a variable and its square. These results indicate that multicollinearity may not always be as significant a problem as expected. However, there are cases where the problems will be severe. For example, in Los Angeles collinearity between air quality and beach proximity may make it impossible to separate the effects.

A final issue concerns whether the environmental effects on properties are fully captured in the effects on property values. It may be that environmental disamenities increase the amount of time a house is expected to be on the market in addition to reducing its expected selling price. This increase in expected selling time could be optimal for sellers if the dispersion of potential offers was increased by the disamenity, but other distributions of potential offers resulted in identical expected selling times (Palmquist 1980b). Most of the empirical work on this question has concerned highway noise (Nelson 1982). The consensus so far is that selling time is not affected significantly by environmental disamenities; therefore, the hedonic effects do fully capture the impacts.

4.4 Estimating the Willingness-To-Pay or Demand Equations

While the theoretical hedonic model developed by Rosen (1974) has proven invaluable to researchers in the field, his suggestions for implementing the estimation of the willingness-to-pay or demand equations have been somewhat misleading. He proposed that the bid and offer functions be estimated as a simultaneous system using shifting offer or supply curves to identify the bid or the demand functions. As will be described in the next paragraph, the use of the offer functions in estimating the bid functions is inappropriate when disaggregate data are used. However, even without the supply side, the problems of identification and endogeneity remain. Rosen's proposal and early attempts by others to implement it are described in the next subsection. The following subsections describe the current understanding of the estimation problems and solutions.

4.4.1 Rosen's Estimation Procedure

After Rosen (1974) discussed the difficulties in deriving closed-form solutions and the amount of structure that must be imposed, he suggested an alternative strategy for estimating marginal bid and offer functions. He had shown the marginal price at a point on the hedonic schedule is equal to the marginal bid of the individual occupying that location. Similarly, the marginal price is equal to the marginal offer of the individual or firm supplying that location. If there are differences among individuals with respect to socioeconomic characteristics and differences among firms with respect to technology and input parameters, then there will be many different bid and offer functions that will determine the equilibrium price schedule. The equilibrium marginal prices will vary along the nonlinear hedonic schedule because of the different bid and offer functions. Rosen suggested that this price variation would allow estimation of the bid and offer functions. Unfortunately for Rosen's suggestion, the bid functions depend on the level of utility that is impossible to observe, and the offer functions depend on profits that are difficult to observe. Rosen simply dropped those two variables from the functions and treated the consumers' equations as uncompensated inverse demands, but as was discussed earlier, this is incorrect.

A few economists tried to implement Rosen's proposed strategy by simultaneously estimating supply and demand for characteristics or by assuming the supply was perfectly inelastic. However, their attempts failed because with disaggregate data there is no need to consider the supply side in the estimation if it is assumed that individual consumers have no market power (Diamond and Smith 1985). This seems like a reasonable assumption in most real estate markets. However, the identification problem to be discussed in the next section is still present. Identification is also a problem when aggregate data are used, whether the supplies are assumed to be fixed or variable. Consideration of the supply side will not necessarily alleviate the problem and may make it worse.

4.4.2 Identification

Rosen's (1974) suggestion for estimating demand functions for characteristics depended on the use of a nonlinear hedonic price function, which generates varying marginal prices. If, instead, he had used a linear hedonic function, which generates only one marginal price, other researchers would not have been tempted to estimate marginal bid functions. However, even with varying marginal prices, all consumers within a market face the same equilibrium price schedule. The observation of a consumer's behavior based on that price schedule provides one point on the consumer's marginal bid function. The other marginal prices are observed only for individuals with other socioeconomic characteristics, and these provide no information on the original consumer's bid for different quantities of the characteristic. Without further structure or data, it is impossible to distinguish between the equilibrium marginal price schedule and the consumers marginal bid functions.

These points can be seen graphically in figure 4.2. The upper diagram reproduces part of figure 4.1. If an economist uses data from a single market, he or she can observe the hedonic price schedule, the bundles of characteristics purchased by the consumers, and the socioeconomic attributes of those consumers. This means that he or she only observes one point on the bid contour for each consumer. There is an infinite number of potential bid contours that are consistent with the observed data. Two contours — one solid line and one broken — are shown for each individual. The lower diagram

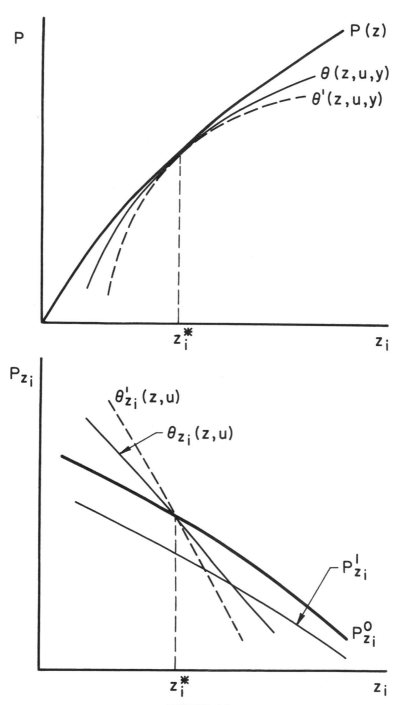

FIGURE 4.2
The identification problem in a single market.

shows the same difficulty for marginal prices and marginal bids. Here it is easy to see that if different marginal price schedules can be observed because of spatially or temporally separated markets, the bid function can be identified.

This point was raised in Brown and Rosen (1982), where they used a quadratic hedonic price equation and linear inverse demand and supply equations to show that coefficients in the second stage would be simple functions of the coefficients in the first stage with no new information.[8] The only ways to identify the second-stage equations would be to impose additional structure on the system of equations or use multiple markets that would generate multiple equilibrium price schedules.

Both McConnell and Phipps (1987) and Epple (1987) emphasized that the system of equations to be estimated included the hedonic equation as well as the demands for the characteristics. However, identification of the hedonic equation is usually provided by exclusion restrictions generated by the theory. Income and the vector of socioeconomic characteristics affect the demand or bid functions but do not enter the equilibrium price equation, so this latter equation is generally identified.[9]

Identification of the demand or bid equations is more difficult. Most of the earlier studies achieved identification (without explicitly considering the issue) by restricting the functional forms or the variables. This can occur if the functional form used in the second stage is markedly different from the functional form of the marginal price equations or if most of the housing characteristics are not included in the bid functions. However, the validity of the results still hinges on the accuracy of the assumptions. Furthermore, identification becomes more difficult if flexible functional forms are introduced in the second stage (e.g., Bender, Gronberg, and Hwang 1980). The exact conditions necessary for identification of the second-stage equations within a single market are considered in McConnell and Phipps (1987), both when the model is linear in the parameters and when it is nonlinear.

Several studies have been devoted to identification issues in a single market setting. Quigley (1982) assumed that the residents had identical generalized constant elasticity of substitution (GCES) utility functions, which is a homothetic functional form. This allowed him to estimate the parameters of the utility function in a single market because the marginal characteristics prices varied. He knew that individuals maximize utility by equating the marginal price to their marginal rate of substitution between the characteristic and the composite good. Homotheticity allowed the consumers' choices to be scaled so that they became observations along a common indifference curve. In

[8] As McConnell and Phipps (1987) correctly noted, these demands could not be integrated back to a quasi-concave utility function.

[9] A different issue was addressed by Epple (1987) in discussing identification of the hedonic equation. If some of the characteristics in that equation are measured with error, an errors-in-variables problem arises. One means of dealing with this is to develop instrumental variables. Epple's results on identification of the hedonic equation require that such instruments be available.

effect, this is how Quigley estimated the parameters of the utility function. Kanemoto and Nakamura (1986) used a similar technique, but instead of specifying the function form of the direct utility function, they specified a particular quadratic form for the bid function. This allowed them to identify the parameters of the utility function. Kanemoto and Nakamura compared their results with Quigley's. Though the mean results were quite similar, the results away from the means differed significantly. This made it impossible to compare the validity of the two results. Horowitz (1987) showed that Quigley's method of using GCES utility functions for identification does not work for all of the possible functional forms of the hedonic equation.

Whereas many of the earlier studies used functional forms for the hedonic equation and for demand or bid equations so that the second-stage estimates were not simply regenerations of the hedonic results, Mendelsohn (1985) developed a possible justification for those functional form restrictions. He pointed out that within a single market, individuals choose different bundles, and therefore different marginal prices, because of differences in their socioeconomic characteristics. Thus, in addition to the direct functional relationship between marginal price and housing characteristics, there is an indirect relationship through the socioeconomic characteristics. He used this relationship to justify including nonlinear terms in the hedonic equation even though they are not present in the bid equations. This may make identification possible. Bartik (1987a) subsequently showed that if there are unobserved socioeconomic characteristics, Mendelsohn's technique does not identify the demand parameters.

Finally, an interesting technique for achieving identification within a single market is used by Cassel and Mendelsohn (not dated). They assumed that every city comprises a single market and thus has a single price schedule. However, some characteristics vary geographically throughout the city. For example, depending on the location of a person's workplace, the cost to the individual of obtaining an amenity may vary. This happens because the cost of the amenity includes the housing price as well as the costs of commuting. Cassel and Mendelsohn's variation in the cost schedule, or total price schedule, between individuals does allow identification without other restrictions;[10] however, the data requirements for this technique are greater than those for typical hedonic studies.

The alternative to attempting to identify the demands for characteristics in a single market is to use data from markets that are separated either spatially or temporally. Interestingly, this technique was suggested by Freeman (1974b). However, since most research in the 1970s was heavily influenced by Rosen's (1974) estimation proposal, Freeman's suggestion was not used until the 1980s.

---

[10] This procedure is closely related to the hedonic travel cost technique developed by Brown and Mendelsohn (1984) and is thus subject to similar potential problems since the price schedule is not the result of equilibrium market forces (see Smith and Kaoru 1987; chapter 8). A method of overcoming these problems is contained in Smith, Palmquist, and Jakus (1989).

The difficulty with identification in a single market is that all consumers face the same equilibrium price schedule. If separate markets are available with separate hedonic equations, then the researcher obtains the necessary variation in the price schedules to which consumers are reacting. It is necessary to assume that consumers are similar between markets if the researcher controls for socioeconomic differences between individuals.

Although many economists have been recommending the use of multiple markets for identification purposes recently, there have only been a limited number of studies implementing this method. Palmquist (1982a, 1983, 1984) has used several data sources to compile a dataset for a number of standard metropolitan statistical areas (SMSA). These data were used to estimate the demand for various characteristics of housing, including air quality. Using the Annual Housing Survey for seven cities and two years, Parsons (1986) has estimated the Almost Ideal Demand System for four attributes. Two of these attributes are based on responses to attitudinal questions on house and neighborhood quality. Bartik (1987b) used data from the Demand Experiment of the Experimental Housing Allowance Program for two cities during a four-year period to estimate the marginal bid for neighborhood quality. Finally, Bajic (1985) used market segmentation within a single city, Toronto, to provide the multiple markets for his study.

Ohsfeldt and Smith (1985) conducted Monte Carlo experiments to determine how much exogenous price variation is necessary to accurately estimate the structural parameters. Not surprisingly, they found that increasing the amount of exogenous price variation increases the reliability of the parameter estimates. Thus, multiple markets are not sufficient for identification unless the hedonic equations differ significantly between markets.

4.4.3 Endogeneity in Hedonic Models

In Rosen's (1974) model, both the marginal prices and quantities of the characteristics are endogenous because they are determined by the interactions of demanders and suppliers. As was discussed earlier, this source of endogeneity can be ignored because the offer or supply functions need not be considered when the data are disaggregate. However, if the equilibrium price equation is nonlinear, then the marginal price and quantity of a characteristic are determined simultaneously. Referring back to figure 4.2, when a consumer selects a quantity of a characteristic to consume, the marginal price paid is simultaneously determined and vice versa because the hedonic price schedule is parametric to the individual. And since price depends on quantity, price is correlated with the error term in the equation explaining quantity demanded. The converse holds for equations explaining prices or bids. Thus, ordinary least squares coefficient estimates will be inconsistent. This point was raised by several economists, including Murray (1983), Mendelsohn (1984), Palmquist (1984), Diamond and Smith (1985), and Epple (1987). The answer economists usually propose is to find instruments for the endogenous variable. The difficulty here is finding instruments that are truly exogenous.

For example, the quantities of other characteristics cannot be used to create instruments because they are also part of the simultaneous decision and thus are correlated with the error term. Socioeconomic characteristics can be used to develop instruments if they are not correlated with unobserved "taste" variables (Palmquist 1984; Epple 1987). However, care is necessary. If the budget constraint has been linearized to allow the estimation (Murray 1983; Palmquist 1984; Mendelsohn 1984), then the adjusted income is no longer exogenous.[11] Instruments for adjusted income as well as characteristics' prices and quantities are necessary. Also, if there are significant unobserved socio-economic characteristics that may be correlated with the observed ones, then the observed characteristics may not used as instruments (Bartik 1987b). Finally, if both supply and demand are endogenous, then only some of the characteristics of demanders and suppliers can be exogenous (Epple 1987). One of the best sources of exogenous variables is the existence of multiple markets. Such markets can provide instruments that are associated with the different locations or time periods (Palmquist 1984; Bartik 1987b; Kahn and Lang 1988).

4.4.4 Sources of Error in the Estimation

There are a variety of possible sources of error in the estimation of a hedonic equation and the demand or bid equations. Many of these have already been discussed, but Epple (1987) provides a useful listing. Three of the potential sources of error arise from problems with measurement and can lead to errors-in-variables problems. The first of these is errors in the measurement of product (housing) price. This measurement can be reliable if the data are disaggregate and record actual sales prices, but if the data are aggregated or use estimates of the value of houses, then this source of error must be considered. One possible solution for this type of error is to develop instruments for those variables. In some cases, the hedonic equation must be estimated simultaneously with the rest of the system (Epple 1987).

A second source of error is measurement error in product characteristics. Data on structural characteristics are usually accurate, but the same is not always true for neighborhood and environmental characteristics, as was discussed in the section on hedonic estimation. The third source of error is

---

[11] Mendelsohn (1984) suggested that there is usually little difference between actual income and adjusted income, and he recommended using actual income. However, housing expenditures make up a large fraction of most household budgets, and thus the nonlinearity adjustment is likely to be necessary (Palmquist 1984).

the measurement of the characteristics of the demanders and suppliers of the product.

Two other sources of error can arise if there are unobserved product or agent characteristics. These errors can not only cause bias because of the variables that were left out, but also can cause the observed socioeconomic characteristics to fail as instruments. A sixth source of error (which is discounted by Epple 1987), may have a significant effect on estimation results. These errors arise because of a difference between the true and the estimated functional form for the hedonic equation. This type of error cannot be ignored because, as was discussed, different functional forms can result in vastly differ results for environmental variables. A final source of error may be asymmetric information on the part of various consumers or consumers and sellers, but as yet this has not been treated within the hedonic framework.

4.5 Benefit Measurement with the Hedonic Model

One of the very earliest works to apply hedonic techniques to valuing the benefits of environmental improvements was Ridker and Henning (1967). This paper was innovative, but it also generated a debate that continued for years. The hedonic equation estimated by Ridker and Henning contained a measure of sulfation levels. In their conclusions, they used the coefficient estimate to calculate the effect of a change in the sulfation level on the value of a house and the value of all properties in the St. Louis area, which they used as a measure of willingness to pay. Freeman (1971) was the first to point out that without a formal model it is difficult to derive conclusions about the welfare effects of an environmental policy. Such models were later provided by Rosen (1974) and Freeman (1974b). As a result of the many contributions on hedonic benefit measurement since that time, there is now a better understanding of the conditions under which the hedonic equation can be used alone for welfare measurement and when additional information is necessary.[12]

A number of conditions determine which technique must be used. These conditions include the size of the environmental improvement, the extent of the area affected by the improvement, and the effect of the improvement on the equilibrium price schedule. Different techniques are used to forecast the benefits of a policy than are used to measure benefits after an environmental improvement has taken place. Benefits also depend on whether or not residents move in response to the environmental change.

[12] A good synthesis of the early debate is contained in Cobb (1977).

4.5.1 Welfare Measurement Using the Hedonic Equation

Freeman (1974a) and Small (1975) were the first to show that the hedonic equation could be used to measure people's *marginal* willingness to pay for an environmental improvement. This is a direct result of the hedonic model because in equilibrium all consumers equate their marginal rate of substitution between the environmental good and the *numeraire,* or marginal willingness to pay for the environmental good, to the marginal price of the environmental good. The hedonic equation reveals the marginal price and thus marginal willingness to pay for environmental quality.

This point is of some theoretical interest, but environmental policy is generally intended to result in nonmarginal improvements in the environment. Even with nonmarginal improvements it is still possible to use the hedonic equation for exact welfare measurement when the equilibrium price schedule is not changed by the policy (Palmquist forthcoming). This may be true when the environmental change affects only a small number of properties relative to the size of the market. In this case the price of properties experiencing the environmental improvement will rise to the level of prices of properties with similar characteristics, but the hedonic equation will remain unchanged. If one thinks of consumers as renting from absentee landlords, it is easy to analyze the welfare effects. The consumers who live in the houses that are to be improved probably move to similar, but unimproved, houses with the same rent. They will reach the original level of satisfaction if their moving costs are zero. Similarly, the improved houses will be occupied by individuals who had lived in comparable houses. These individuals also will be as well off as before, so that the effect on these consumers also will be zero. Landlords of the affected houses, on the other hand, will receive an increase in rent because of the improvement. The amount of this increase can be determined from the original hedonic equation. The change in rents captures all of the benefits under these circumstances.

The assumption of absentee landlords can easily be relaxed. If the houses are owner-occupied, then the improvement provides a capital gain for the individuals living in the improved houses. However, since they are a small part of the market, the hedonic equation is still uninfluenced and the analysis proceeds as before.

It is more complex to relax the assumption that moving costs are zero. If the hedonic equation remains unchanged, the landlords will gain as before, whereas the residents will be made worse off. This happens because the residents must pay for an improvement they had previously chosen to forfeit. They could have rented a house in an area where the environmental conditions were better, but they did not because of the higher rent. When they are forced to do so, they are made worse off. Therefore, their loss in welfare must be

subtracted from the welfare gained by the landlords.[13] If the price schedule
changes because moving costs prevent some movement to and from other
areas, then welfare measurement is more complex. The techniques discussed
in the next section will be relevant here.

Another assumption that could be made in estimating the benefits of
environmental improvements from the hedonic equation is that the consumers
are identical. If there are no socioeconomic differences among the residents,
everyone will have an identical bid function. The equilibrium price schedule
will coincide with these common bid functions, so estimating the hedonic
equation reveals the bid equation. Discussion of this possibility will be
postponed until the next section because it is common in models that have
evolved from urban location models. The practical application of such models
to benefit measurement may be open to question.

4.5.2 *Welfare Measurement Using Bid or Demand Functions*

When an environmental change affects a large number of people, the hedonic
price function will change and the level of utility of individuals may change.
In this case, the hedonic equation alone is inadequate for estimating the
benefits of the improvement. Some knowledge of the consumers' preferences
is necessary for measuring willingness to pay. In addition, a knowledge or a
forecast of the change in the hedonic price equation is often required for
exact measurements of welfare.

Early studies (Harrison and Rubinfeld 1978; Blomquist and Worley 1981)
sought to estimate a marginal willingness-to-pay function for environmental
quality and then integrate this function between two levels of environmental
quality to derive benefits. However, if the problems in estimating the marginal
willingness-to-pay functions are ignored, at least two problems with this
technique remain. First, for the technique to work, economists must assume
that consumers do not move in response to the environmental change. Second,
they must assume that the welfare measures derived from the uncompensated
willingness-to-pay functions provide a close approximation to the true measures
that would be derived from compensated functions. Relaxing these assumptions frequently requires using the change in the price schedule.

Parsons (1986) gave examples of calculations of welfare change in the
special case when the hedonic equation is linear and the price change is
known. The environmental change has two effects: the environmental quality

---

[13] Palmquist (1990) showed that the transaction and the moving costs, which are relatively
easy to quantify for localized externalities, always provide an upper bound and often provide
an exact measure for the loss to tenants from remaining in a suboptimal location after the
environmental improvement. This measure must be combined with the welfare effect on the
landlord to get the total effect. Bartik (1986) considered a related issue having to do with
benefit-cost analysis of neighborhood revitalization when the residents develop strong ties to
their existing neighborhoods and thus face "psychological" moving costs.

at some houses is improved and the prices of houses are changed. It is reasonable to assume that the improvement in environmental quality reduces the price of environmental quality. The residents as consumers of housing services gain because they pay lower prices for the original units of environmental quality; furthermore, they would have been willing to pay more than they had to for the additional units. Given constant marginal prices because of the linear hedonic, their gains can be represented by the area to the left of the compensated demand for environmental quality between the two prices. However, the residents as the owners of the houses are also affected. The implicit rent is lower on the original units of quality, but rent is gained on the new units of quality. If the residents choose to stay in their same houses after the environmental change, then their net welfare measure is the area beneath the compensated demand between the two quality levels.

In general, the change in the price of environmental quality will cause individuals to desire more or less quality than they receive at their original houses after the change. If moving costs are low enough, they will relocate to gain additional benefits. Thus, the welfare gains will exceed the area beneath the compensated demand between the two quality levels at the original houses. If the compensated demand has been estimated and the price change correctly forecast, it is possible to measure this welfare effect exactly.

Since the hedonic equation is generally nonlinear, forecasting the change in the price schedule is quite difficult.[14] For this reason it is useful to separate the analysis into *ex post* welfare measurement, which is retrospective, and *ex ante* welfare measurement, which is prospective. *Ex post* welfare analysis assumes that the hedonic price schedule is known both before and after the environmental change. However, there is still a problem in welfare measurement since marginal prices are not parameters in the demand equations (McConnell and Phipps 1987). The individual takes the parameters of the equilibrium price schedule, rather than the marginal prices, as given. Standard duality results are no longer available for welfare measurement.

Fortunately, Palmquist (1988) demonstrated that if the budget constraint is linearized at the chosen bundle, it is possible to derive pseudo-Marshallian demands that depend on marginal prices. These are not equivalent to the Marshallian demands that depend on the parameters of the hedonic equation, but they can be estimated from observed data. The usual duality techniques can be used to transform the estimated pseudo-Marshallian demands to a pseudoexpenditure function for welfare measurement. However, while the pseudoexpenditure function accurately represents preferences, the resident's willingness to pay will depend not only on the difference in the pseudoex-

[14] In the usual *ex ante* welfare analysis with parametric marginal prices, it is necessary to forecast the prices after a policy change. While this may be difficult, forecasting a single price is considerably easier than forecasting an entire schedule, as is required here.

penditure function but also on the differences in the inframarginal characteristics prices. The equations for compensating and equivalent variation are:

$$CV = y^1 - y^0 + e(p_1^0, \ldots, p_n^0, p_x^0, u^0) + P^0(z^0) - \Sigma \, p_i^0 z_i^0$$
$$- e(p_1^1, \ldots, p_n^1, p_x^1, u^0) - P^1(z^1) + \Sigma \, p_i^1 z_i^1$$
$$EV = y^1 - y^0 + e(p_1^0, \ldots, p_n^0, p_x^0, u^1) + P^0(z^0) - \Sigma \, p_i^0 z_i^0$$
$$- e(p_1^1, \ldots, p_n^1, p_x^1, u^1) - P^1(z^1) + \Sigma \, p_i^1 z_i^1,$$

where superscripts 0 and 1 denote the original and subsequent time periods respectively and $e(\cdot)$ is the psuedoexpenditure function. If the owners of the houses are unable to vary any of the characteristics of houses in the response to the environmental change, then there is simply a change in rents that can be combined with the consumer's willingness to pay to derive the total welfare change. If, more realistically, the owners can vary other characteristics when the environment changes, then the nonlinear price schedule will affect producer welfare measurement. A similar process of linearization enables a psuedovariable profit function to be estimated, and the results can be used for exact welfare measurement here as well (Palmquist 1988).

Ex ante welfare measurement is commonly required. But an exact welfare analysis is difficult if the equilibrium price schedule after the environmental change is unavailable. A lower bound on the benefits was developed by Bartik (1988b) and Palmquist (1988). Bartik (1988b) showed that the true benefits equal the sum of four factors: (1) residents' willingness to pay, by households, for an improved environment if they do not move; (2) the cost savings, if any, to landlords at improved sites; (3) the profit gains, if any, to landlords if they alter housing supply; and (4) the household utility gains, if any, from moving. Because the last three elements are all greater than or equal to zero, the sum of the households' willingness to pay for improvements at their original location is a lower bound for the total of the four sources of willingness to pay.

Palmquist (1988) followed a different path in developing this result. His technique may be useful in the second-stage hedonic estimation. Palmquist reasoned that just as preferences that are continuous, quasiconcave, and increasing can be represented by either the direct utility function, the indirect utility function, or the expenditure function, they can also be represented by the distance function. The distance function depends on quantities of the characteristics and the level of utility, and can be thought of as representing people's willingness to pay for a vector of characteristics given a level of satisfaction. The parameters of the distance function can be estimated by transforming it to the direct utility function and using the Hotelling-Wold theorem to derive the uncompensated inverse demands. The estimates for these demands can be used to calculate the distance function parameters. The welfare measures, quantity compensating variation and quantity equivalent variation, are easily calculated from the distance function. It also is possible

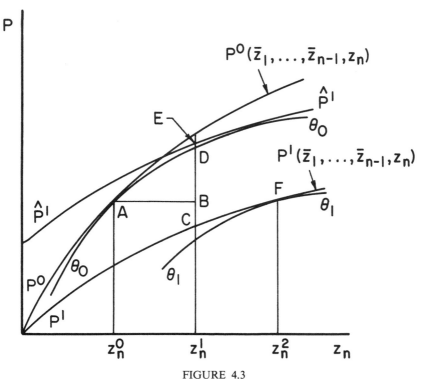

FIGURE 4.3
Welfare measurement when the hedonic schedule changes.

to derive the compensated inverse demands for characteristics by differentiating the distance function using the Shephard-Hanoch theorem. Since the quantity compensating and equivalent variations hold the quantities of other housing characteristics fixed, they represent a lower bound for willingness to pay when moving is possible.[15] If moving costs prevent individuals from moving in response to environmental changes, then these measures are exact.[16]

It is also possible to develop this lower bound in terms of the bid function. Figure 4.3, which is reproduced from Palmquist (1988), shows an initial hedonic price schedule $P^0(z)$ with respect to an environmental characteristic z_n. An individual maximizes satisfaction by choosing to consume z_n^0 units of the characteristic at point A, where the bid function θ_0 is tangent to the price schedule. An environmental improvement increases the exogenously determined level of z_n to z_n^1 in the original house and shifts the price schedule to $P^1(z)$. If the consumer can change houses, he or she will move to F and

[15] Similar techniques can be applied with a differentiated factor of production, such as land, using the variable profit function (Palmquist 1989).

[16] This assumes that the landlords do not make changes in other characteristics, but because they cannot attract new tenants, this may not be unreasonable.

consume z_n^2 units. The compensating variation can be found by constructing price schedule $\hat{P}_1$, which is parallel to P^1 and tangent to θ_0. The vertical distance between $\hat{P}_1$ and P^1 (for example, CE) is the compensating variation. The change in rents is the negative of distance BC, so the change in welfare is distance BE. However, the quantity compensating variation for the change from z_n^0 to z_n^1 is distance BD. It is clear that BD is a lower bound for BE. As discussed in Palmquist (1988), this result is robust.

Bartik (1988b) discussed the possibility of developing an upper bound for the true welfare measure by using the hedonic equation before the environmental change. He found that the hedonic equation sometimes did not provide an upper bound when landlords could adjust their housing supply. However, Bartik argued that under most circumstances the hedonic would provide an upper bound on benefits. This technique would be very useful because it would eliminate the second-stage estimation, which is demanding with respect to data and estimation. This issue deserves further attention.

4.6 Environmental Benefit Measurement in Related Land Value Models

Various theoretical contributions on land values and benefit measurement do not use the usual hedonic methodology. Nonetheless, this research is closely related and reaches many of the same conclusions as did those described in the preceding section. For the purposes of this chapter, these diverse contributions have been loosely divided into two lines of research. The first had its origin in the essay by Strotz (1968) and was greatly influenced by Lind (1973). This research sought to develop techniques for using changes in land rents to measure or approximate the benefits of public programs. The second approach modified urban economics models in an effort to discover whether land rent changes due to public policies can be predicted and used to measure willingness to pay, and whether the demand for amenities can be calculated from rent gradients. Much of the early work in this area was by Polinsky, Shavell, and Rubinfeld.

4.6.1 Rent and Benefit Measurement

Strotz (1968) apparently was the first to attempt to develop a model of how land rent changes can be used to measure the benefits of land improvements. He considered a polluted urban area in which an abatement program cleaned up the north half of the city. The marginal benefit measure that he derived indicates that one must add the rent increases that landlords receive in the north and the rent reductions that tenants enjoy in the south. However, this depends crucially on the equality of the number of parcels of land in the north and the south. This necessary assumption and several others limit the applicability of the Strotz model.

Lind (1973) addressed similar questions in a more general framework using an assignment model. Lind's model has a fixed number of parcels and many production and consumption activities, each of which has a bid for each of the parcels. The parcels are occupied by the activity with the winning bid. Improvements occur on some of the parcels. In response, the activity with the winning bid for any parcel may change, so there may be a series of activity relocations. One of Lind's important results was showing that the benefits of an improvement net out on parcels that are not directly affected by the improvement. All that is required is to consider the change in the profits to both the activities that stay on the improved land and to those that make moves involving the improved land. This change in profits calculated at the old prices (new prices) is an upper bound (lower bound) on the benefits of the land improvement. The other important result from Lind (1973) was that the net change in rents on the land directly affected by the improvement provides an upper bound for the benefits if there is no change in profits or consumer surplus.

Freeman (1975) pointed out that the zero-surplus assumption generally cannot hold if there is a change in the rent schedule; therefore, the change in rents on the improved land may not provide the upper bound. However, a suggestion by Freeman, which is proved for the general case by Lind (1975), is quite useful. A comparison of the rent for an unimproved site with that for a comparable site that has already been improved provides an upper bound for the benefits of the improvement. The hedonic analog of this would use the hedonic equation to provide an upper bound on benefits. However, the Lind model does not allow for such adjustments as the landlords varying the characteristics within their control in response to the environmental improvements (Bartik 1988b).

Pines and Weiss (1976) used a model similar to Strotz's in dealing with the question of valuing land improvements. As in Strotz (1968), they divided area into two parts with everyone owning land in both parts. Pines and Weiss assumed individuals live in one area only and are endowed with equal amounts of the composite good. Their *ex post* measure of benefits combined the changes in people's incomes due to changes in rents in the two areas and the changes in people's housing costs due to the changes. Their measure is only approximate and can be an over- or underestimate of the true benefits. It is exact only when the improvement is purely land augmenting. Pines and Weiss (1976) also discussed *ex ante* benefit measurement, but they only considered marginal changes. They showed, as did Freeman (1974a,b) and Small (1975), that the marginal benefits of an improvement equals the marginal change in the rent gradient.

A recent study (Kanemoto 1988) also fits with this group even though the term "hedonic prices" is in the title. The identical consumers again derive

utility from a composite good, lot size, and an amenity. The various characteristics of housing can only be incorporated in the composite good if the
hedonic equation is linear. This may be true in a long-run equilibrium, but
certainly not in the more relevant short run. Kanemoto's model is similar to
those in Strotz (1968) and Pines and Weiss (1976). His main result is that
net benefits estimated using preimprovement land rents will exceed an
equivalent variation measure of the true benefits. The "hedonic" benefit
measure used is the amount of land to be improved multiplied by the
difference before the improvement in rents on already improved land and the
land to be improved. This is similar to the upper bound developed by Freeman
(1975) and Lind (1975).

4.6.2 Urban Spatial Models

A related and concurrent line of research uses a typical model of an urban
area with a central business district (Polinsky, Rubinfeld, and Shavell 1974;
Polinsky and Shavell 1976; Polinsky and Rubinfeld 1977). The amenity —
air quality — improves with distance from the central business district in a
city with a fixed boundary. The identical individuals receive utility from a
composite good, land (housing), and the amenity, while they spend their
exogenously given income on the first two items and transportation. The
researchers consider four questions when there is a change in the amenity:
How will land rents be affected by the change? Can the new rent schedule be
predicted? Can willingness to pay for the amenity be estimated? and, Can the
demand for the amenity be estimated? The answers to these questions depend
on the type of city.

If the city is "open," so people can migrate into and out of the city, and
"small," so prices are exogenously fixed elsewhere in the system, then an
increase in the amenity will cause rents to increase for affected land but rents
elsewhere will not change. The new rents can be predicted from the initial
rent gradient and the benefits, which all accrue to the absentee landlords, are
measured by the increased rents. The demand for the amenity can only be
estimated if additional structure is provided. For example, if the common
utility function is assumed to be Cobb-Douglas, the amenity demand function
can be derived from the rent gradient.

If the city is "closed," so the population is fixed, then the rent schedule
will change. Aggregate rents may increase, but not necessarily. For example,
with the Cobb-Douglas utility function and a uniform improvement in
amenities, rents will be unchanged. In general, it will not be possible to predict
the new rent schedule. Because there may be a change in the utility level of
the residents, it is impossible to measure willingness to pay even if the new

rent schedule is known. The demand for the amenity can be predicted under exactly the same conditions as in the small, open city.

In Polinsky, Rubinfeld, and Shavell (1974) and Polinsky and Rubinfeld (1977), they also considered cases in which there are various classes of consumers. This makes their model more general. However, since they derive the rent schedule from the indirect utility function of a typical consumer, they can only derive a rent schedule that applies to properties occupied by the same class of consumers. Since a given class will occupy only a small fraction of the land in the city, this may be a significant restriction.

Cobb, Barkume, and Shapiro (1978) in a comment on Polinsky and Shavell (1976) suggested that the rent gradient contained more information on consumer preferences than Polinsky and Shavell suggested with their Cobb-Douglas example. They said this happens because the slope of the rent schedule with respect to the amenity is equal to the marginal rate of substitution of the amenity for the composite good. And since all residents are identical and achieve the same level of satisfaction, the rent schedule can be used to recover an indifference surface for a typical consumer. However, as Polinsky and Shavell (1978) pointed out in their reply, the knowledge of one indifference surface is inadequate to measure benefits in a closed city where the residents' levels of satisfaction can change. They said that some structure, such as the assumption of homotheticity, would still be necessary.

The papers by Scotchmer (1985, 1986) can be considered a part of this line of research, particularly the latter paper, although they are clearly related to the "Land Rent and Benefit Measurement" studies as well. Scotchmer considers her work to be in the hedonic tradition, although she considers only amenities, lot size, and a composite of other goods. Her generally long-run perspective is used to aggregate all of the structural characteristics of the houses into a single measure. This housing aggregate has an exogenous price determined by the hedonic equation, which is linear in the long run when costless repackaging is possible. The housing measure can then be aggregated with all other goods to obtain the single composite commodity. Scotchmer's link to the Rosen (1974) model is through lot size. Rosen considered the case in which the individual consumes more than one unit of the differentiated product. This modification is straightforward if all units consumed are the same (Houthakker 1952). Scotchmer considered land to be the differentiated product (differentiated by the amenity) and the number of units of land consumed to vary at a constant price.

Her contention was that a cross-section of land prices within a single market cannot capture people's willingness to pay for differences in amenities when lot size varies with amenities, even if there is a homogeneous population. If the lot price is per unit of land, the problem is that the number of people per unit of land varies with location. If the lot price is per dwelling, then the difference in price represents both the difference in the amenity and the difference in lot size. In the short run of her model (maintaining the linear hedonic equation assumption), if the amenity changes without lot sizes being varied, it may not be possible to observe that combination of lot size and amenity. However, if the expenditure function is separable in the amenity and lot size, then it is possible to estimate willingness to pay by estimating both the lot price gradient and the population density gradient with respect to the amenity. Scotchmer's model cannot allow for demographic differences among residents.

Many of the issues Scotchmer raised are closely related to issues in a more traditional hedonic framework. In the latter framework, lot size and the amenity are simply additional characteristics of housing that influence the equilibrium price. Combinations of lot size and amenity may vary because the other characteristics vary as well. As Scotchmer noted, the rent gradient, but not the willingness-to-pay function, can be observed within a single city (unless further assumptions are made). This is analogous to the identification problem in hedonic models, where there are methods, such as multiple markets, for solving it. Whether or not the long run is the correct framework for modeling real estate market responses to environmental changes is an important question.

4.7 Benefit Measurement with Hedonic Wage and Property Value Models

Recently, there have been several models developed that attempt to combine the insights of hedonic wage models and hedonic property value models (Cropper and Arriaga-Salinas 1980; Cropper 1981b; Roback 1982, 1988; Hoehn, Berger, and Blomquist 1987; Blomquist, Berger, and Hoehn 1988). These models are an outgrowth of Rosen (1979). In that article Rosen pointed out that it is necessary to consider both the supply and demand for labor services in explaining the outcomes of hedonic wage studies. Previous studies had focused on the disamenities of urbanization that required higher wages to attract workers to urban areas. But these models did not explain the trends toward urbanization that have dominated in this century because they gave no reasons why firms prefer to locate in urban areas nor how they can pay the higher wages. Rosen also indicated that an individual's work decisions will be influenced by land rents in an area. In the simple model Rosen chose to use, he could focus on either the real wage gradient or the rent gradient to obtain the same information.

The later authors generalized Rosen's model in several directions with the result that both gradients are relevant. Individuals' willingness to pay for an increase in an amenity will be partially reflected in the labor market and partially in the land or housing market. Assume that individuals derive satisfaction from the consumption of a composite commodity x, land or housing L, and an amenity A according to a utility function $U(x,L,A)$.[17] They receive a wage per unit time w. By adjusting the time units, w is an individual's income per period.[18] This is spent on the composite good (the price of which is normalized to one) and on L (the price of which is r). The indirect utility function that results from the constrained utility maximization problem is $V(w,r;A)$. The models almost universally assume that all individuals are identical, and thus all achieve the same level of satisfaction, $u^0 = V(w,r;A)$.

The composite good is produced by identical firms using a constant returns-to-scale technology that uses labor and sometimes other factors of production. The firms may be influenced directly and indirectly by the amenity through factor prices. In some of the models, the technology also uses land or capital, and there may be agglomeration economies. These modifications do not affect the basic conclusions. The resulting unit cost function is, for example, $C(w,r,A)$, which is equated to the product price that can be normalized to unity. This equation and the indirect utility function equation are central to the models, which are then closed by making various assumptions about whether or not the boundaries of the cities are fixed or are determined endogenously. This system of equations is differentiated with respect to the amenity, and the matrix equation is solved for dw/dA and dr/dA. Each of these depends on the effects of w, r, and A on both utility and costs.

The models can be used for measuring marginal willingness to pay for improvements in the amenities by totally differentiating the indirect utility function, solving for V_A/V_w, and then using Roy's identity,

$$V_A/V_w = (-V_r/V_w)dr/dA - dw/dA$$
$$= L\, dr/dA - dw/dA.$$

Hedonic equations for housing and wages are estimated and used to estimate the derivatives in the equation above. The theoretical model could help in determining the specification of the estimating equations, although this would be quite complex and has not been done.

This model shows that welfare measures may have to consider both wages and property values. An interesting question becomes, what is the relative importance of the labor and land markets in the measure? Bartik and Smith

[17] In Cropper and Arriaga-Salinas (1980) and Cropper (1981a) the utility function and the production function are given specific Cobb-Douglas functional forms. This model also allows the amenity to vary linearly with distance from the central business district within an urban area. The model is related to that in Polinsky and Shavell (1976). The other authors use unspecified functional forms until the empirical applications and hold constant the amenity within urban areas or counties.

[18] In Hoehn, Berger, and Blomquist (1987) and Blomquist, Berger, and Hoehn (1988) commuting costs are incorporated, so income is reduced by these costs, which are dependent on location. The others do not include these costs.

(1987) showed that under plausible circumstances the relative impact of a change in amenities on wages and rents depends on the relative share of land and labor in production costs. When labor's share is larger (as one would expect), most of the amenity effects will be on land rents.

While these models are interesting, they still require some restrictive assumptions. One of the most questionable is that all individuals are identical. As Rosen (1979) pointed out, if the workers differ, one cannot use the equilibrium hedonic schedules to measure an individual's willingness to pay for nonmarginal amenity improvements. This problem can be avoided only if all consumers are identical or if the researcher's interest is only in measuring marginal willingness to pay. In hedonic property value work there has been interest in estimating the underlying structural equations, but this has not been attempted in the wage and property value hedonic literature. This will be necessary before it is possible to measure the benefits of nonmarginal environmental improvements with differing individuals in the economy. Roback (1988) modified the model to allow for two types of workers rather than assuming all individuals are identical. Both types of workers are necessary for production and thus both live in all locations. Although Roback (1988) focused on cost of living differences between cities, differences in workers will be an important consideration for benefit measurements using wage differences.

Blomquist, Berger, and Hoehn (1988) are interested in estimating the quality of life in various urban areas — a goal of many studies in this area — rather than in measuring the value of amenity improvements. In their studies they used the empirical results to generate quality of life indexes based on amenities for the urban areas. However, since all individuals in their model were identical and moved freely between cities, they all attained the same level of utility. Thus, it is difficult to say that the quality of life is better in one city than in another. The cities could be ranked by how high the wages are and how low the cost of housing is and generate the opposite rankings. Hedonic wage studies, it seems, are more valuable for estimating people's willingness to pay for amenities rather than for ranking cities.

A number of questions still must be addressed by hedonic wage studies. One of the most important is how to incorporate the diversity of individuals into the models. This will require estimation of the structural equations of the models as well as the equilibrium wage schedule. The extent of the various markets must also be addressed. For example, are real estate markets and labor markets truly national in scope? Are there moving costs between cities that must be considered? Are firms also required to move for an equilibrium to be attained. Firm size is indeterminant when returns to scale are constant; therefore, one firm in each urban area is adequate. But this would not be true for other production technologies. Do workers simultaneously choose a job and a house? These and other issues will have to be resolved as the promising early results with hedonic wage studies are refined.

4.8 Using Resale Indexes as an Alternative to Hedonic Regressions

Sometimes the effects of a substantial change in an environmental variable is of interest to the economist. For these cases it is possible to estimate hedonic regressions before and after the change, but it requires substantial data. An alternative is to use information on houses that have sold more than once during the time in question. Intuitively, the economist seeks to discover whether price differences for resales differ depending on the property's environmental change. For this technique to work, three conditions have to be met: a substantial number of properties have to have sold more than once; the environmental change has to differ among properties; and the researcher has to control for other major changes that took place in the neighborhood or houses during that time. Most neighborhoods probably have sufficient turnover to meet the first condition. As building permits are required for major additions or renovations, the researcher can control for such changes or exclude such observations, satisfying the third condition.

A technique developed in Palmquist (1982b) estimated the environmental effect while controlling for general changes in the real estate price level in the area and for depreciation of houses. The only data his technique requires are the dates and prices of sales and the environmental variable in the simplest case. His model assumes that the typical hedonic relationship holds so that the price for which a house sells depends on a variety of characteristics of the structure and neighborhood, including the age of the structure, the environmental quality at that location, and the general real estate price level in the neighborhood. The characteristics can enter with any functional form except for depreciation, which is assumed to be geometric, and the price changes due to the environmental change, which are percentages. For each pair of sales of a house, a depreciation-adjusted price ratio is formed that depends on the change in the real estate price index and the change in the environmental variable.[19] The other characteristics, which did not change between sales, cancel out. Taking the natural logarithm of both sides of the equation yields the equation to be estimated,

$$r_{itt'} = -\beta_t + \beta'_t + \gamma N_{itt'} + v_{itt'},$$

where $r_{itt'}$ is the log of the ratio of the sales prices of house i at time t' and t, β_t and β'_t are the values of the real estate price index at time t and t', and $N_{itt'}$ is the environmental change between the sales, for example, the difference in the noise levels at the house.

This equation can be estimated once the error structure has been considered. If the errors in the hedonic regression have the usual distribution properties, for example, $E(\epsilon) = 0$ and $E(\epsilon'\epsilon) = \sigma^2 I$, then the covariance of the estimated equation will be nonzero when there are more than two sales for a given

---

[19] If other variables have also changed, they can be incorporated.

house (see Palmquist 1982b). The estimator must take this covariance structure into account. It then becomes possible to estimate the coefficient of the environmental variable in a hedonic regression without collecting the data necessary for hedonic estimation. This estimate can be used in exactly the same way as are the hedonic estimates discussed above.

In addition to studies of highway noise, for which this technique was developed, it recently has been used by Mendelsohn (1987) to measure the impact of PCB dumping on adjacent houses. He used the same basic model, except that he treated the real estate price index differently. Instead of estimating the index that applies to the neighborhood, he used variables based on the GNP nonfarm residential deflator or the average sales price of home in Massachusetts. He also found that the environmental change has significant effects on property values.

4.9 Discrete Choice Models

The premise of hedonic methodology is that individuals' utility depends on the characteristics of housing they consume. It is generally assumed that although individuals consume only one house, they can continuously vary most of the characteristics of houses by changing dwellings. Thus, the hedonic methodology avoids the discreteness of individual houses by concentrating on the continuous characteristics. However, some research has been done on modeling housing decisions as discrete choices. These studies can be classified as either random utility models or random bidding models. The random bidding has been made more realistic through the incorporation of transactions costs.

4.9.1 Random Utility Models

One of the earliest papers using a random utility model to describe housing choices was Quigley (1976). Quigley used what he called a mixed direct-indirect utility function (simply called an indirect utility function below) that incorporated the budget constraint and depended on the characteristics of the house and the price of the house. According to this model, individuals choose the dwelling that yields the highest utility from among the houses available to them. The researcher is unable to observe all of the characteristics that are considered by the individuals when making their decisions; therefore, the indirect utility function is stochastic rather than deterministic. Quigley considered the probability that a particular type of consumer will occupy a house with certain characteristics and price. If the random elements in the indirect utility functions have the extreme value distribution, then conditional logit can be applied in estimating the probabilities.

Conditional logit implies certain restrictions on the behavior of the consumers. If one attempts to include socioeconomic variables in the indirect utility function to allow for differences in the behavior of different types of individuals, the socioeconomic variables cancel out in the estimating equation. For this reason, Quigley and many that followed this line of research estimated separate equations for different population segments when the population was grouped by variables such as income and family size. This allows for differences in socioeconomic characteristics, but makes it difficult to interpret the effects of changes in these variables.[20] The technique also sacrifices information by converting continuous variables, such as income, into discrete classes. Another restriction implied by conditional logit is the well-known independence from irrelevant alternatives (IIA) condition. It seems unlikely that the odds in choosing between two houses will always be uninfluenced by the introduction of a third house. For example, the new alternative may be virtually identical to the second of the two houses. The odds of a person choosing the first house over either the second or the third can be expected to be the same as the previous odds of choosing the first over the second. Thus, the new odds of a person choosing the first over the second will have increased.

A theoretical paper by McFadden (1978) on housing choices has received only limited application in housing but has been influential in a number of other fields. McFadden showed that the assumption of IIA could be relaxed by using nested logit or the General Extreme Value model. Nested logit introduces a hierarchical decision process in which decisions at different levels are linked through the inclusive values in the estimation. Such a decision process is probably a reasonable representation of the way housing choices are made. However, the exact nature of the appropriate nesting may not be determined theoretically. Alternative plausible hierarchies should be considered.

A second important contribution of the McFadden paper to the discrete analysis of housing choices was his demonstration that it may be unnecessary to consider all alternatives in estimating the choice probabilities. For example, a resident may choose one house over thousands of other houses, but not all of these other houses have to be considered in the estimation. Instead, the houses can be divided into a subset of all the choices containing the chosen house and a randomly selected subset of the houses not chosen. Several sampling methods are acceptable, but the easiest is probably when the probability of choosing an alternative depends only on the number of alternatives. This sampling rule satisfies McFadden's uniform conditioning property. Maximizing the likelihood function with this sample of observations yields consistent estimates of the parameters. This satisfies the constraints imposed by data processing equipment.

---

[20] Socioeconomic variables can be entered in the estimated equation if they only enter as interaction terms with housing characteristics. This is somewhat restrictive but is still useful.

In a recent article, Quigley (1985) used both of the techniques suggested by McFadden to estimate a consumer's choice of dwelling by hierarchically considering the characteristics of the house, the neighborhood, and the public services. He determined that the independence of irrelevant alternatives seemed to impose an inappropriate restriction on the estimation. This paper represents an important step in this type of estimation. However, Quigley did not consider environmental quality, and he did not consider alternative nesting procedures or the effects of socioeconomic characteristics on the parameter estimates. These steps will be necessary before this technique can be used for environmental benefit measurement.

4.9.2 Random Bidding Models

A closely related alternative to the random utility models of housing choice was developed by Ellickson (Ellickson, Fishman, and Morrison 1977; Ellickson 1981). His random bidding model seeks to predict the type of household that will have the winning bid for a particular house. The random utility model, on the other hand, seeks to predict the type of house that will be chosen by a given type of household. Instead of using the indirect utility function, Ellickson used the bid function from hedonic theory. In his model, a resident occupies the house for which he has the winning bid. The bids depend on the characteristics of the house,[21] and consequently, contain a stochastic element because the researcher cannot observe all housing characteristics. This error term again follows an extreme value distribution.

Ellickson suggested that since the parameters were only identifiable after normalization, it would be impossible to measure consumer's willingness to pay from his model. If this were true, it would certainly limit the usefulness of the model for environmental benefit measurement. Fortunately, Lerman and Kern (1983) showed that it is possible to estimate willingness to pay in a random bidding model if Ellickson's estimation procedure is modified. Because the price paid in the winning bid is observed, this additional information allows identification.

The empirical applications of the Ellickson model (Ellickson 1981) and the Lerman and Kern model (Gross 1988) have not included environmental variables. Gross did compare his results with the hedonic results of others who used the same data set. But because of shortcomings in the results from both models, no conclusions were possible.

---

[21] The bids should also depend on income and utility. Ellickson focuses on particular types of individuals and assumes that these variables are the same for individuals of a particular type.

4.9.3 Transaction Costs in the Random Bidding Model

Horowitz (1986) made an important advance in the random bidding models when he incorporated several realistic elements into the bidding process. Bids on houses are received sequentially over a period of time, and sellers cannot be assured that they can return to a bid rejected earlier if it proves to be the highest bid. Thus, the seller will accept a bid that may be below the maximum possible bid. Buyers also have limited information, so their actual bid may differ from the bid they would make if they were perfectly informed. Finally, the distribution of bids on a house may be truncated at the asking price. Horowitz generalizes the Ellickson-Lerman-Kern model to incorporate these possibilities. Horowitz's paper is difficult to interpret for specific variables because he used principal components to aggregate the housing characteristics. When Horowitz's prediction results were compared with the results of Lerman-Kern model using the same data, his model performed considerably better. Horowitz (1985) used his model to consider air quality in Baltimore. The air quality variables were appended to the principal components, but they were not significant, and in one case, had the wrong sign.

4.9.4 Summary

Discrete modeling may provide useful results in environmental benefit measurement, although such applications have not yet been made. Rapid advances in the available techniques are encouraging. The results with discrete choice modeling can be used as a check on the standard hedonic model. However, the apparent ease of measuring willingness to pay in the discrete choice models is not as costless as it appears. Strong assumptions about the functional forms of the indirect utility function or bid function are necessary. If comparable assumptions are made in the hedonic model, willingness to pay is identified in that model as well (e.g., Quigley 1982). Also, the effects of socioeconomic variables are more difficult to incorporate in the discrete choice models. At the moment, it appears that there are strengths and weaknesses with each technique.

4.10 Conclusions and Future Directions

As this chapter has shown, there has been rapid progress on many theoretical and econometric issues in hedonic analysis as well on many related techniques. However, it is not surprising that these advances are not as yet completely implemented. There should be innovative applications of these techniques in the coming years.

There remain a variety of areas where further theoretical and econometric research should yield valuable results. Several examples are the following.

1. Currently the discrete choice models are viewed as alternatives to hedonic models. For some questions, one type of model is more appropriate, while for other questions the alternative models seem more promising. It is likely that an integration of the two methodologies will be fruitful.

2. In the benefits measurement area the differences in the assumptions and conclusions between the long-run models and the short-run models requires further attention (see section 4.6). Which framework is more appropriate for measuring the benefits of environmental policies?

3. Can the long-run models be modified to consider many consumer types and will this modify the results?

4. The appropriate size for various hedonic markets should receive further consideration. For real estate markets, a majority of researchers seem to favor urban area markets. However, recently there are those who have argued that the markets are more limited, while others have used national markets for real estate as well as labor markets. Indeed, should these two markets have the same scope?

5. Lower and upper bounds have now been derived for benefit measures in hedonic models. However, the size of the region between those bounds remains an important question.

6. The accuracy of environmental benefit measurement depends crucially on the appropriateness of the variables representing environmental quality. Further research on how such quality is perceived will be valuable in designing correct measures.

This list is certainly not exhaustive, but it does indicate the breadth of important research topics in hedonic methodology that remain.

Measuring the Demand for Environmental Quality
John B. Braden & Charles D. Kolstad (Editors)
© Elsevier Science Publishers B.V. (North-Holland), 1991

Chapter V

CONSTRUCTED MARKETS

RICHARD T. CARSON

University of California, San Diego

5.1 Introduction[1]

Markets where environmental commodities may be directly bought and sold are scarce. This has led economists to develop techniques such as household production-travel cost analysis (see chapter 3) and hedonic pricing (see chapter 4) in order to infer the value of environmental commodities from transactions for other goods. The alternative approach is to construct markets where environmental amenities may be bought and sold. These markets may be either hypothetical or real. The objective in either type of market is to measure the consumer's willingness to pay or willingness to accept compensation for the environmental amenity of interest.

While hypothetical markets are most often created during the course of a survey interview, the creation of real markets can take several routes. For instance, a city government creates a market for a park when it holds a public referendum to decide whether the community should establish the public park, and a developer creates a market for units with an ocean view when he or she sells otherwise identical units for different prices depending upon whether they do or do not have views.[2] However, most often economists create these markets using groups of test subjects, and for that reason they are sometimes referred to as experimental markets. In this chapter, a term coined by Richard Bishop, "simulated market," will be used to refer to any market in which real money actually exchanges hands for the usually un-

[1] The author wishes to thank W. Michael Hanemann, Kerry M. Martin, Robert Cameron Mitchell, and the editors for their helpful comments. The remaining errors, of course, are those of the author. The author also wishes to acknowledge the financial support of the University of California Water Resources Center, grant W-722, in writing this chapter.
    [2] Offering the ocean view as an option with a known price effectively unbundles the ocean view from the structure. The hedonic pricing method is essentially a theoretical and statistical approach to unbundling and pricing a commodities characteristic.

marketed commodity. Perhaps the key characteristics of any constructed market, hypothetical or simulated, is that initially the market is unfamiliar to its participants.

The historic antecedents for using created markets to value commodities date back to at least the 1940s. Ciracy-Wantrup (1947, 1952) advocated the use of survey techniques to determine the demand for environmental commodities, and Bowen (1943) showed how to determine demand for public goods using the results of referenda. The history of test markets in marketing, a close cousin of our simulated markets, is even older. The strongest influences on current work are, however, much more recent. The most well-developed variant of the hypothetical market approach, known as contingent valuation, stems largely from papers by Davis (1963, 1964) and Randall, Ives, and Eastman (1974), while current work on simulated markets derives largely from work in experimental economics by Charles Plott, Vernon Smith, and their associates, and from a paper by Bishop and Heberlein (1979).[3]

Working with constructed markets often makes economists uncomfortable because in doing so they move beyond the usual purview of economics into the realm of other disciplines such as experimental design, marketing, political science, psychology, sociology, and survey research. What has driven economists to use constructed markets is the market's great flexibility, particularly in valuing environmental commodities or aspects of environmental commodities which are difficult, if not impossible, to value using other benefit estimation techniques. In spite of strong attacks by some economists, constructed markets are becoming more and more widely accepted. For instance, contingent valuation, the most frequently used of the constructed market techniques, is endorsed as a benefits estimation technique in the Water Resources Council (1983) guidelines and to a lesser degree by the U.S. Department of Interior (1986) rules for natural resource damage assessment. Contingent valuation is used by a number of federal agencies, such as the Environmental Protection Agency, the Forest Service, the Department of Interior, the National Marine Fisheries Service, and the Army Corp of Engineers; by various state agencies, such as the Alaska Department of Fish and Game, the Colorado Attorney General's Office, and the Metropolitan Water District of Southern California; by major research organizations, such as the Electric Power Research Institute and Resources for the Future; by government agencies in other countries, such as Australia, Canada, and Norway; and by international organizations, such as the World Bank. The number of resource valuation studies based on constructed markets is growing at a rapid rate.

In terms of specific program areas, contingent valuation has been used most extensively to value changes in air quality (e.g., Tolley and Fabian 1988), water quality (e.g., Smith and Desvousges 1986b), and recreation (e.g., Sellar,

---

[3] See Plott (1982) for a discussion of the history of experimental economics and Mitchell and Carson (1989) for a discussion of the historical development of the hypothetical approaches to valuing nonmarket goods.

Stoll, and Chavas 1985). The technique is also receiving a great deal of attention in the valuation of risk reductions (e.g., Jones-Lee, Hammerton, and Philips 1985). While these are the main application areas to date, a remarkable range of both environmental and nonenvironmental goods have been valued using constructed markets.

Simulated markets for environmental goods have been primarily used to assess the performance of hypothetical markets, with the best examples being the work of Richard Bishop and his colleagues at the University of Wisconsin and that of William Schulze and his colleagues at the University of Colorado. These economists have also focused on comparing the differences between people's willingness to pay (WTP) for welfare changes and their willingness to accept compensation (WTA) measures. Perhaps the largest body of work in experimental economics looks at free-riding behavior (Marwell and Ames 1981; Bohm 1972). Coursey and Schulze (1986) described how the results from laboratory experiments could be used to help develop better contingent valuation methods. Table 5.1 briefly describes a number of representative contingent valuation and simulated market studies.

5.2 Theoretical Foundation

Constructed markets enjoy a very strong theoretical foundation. Depending on the property right assigned, the preferred Hicksian welfare measure can be expressed in terms of either willingness to pay or willingness to accept compensation. Constructed markets, in principle and in contrast to other benefit measurement techniques, can directly obtain WTP or WTA. The other benefit measurement techniques obtain measures of Marshallian consumer surplus that, in many instances, are good approximations of WTP or WTA.[4] Assume, for instance, that an organization or institution is considering an improvement in environmental quality and desires a measurement of WTP (i.e., the Hicksian compensating surplus — see chapter 2). A participant is asked to respond by giving the difference between two expenditure functions:

$$e(p,q_0;U_0,Q,T) - e(p,q_i;U_0,Q,T), \qquad (5.1)$$

where p is the vector of prices for the marketed goods, q_i is the environmental amenity being changed, U_0 is the initial, or status quo, level of utility to which the respondent is assumed to be entitled, Q is a vector of the other public goods that are assumed not to change, and T is a vector of the participant's taste parameters (Deaton and Muellbauer 1980). The value of the first

---

[4] Exact measures of WTP or WTA can be obtained using the travel cost or hedonic pricing methods if very strong assumptions can be made about the specification of the utility function (e.g., Hausman 1981). One of the major advantages of using constructed markets is that in many instances it is possible to avoid making specific assumptions about the form of the utility function.

TABLE 5.1
Representative contingent valuation and simulated market studies.

Authors (year)	Good Valued	Research procedure(s)	Elicitation method
Partial list of contingent valuation studies			
Water quality studies			
Carson, Hanemann, and Mitchell (1986)	Water quality bond issue	Telephone	Take-it-or-leave-it
Carson and Mitchell (1988)	National water quality	Personal interview	Payment card
Davis (1980)	Potomac River	Personal interview	Direct question
Gramlich (1977)	Charles River and national water quality	Telephone, personal interview	Take-it-or-leave-it, direct question
Greenley, Walsh, and Young (1981)	Colorado River	Personal interview	Bidding game
Hanemann (1978)	Boston beaches	Personal interview	Bidding game
Loomis (1987)	Mono Lake	Mail	Take-it-or-leave-it, direct question
Oster (1977)	Merrimack River	Telephone	Direct question
Smith and Desvousges (1986b)	Monangahela River	Personal interview	Bidding game, direct question, payment card, contingent ranking
Sutherland and Walsh (1985)	Flathead Lake, Montana	Mail	Direct question
Air quality studies			
Brookshire, Ives, and Schulze (1976)	Siting of plant and visibility	Personal interview	Bidding game
Loehman (1984)	Visibility in San Francisco	Personal interview	Payment card
Loehman and De (1982)	Air pollution control	Mail	Payment card
Rae (1983)	Visibility at national parks	Personal interview	Contingent ranking
Randall, Ives, and Eastman (1974)	Visibility and environmental damage	Personal interview	Bidding game
Ridker (1967)	Air pollution	Personal interview	Direct question
Rowe, d'Arge, and Brookshire (1980)	Visibility in Four Corners Region	Personal interview	Bidding game
Rowe and Chestnut (1989)	Visibility in national parks	Mail	Payment card
Schulze, Brookshire, et al. (1983)	Visibility in Grand Canyon	Personal interview	Bidding game
Tolley and Fabian (1988)	Visibility in Eastern U.S.	Personal interview	Bidding game, direct question

TABLE 5.1 *Continued*

Authors (year)	Good Valued	Research procedure(s)	Elicitation method
Risk studies			
Acton (1973)	Heart attack programs	Mail, personal interview	Direct question
Frankel (1979)	Value of life (airline crash)	Personal interview	Direct question
Hammerton, Jones-Lee, and Abbott (1982)	Statistical life	Personal interview	Direct question
Hammitt (1986)	Food-borne risks	Focus group	Direct question
Jones-Lee (1976)	Value of life	Mail	Direct question
Jones-Lee, Hammerton, and Philips (1985)	Safety	Personal interview	Direct question, bidding game
Mitchell and Carson (1986b)	Trihalomethanes	Personal interview	Direct question
Mulligan (1978)	Nuclear plant accidents	Personal interview	Bidding game
Smith and Desvousges (1986b)	Hazardous waste disposal sites	Personal interview	Direct question
Tolley and Babcock (1986)	Health risks	Mail, personal interview	Bidding game
Land/recreation facilities studies			
Bergstrom, Dillman, and Stoll (1985)	Agricultural land preservation	Mail	Payment card
Bishop and Boyle (1985)	Illinois State Beach	Mail	Take-it-or-leave-it
Daubert and Young (1981)	Instream flows	Personal interview	Bidding game
Majid, Sinden, and Randall (1983)	Public parks	Personal interview	Bidding game
McConnell (1977)	Day at beach	Personal interview	Bidding game
Randall et al. (1978)	Surface coal mine reclamation	Personal interview	Bidding game
Roberts, Thompson, and Pawlyk (1985)	Offshore diving platforms	Mail, personal interview, telephone	Bidding game
Thayer (1981)	Environmental damage	Personal interview	Bidding game
Walsh, Miller, and Gillman (1983)	Ski capacity	Personal interview	Bidding game
Walsh, Loomis, and Gillman (1984)	Wilderness protection	Mail	Direct question

TABLE 5.1 *Continued*

Authors (year)	Good Valued	Research procedure(s)	Elicitation method
Wildlife, hunting, and fishing			
Brookshire, Eubanks, and Randall (1983)	Grizzly bears, bighorn sheep	Mail	Direct question
Brookshire, Randall, and Stoll (1980)	Elk hunting	Personal interview	Bidding game
Cameron and James (1987)	Recreational fishing	Personal interview	Take-it-or-leave-it
Cocheba and Langford (1978)	Waterfowl hunting	Mail	Payment card
Hageman (1985)	Marine mammals	Mail	Payment card
Hammack and Brown (1974)	Migratory waterfowl	Mail	Payment card
Samples, Dixon, and Gower (1986)	Humpback	Focus group	Direct question
Sorg and Nelson (1986)	Elk hunting	Telephone	Bidding game, direct question
Stoll and Johnson (1985)	Whooping crane	Mail, personal interview	Bidding game
Wegge, Hanemann, and Strand (1985)	Recreational fishing	Mail	Take-it-or-leave-it
Partial list of simulated market studies			
Bishop and Heberlein (1980)	Goose permits	Mail	Take-it-or-leave-it
Bishop and Heberlein (1986)	Deer permits	Mail	Take-it-or-leave-it
Bohm (1972)	Free-riding behavior	Laboratory	Direct question
Bohm (1984)	Government	Mail	Direct question
Coursey, Hovis, and Schulze (1987)	WTP vs. WTA	Laboratory	Bidding game
Ferejohn and Noll (1976)	PBS programming	Mail	Iterative ranking of programs
Hoffman and Spitzer (1982)	Coase Theorem	Laboratory	Payoff chart
Knetsch and Sinden (1984)	WTP vs. WTA for lottery tickets	Laboratory	Direct question
Knez and Smith (1989)	WTP vs. WTA for asset units	Laboratory	Direct question
Marwell and Ames (1981)	Free-riding behavior	Mail, telephone	Payoff chart

expenditure function is Y_0, the participant's current income; the value of the second expenditure function is the level of income that solves for U_0 given p, q_i, Q, and T. WTP is defined as the difference between Y_0 and Y_i. Willig (1976) has shown that equation (5.1) can be expressed in an equivalent form known as the *income compensation function*. If WTP is the desired benefit measure, this function, sometimes referred to as the WTP function, is given by

$$\text{WTP}(q_i) = f(p, q_i, q_0, Q, Y_0, T), \tag{5.2}$$

where q_0 is now taken explicitly to be the baseline level of the public good of interest, and the functional form chosen for $e(\cdot)$ or $f(\cdot)$ imposes restrictions on the other. Equation (5.2) forms the basis for estimating a valuation function that depicts the monetary value of a change in economic welfare that occurs for any change in q_i.

Four additional theoretical questions have occupied the attention of contingent valuation researchers. Two of these, the treatment of uncertainty and the decomposition of an agent's benefit from a change in q_i, can be handled in a straightforward manner in a constructed market framework. The other two — should WTP or WTA be used as the measure of economic welfare and how should individual WTP or WTA be aggregated — are not easily resolved because they involve fundamental philosophical issues. Each of these questions is taken up in turn.

Smith (1987b) has shown that uncertainty can be introduced into this framework in a very natural way by replacing the standard expenditure function in equation (5.1) with the concept of a planned expenditure function in order to obtain the desired *ex ante* welfare measure (also see chapter 2). In the simplest sense, the planned expenditure function returns the amount of money just needed *ex ante* to preserve the perceived status quo of expected utility. Because participants in a constructed market naturally take into account both the uncertainty in their demand and any revealed uncertainty of supply when they make their decisions, their responses are consistent with *ex ante* decision making and welfare measures. In contrast, the other benefit estimation techniques must now contend with the need for a technical correction factor known as *option value* (Chavas, Bishop, and Segerson 1986) because they measure *ex post* rather than *ex ante* economic welfare.

Often inspired by the way that various environmental laws are written and by the limitations of the other benefit measurement techniques, researchers who use constructed markets often attempt to disaggregate (or aggregate) WTP/WTA measures obtained from asking the participant to evaluate equation (5.1).[5] The most popular decomposition is between use and existence values. This happens because existence values typically are not measured by other

---

[5] For example, the Clean Air Act does not allow a monetary value to be placed on health benefits but calls for consideration of economic values for "secondary" benefits such as visibility improvements.

benefit measurement techniques, such as travel cost analysis. The exclusion of existence values creates a bias in the travel cost analysis; the question, of course, is how "big" is the bias. [In chapter 10, Randall uses the expenditure function representation from equation (5.1) to investigate this issue.] Closely related is the issue of how to aggregate or disaggregate benefits over different geographical areas or different policies. This question, too, has an expenditure function representation (see chapter 10 and Hoehn and Randall 1989) and turns crucially on substitution elasticities. One of the key results of the Hoehn and Randall formulation is that it demonstrated the importance of sequence in valuing environmental amenities or disaggregating total value. This is a disturbing finding for policy makers because it means that an environmental amenity does not have a "context independent" value.

The essential problem is that a particular policy change is not well specified with regard to another policy change unless the sequence of the two changes is known by the participant. Individuals living in an area that has several polluted lakes will place a greater value on the first lake that is cleaned up in their area than on the second. They do this for several reasons. First, each cleaned lake becomes a substitute for subsequent lakes that require cleaning. Second, the individual's allocation of money for the first lake cleaned up reduces the money he or she has available for cleaning up another lake. If separate studies value the lakes individually, however, participants will treat whichever lake they are asked to value as if it is the only lake to be cleaned up. An overvaluation of the benefits of a combined cleanup will occur if the separate values are added up. If the lakes are valued in sequence in a single study, the benefit estimates for the individual lakes — but not for the entire set of lakes — will be biased unless the valuation sequence replicates the actual sequence in which the cleanup will occur. It should be clear that any good being valued has a place in a sequence relative to some other good — either the other good will be provided before, at the same time, or later than the good of interest.[6]

One of the most enduring controversies in constructed markets is whether WTP or WTA should be used as the welfare measure. Many economists thought that this controversy had largely ended with Willig's (1976) results that showed that for a price change, the difference between WTP and WTA was a function of the income elasticity, and that for reasonable values of the income elasticity, the difference between WTP and WTA had to be small. The other benefit measurement techniques, because they were based on estimated Marshallian demand curves, were incapable of directly providing evidence on the difference between the two Hicksian welfare measures. WTP

---

[6] Although efforts to decompose a WTP response into use value and existence value have probably received too much attention given its policy relevance (because total WTP is already the desired welfare measure) and determining the substitution relationships between environmental amenities has received far too little attention given its large potential policy relevance.

and WTA could, however, be directly measured using constructed markets and the empirical results consistently showed large differences.

These differences helped spawn a great deal of research. Psychologists such as Kahneman and Tversky (1979) put forth theories of why people treated gains and losses asymmetrically, while economists such as Randall and Stoll (1980) extended Willig's work to quantity changes, and Bockstael and McConnell (1980) looked at corner solutions. Bishop and Heberlein (1979) undertook a major experiment to see if the differences were related to the hypothetical nature of contingent valuation, and Coursey, Hovis, and Schulze (1987) looked at how the two measures of value behaved in repeated trials of the simulated market. The number of papers that have attempted to measure both WTP and WTA or rationalize the differences between the two has become quite large.[7]

The most noteworthy recent paper on this topic is Hanemann's forthcoming paper. Hanemann shows that with imposed quantity changes, the theoretical difference between WTP and WTA is governed by the ratio of an income elasticity to a substitution elasticity rather than by an income elasticity alone, as is the case with Willig's price changes. Substitution elasticity refers to the ease with which other market commodities can be substituted for the given public good while maintaining an individual at a constant level of utility. This elasticity of substitution takes a value of zero if no amount of increment in any market goods can substitute for the change in the public good, and a value of infinity if at least one market good is a perfect substitute for the pubic good. It can be shown that the *smaller* the substitution effect (that is, the fewer substitutes available for the public good) and the *larger* the income effect (that is, the greater the income elasticity of demand for the public good) the *greater* the disparity between WTP and WTA. Conversely, if *either* the income effect is zero *or* the substitution effect is infinite, then WTP and WTA must coincide. If the public good in question is unique and the income elasticity of ordinary magnitude, then the difference between WTP and WTA can be quite large. Hanemann's results appear to encompass many of the previous empirical findings. The largest differences between WTP and WTA tend to be observed when the good being valued is unique; repeated "sales" of the good in question, of course, make that good more commonplace.

Hanemann's work is unsettling because it implies that, in contrast to Willig's results, there may be large real differences between WTP and WTA for unique environmental goods. This suggests that the property right chosen is important. While there are some researchers who are hopeful that contingent valuation might one day be able to measure WTA, the current consensus is that WTA cannot now be reliably measured using a contingent valuation survey. The problem in a contingent valuation market is creating either a plausible situation in which the implicit agent who will purchase the good is likely to convey

---

[7] See Mitchell and Carson (1989) for a review of this literature.

the money to the participant who can sell the good so that the seller's rational response is to set the price so high that the good will not be sold or a situation in which the purchaser has no choice but to purchase the good so that the seller's rational response is to ask for the highest feasible amount and not the minimum WTA.[8]

Sometimes WTP is obviously the correct welfare measure, in which case the task of the designer of a constructed market is simplified. Sometimes, however, WTA appears to be the correct welfare measure. When this is the case, the debate on WTP versus WTA is sometimes decided in favor of WTP based on questionable logic, such as the following: WTA is the correct measure but since it cannot be measured, the researcher should measure WTP instead. This logic was adopted, for instance, in the U.S. Department of Interior (1986) natural resource damage assessment guidelines and was certainly easier to defend before Hanemann's result.

Mitchell and Carson (1989) have argued that perhaps WTP is the correct property rights assignment in many instances where WTA at first appears to be the correct assignment. For instance, the WTA property right may appear to be correct when an electric utility is responsible for an air quality problem in the city where it is located and the people in the city are assumed to have a right to clean air. The WTA question would inquire how much the city's residents would have to be paid to voluntarily accept the poorer quality air. However, the utility is either publicly owned or regulated, so that residents can have better air quality and higher electricity prices or lower electricity prices and poorer air quality. In such an instance, the residents may possess the right to clean air but they have to pay for it through higher electricity prices. Thus, the effective property right is WTP not WTA. Participants in constructed markets appear to have little problem with this concept if they are told how their money will be used to solve the problem. The key property that makes WTP rather than WTA appropriate is that the same group of agents effectively form both sides of the transaction.

Assume that the desired property right specification leads one to choose the ith agent's willingness to pay, WTP_i, as the welfare measure of choice for that agent. Should the aggregate welfare measure used be N, the population size, times the mean WTP or N times the median WTP, $M(WTP)$? The standard economic welfare, benefit-cost framework (Just, Hueth, and Schmitz 1982) favors N times mean WTP as the measure that is consistent with the potential Pareto improvement criteria. The public choice literature, however, places much more emphasis on a voting criteria in making decisions about public goods. Constructed markets have the good or bad property, depending

[8] Garbacz and Thayer (1983) provide one instance where WTA seems to have been accurately measured. They asked seniors how much they were willing to accept in the form of higher benefit payments in order to voluntarily give up a senior companion program. This paper seems to succeed in measuring WTA because of the credibility of the option of the government maintaining the senior companion program.

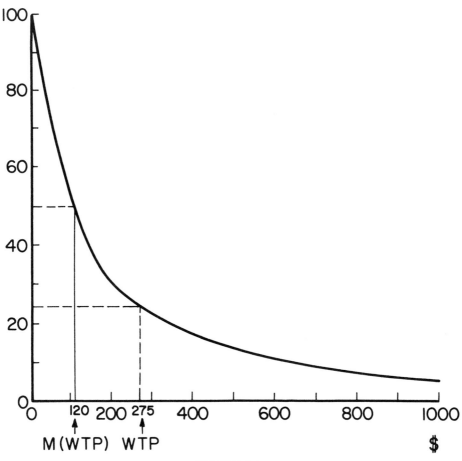

FIGURE 5.1
Percent willingness to pay specified amounts for a fixed quantity of public good.

upon one's perspective, of illuminating the potential divergence between these two criteria because one of the most succinct ways of displaying the results from a constructed market exercise is to display a graph of the distribution of the WTP_i's. Figure 5.1, taken from Carson and Mitchell's (1988) study of WTP for a national clean water program, is typical of the difference between mean WTP and M(WTP) that is often observed. A program that is justified using mean WTP may not be justified using M(WTP). A family of estimators that includes both the mean and the median as special cases is the α-trimmed, where the α largest and smallest observations are given zero weights in calculating the estimate.[9] The statistical properties of the family of the α-trimmed mean estimator are discussed later in this chapter.

[9] It should be emphasized that the observations are not being "thrown way" in calculating

5.3 Designing Constructed Markets

In an ordinary private goods market, a commodity can be bought or sold on a regular basis. Constructed markets have the opposite property. A commodity can only be bought or sold in a constructed market on the terms, including times, defined by whomever set up the constructed market. Constructed markets are of two types: simulated and hypothetical. In a simulated market, the participant makes an "actual" transaction for the good in question. In a hypothetical market, the participant states preferences or makes a pledge about the transaction for the commodity in question. For most purposes, there is no need to distinguish between simulated or hypothetical markets.

Constructed markets may or may not involve experiments, that is, the random assignment of different participants to different treatments, such as different market rules, different market prices, or different commodity characteristics. The term *experimental market* is somewhat of a misnomer as it implies nothing about random assignment of participants to different treatments. This principle of random assignment forms the basis of experimental economics, experimental psychology, and much of statistics. The random assignment of respondents to different treatments within surveys has a long history (Fienberg and Tanur 1985). The topic of experiment design as it relates to constructed markets is taken up in a later section.

A constructed market explicitly or implicitly defines both the payment mechanism and the agent on the other side of the transaction who will deliver or receive the commodity being traded. Three problems are common to the design of all constructed markets: first, structuring the rules of the market in which the good is to be bought or sold; (2) describing the good being valued; and (3) eliciting values or indicators of value in that market. The first two are closely related and are often referred to as the market scenario, which is discussed in this section. The third will be taken up later in this chapter along with other issues, such as market administration, sample design, and estimation of valuation functions.

5.3.1 Market Scenario

How do you tell participants in a constructed market what they are actually buying? Unfortunately, environmental goods such as air quality, water quality, and the risk of toxic chemicals tend to be intangible. In large part, the art of designing constructed markets lies in the description of such goods. The designers of constructed markets have become quite clever in doing this. They

---

the α-trimmed. The α-trimmed mean is an estimator based on order statistics where the α largest observations are assumed to be offset by the α smallest observations. In doing the trimming, only the rank of the observation is considered and not its absolute value. That is why this estimator becomes more and more resistant to outliers as α is increased.

use photographs to depict different visibility levels due to air quality changes; they denote changes in water quality by what types of water-based recreation are feasible; and they use risk ladders that include familiar activities to inform participants about the effects that changing drinking water standards might have.

Successfully describing the good to be sold is only half of the problem. The other half is to successfully describe a market mechanism under which the good can be sold. The major choice facing the researcher is whether to emulate a private goods market or a public goods market, specifically a referendum situation. The private goods market seems to work well for quasi-public goods, such as duck permits, where exclusion is possible and likely to be desirable. For goods that closely resemble pure public goods, a referendum may be the more logical choice. This choice, however, is not at all neutral. Participants presume the aggregation rule is being used and that other individuals are possibly free riding. Their perception of whether the good can actually be delivered as described is also influenced by the market mechanism used and the description of the agent on the other side of the market.

The wording of the constructed market scenario is critical because it provides the stimulus to which the participants respond. The researcher who designs a constructed market creates a scenario for the participant of which some features, such as the quality of the good, are intended to be taken into account by the participant when he or she assesses the value of the amenity. Other features, which may include the provider of the good or the sequence of questions, are intended to provide a plausible background for the valuation situations without themselves influencing the valuation outcome.

One of the difficulties in designing a constructed market is that it must meet the dual criteria of satisfying the requirements imposed by economic theory and the need of the respondents for a meaningful and understandable set of questions. Someone who wishes to evaluate a study must have access to the complete text of the questionnaire as administered. Table 5.2 shows a set of design criteria that must be met by any constructed market attempting to value an environmental good for policy purposes and the consequences of not meeting them. Each of the five criteria is a necessary, but not sufficient, condition for a valid scenario; together they may be regarded as necessary and sufficient.[10]

The first two criteria concern the fit between the subject matter of the scenario and the requirements of theory and policy. If, for example, the scenario describes the wrong property right or budget constraint, the data will be incompatible with economic theory. From a policy perspective, perhaps the most crucial aspect is that the scenario adequately describes the amenity change that the policy maker wishes to value. If the findings of a constructed

---

[10] Even if the scenario is designed correctly, there are other ways in which a constructed market study can fail to obtain valid and reliable data such as from a bad sampling design or faulty execution of the questionnaire.

TABLE 5.2
Scenario design criteria and contingent valuation measurement outcomes.

Is the scenario . . .	If not, respondent will . . .	Measurement consequence
Theoretically accurate?	Value wrong thing (Theoretical misspecification)	Measure wrong thing
Policy relevant?	Value wrong thing (Policy misspecification)	Measure wrong thing
Understandable by respondent as intended?	Value wrong thing (Conceptual misspecification)	Measure wrong thing
Plausible to the respondent?	Substitute another condition or	Measure wrong thing
	Not take seriously	Unreliable, bias-susceptible DK, or protest zero
Meaningful to respondent?	Not take seriously	Unreliable, bias-susceptible DK, or protest zero

market study of risk benefits was intended to apply to low-level risk reductions, such as from two in one million to one in one million, a scenario which describes risks of one in a thousand or even one in a hundred thousand would be misspecified. Similarly, the description of a new recreational area should include all its salient features if the WTP amounts are to represent its true value. It is important, in this context, to be aware of the trade-off between generality and specificity in the descriptions of amenities in constructed market studies. The researcher often wishes to apply his or her results to a variety of settings that require findings that are insensitive to the details of a particular scenario, such as the location of a recreational area in Ohio rather than Indiana or the use of a utility bill payment vehicle instead of a "higher prices and taxes" vehicle. However, sometimes what seems to be minor changes in the description of an amenity have large effects on the elicited WTP amounts. Therefore, the closer the fit between the amenity valued in a constructed market study and the amenity a policy analyst wishes to value, the greater the confidence the analyst can have that the findings are relevant to the policy decisions.

Presuming that the scenario is properly specified from the standpoints of theory and policy, it is necessary to communicate the scenario accurately to the respondents. Conceptual misspecification occurs when respondents understand the scenario in a different way than the researcher intended. This problem tends to be underestimated by researchers untrained in survey research techniques. As Sudman and Bradburn (1982) observe:

The fact that seemingly small changes in wording can cause large differences in responses has been well known to survey practitioners since the early days of surveys. Yet, typically, the formulation of the questionnaire is thought to be the easiest part of the design of surveys — so that, all too often, little effort is expended on it.

For example, some respondents think of "environmental problems" as including trash on city streets and local crime. Their definition encompasses a broader range of concerns than was most likely intended by the individual who used the term in the survey instrument. Comprehension problems can seriously distort WTP estimates. The researcher will measure the wrong thing if, for instance, respondents think they are being asked about drinking water in a study that was intended to inquire about surface-water quality in lakes, rivers, and streams; or if they think they are being asked to define a "fair" price for an amenity instead of the highest amount they would pay for it before doing without it; or if they think they are being asked to value a risk reduction that will reduce the risk from a contaminant to zero when, in fact, some risk will remain. This places an unusually heavy burden on the designer of a constructed market study to undertake a careful, and if necessary, extensive program to try out the instrument under various conditions. Converse and Presser (1986) provide one description of this process.

Just because a respondent does understand or can understand the scenario, does not mean that he or she will be sufficiently motivated to take the hypothetical situation into account and determine the value of the amenity to him or her. Two factors, plausibility and relevancy, are particularly important in motivating valid responses to scenarios. *Plausibility* involves a variety of factors, all of which enhance the realism of the hypothetical market. Is the hypothetical market sufficiently believable to the respondent that he or she will take it seriously? If a good, such as a hunting license or the use of a state park, is currently provided at a relatively nominal cost, respondents may find it difficult to believe that the good can have a value that is significantly higher than these reference amounts even if, in fact, it does. Is it conceivable to the respondent that the outcomes described in the scenario could occur? Respondents who do not believe, for example, that nuclear power can be made "safe" will be incredulous if a scenario asked them how much they would pay for programs to reduce the risk from a given nuclear power plant to close to zero. Is the choice situation one that makes sense to the respondent? An electric utility bill will be a more plausible payment vehicle than will be a sales tax for an air visibility scenario because the former has a more understandable connection to the cause of the visibility changes than does the latter. A hypothetical referendum often makes more sense to respondents than does a hypothetical private goods market for nonmarketed goods. In all these ways, plausibility reduces the uncertainty in the respondent's mind about the choice situation.

There are two undesirable outcomes that may occur if the respondent perceives the scenario as implausible. One is that respondents may substitute what they believe to be a more plausible condition for the one described in the scenario. When asked to value a recreational area via a scenario that has the users paying for it, the respondents may (consciously or unconsciously) assume that the government will pay for it out of taxes, and as a result, undervalue it in their WTP amounts. The result would be a WTP amount for the appropriate good under conditions other than those intended by the researcher. The second outcome is that the respondent will not be motivated to take the valuation exercise seriously. To the extent that this occurs, a variety of measurement consequences may result, none of them desirable and some subversive of accurate benefit estimates. The respondent might take a wild guess at an amount, which would affect the reliability of the WTP estimate, or the respondent might be motivated to minimize the effort involved in answering the valuation question by saying "don't know," by giving a protest zero (a $0 willingness-to-pay amount offered to appease the interviewer which does not represent a true $0 valuation), or by giving a biased WTP amount. A classic example of bias is when respondents' WTP amounts vary systematically according to whether a $1 or a $10 amount is used as a starting point for a bidding game elicitation framework.

Bias, in the sense that it is used here, refers to systematic errors. Unlike random error, which is amenable to assessment by sampling and replicating the survey, no applicable body of theory exists by which validity can be assessed (Carmines and Zeller 1979; Bradburn 1982) because there are no explanatory models of the cognitive processes that underlie respondents' verbal self reports (Bishop 1981). In these circumstances, the prevention of systematic error necessarily has an ad hoc character about it, although survey researchers have developed rules of thumb, based on experience and a growing body of survey experiments, which serve to minimize bias.[11]

It is difficult to make a general statement about the likely magnitude of potential biases. The reason is that the threat of various biases is quite specific to the contingent valuation scenario being valued. Most biases in contingent valuation surveys are avoidable; however, some biases, such as starting point bias in a bidding game (which is explained later in the chapter) and sample selection bias in a mail survey, will almost always be present. Typically, most other problems in contingent valuation surveys relate to the people being given inadequate descriptions of what the researchers actually want to value. This can result in large differences between what the researchers actually value and what they intended to value.

The question of bias is complicated in CV surveys by the general absence of a measurable true WTP value for public goods that can be used to assess the validity of a given study. This means that bias must be inferred from the researchers' partial understanding of respondent behavior; for example, re-

---

[11] See Mitchell and Carson (1989) for a further discussion of this issue and a preliminary framework for understanding respondent behavior in CV surveys.

searchers know that questions asked in certain ways will likely cause people to distort their answers. Or bias must be inferred from evidence in the survey that shows that changing the wording of the scenario in ways that are not expected to affect the WTP amounts does, in fact, do so. "Not expected" is a key phrase here because some differences may be legitimate contingent effects. The possibility of starting point bias was indicated by theories that suggest that under conditions of uncertainty, respondents might take initial amounts as information about the "correct" value for the good. The effect was demonstrated in several experiments.

This observation requires some explanation because until recently there was some confusion in the literature on this point. Earlier researchers assumed that only the nature and the amount of the amenity being valued should influence the WTP amounts; all other scenario components, such as the payment vehicle and method of provision, should be neutral in effect (Rowe, d'Arge, and Brookshire 1980). Therefore, according to this view, an experimental finding that the WTP amounts for a given study differ according to whether a utility bill or a sales tax payment vehicle is used was evidence of "information bias." More recently, Arrow (1986), Kahneman (1986), and Randall (1986) have argued against this view, holding that important conditions of a scenario, such as the payment vehicle, should be expected to affect the WTP amounts. According to their view, respondents in a CV study are not valuing abstract levels of provision of an amenity; instead, they are valuing a policy that includes the conditions under which it will be provided and the way the public is likely to be asked to pay for it. This notion that a public good does not have a value independent of its method of financing goes back at least to Wicksell's (1967) studies and is fully consistent with economic theory.

The uncertainty induced by implausible scenarios promotes bias because the respondents are susceptible to treating supposedly neutral elements of the scenario, such as the starting points, as clues to what the value of the amenity should be. Table 5.3 summarizes several types of bias that result from the respondents being influenced by the interview or treating elements of the contingent market as providing information about the "correct" value for the good. In each case, the respondent's WTP amount is distorted directionally by the scenario feature. For example, the undermotivated respondent may assume the amenity is important because an interviewer has gone to the trouble of asking him or her about it. As a result, the respondent will give a higher amount than he or she would if they were properly motivated to express its true value to them (importance bias).

Finally, the *relevance* of the amenity to the respondent can also play a role in motivating thoughtful responses. If the CV study interviews Colorado residents about an expansion in skiing opportunities, it's likely that the interviewers will have more difficulty motivating those residents who do not ski to take the study seriously. If so, the same array of measurement

TABLE 5.3
Typology of potential response effect biases in CV studies.

Incentives to misrepresent responses

Biases in this class occur when a respondent misrepresents his or her true willingness to pay (WTP).

Strategic bias	Where a respondent gives a WTP amount that differs from his or her true WTP amount (conditional on the perceived information) in an attempt to influence the provision of the good and/or the respondent's level of payment for the good.
Compliance bias	
Sponsor bias	Where a respondent gives a WTP amount that differs from his or her true WTP amount in an attempt to comply with the presumed expectations of the sponsor (or assumed sponsor).
Interviewer bias	Where a respondent gives a WTP amount that differs from his or her true WTP amount in an attempt to either please or gain status in the eyes of a particular interviewer.

Implies value cues

These biases occur when elements of the contingent market are treated by respondents as providing information about the "correct" value for the good.

Starting point bias	Where the elicitation method or payment vehicle directly or indirectly introduces a potential WTP amount that influences the WTP amount given by a respondent. This bias may be accentuated by a tendency to yea-saying.
Range bias	Where the elicitation method presents a range of potential WTP amounts that influences a respondent's WTP amount.
Relational bias	Where the description of the good presents information about its relationship to other public or private commodities that influences a respondent's WTP amount.
Importance bias	Where the act of being interviewed or some feature of the instrument suggests to the respondent that one or more levels of the amenity has value.
Position bias	Where the position or order in which valuation questions for different levels of a good (or different goods) suggest to respondents how those levels should be valued.

consequences described earlier for implausible scenarios are likely to occur, and since even in Colorado the number of nonskiers is likely to be large, the results could seriously distort the benefit estimates. Interviewer bias, for example, might induce many of these people to say they would be willing to pay a nominal amount in order to avoid appearing "cheap" in the eyes of the interviewer.[12] Aggregated over a large number of nonskiers, annual WTP amounts of one or two dollars, offered by people who really, if they considered the matter, would value the amenity at $0, could substantially bias the estimate upwards.

[12] The best way to avoid interviewer bias, of course, is to get nonthreatening interviewers who have little interest in the actual responses. Graduate students working on the project do not tend to meet these criteria.

TABLE 5.3 *Continued*

Scenario misspecification

Biases in this category occur when a respondent does not respond to the correct contingent scenario. Except in theoretical misspecification bias, in the outline that follows it is presumed that the intended scenario is correct and that the errors occur because the respondent does not understand the scenario as the researcher intents it to be understood.

Theoretical misspecification bias	Where the scenario specified by the research is incorrect in terms of economic theory of the major policy elements.
Amenity misspecification bias	Where the perceived good being valued differs from the intended good.
Symbolic	Where a respondent values a symbolic entity instead of the researcher's intended good.
Part-whole	Where a respondent values a larger or a smaller entity than the researcher's intended good.
Geographical part-whole	Where a respondent values a good whose spatial attributes are larger or smaller than the spatial attributes of the researcher's intended good.
Benefit part-whole	Where a respondent includes a broader or a narrower range of benefits in valuing a good than intended by the researcher.
Policy-package part-whole	Where a respondent includes a broader or a narrower policy package than the one intended by the researcher.
Metric	Where a respondent values the amenity on a different (and usually less precise) metric or scale than the one intended by the researcher.
Probability of provision	Where a respondent values a good whose probability of provision differs from that intended by the researcher.
Context misspecification bias	Where the perceived context of the market differs from the intended context.
Payment vehicle	Where the payment vehicle is either misperceived or is itself valued in a way not intended by the researcher.
Property right	Where the property right perceived for the good differs from that intended by the researcher.
Method of provision	Where the intended method of provision is either misperceived or is itself valued in a way not intended by the researcher.
Budget constraint	Where the perceived budget constraint differs from the budget constraint the researcher intended to invoke.
Elicitation question	Where the perceived elicitation question fails to convey a request for a firm commitment to pay the highest amount the respondent will realistically pay before preferring to do without the amenity. (In the discrete-choice framework, the commitment is to pay the specified amount.)
Instrument context	Where the intended context or reference frame conveyed by the preliminary nonscenario material differs from that perceived by the respondent.
Question order	Where a sequence of questions, which should not have an effect, does have an effect on a respondent's WTP amount.

The preceding paragraphs should have clarified that the frequently used term *hypothetical bias* is a misnomer. It's a misnomer because even though the hypothetical nature of the situation may increase the variance of the responses and may make the responses more susceptible to other potentially biasing influences, no evidence exists from WTP studies to suggest a systematic direction for the results of a hypothetical as opposed to a simulated market. Likewise, the frequently used term *information bias* is a misnomer. Participants take into consideration the information available to them in formulating their responses. The problem is that most information likely to be provided to a participant in a constructed market is unlikely to be neutral with respect to willingness to pay for a particular good. In particular, participants have preferences over who provides the good, how it will be provided, and who else will have to pay for it. Therefore, the terms hypothetical bias and information bias should be banished from the vocabulary of constructed market discussions.

5.4 Elicitation Methods

For those who have not actually worked with constructed markets, avoiding strategic behavior and problems with question wording most often appears to be the primary issue in using constructed markets. For practitioners, the central issue is often "how is the valuation response actually going to be elicited." This choice of the elicitation method tends to encompass many of the same issues surrounding threats to reliability and validity.

The most obvious elicitation method is to simply ask someone "What is the most you are willing to pay for this environmental good?" This approach is known as the *direct question method* and it has a number of problems. The major problem is the difficulty that people have answering questions of this type. Difficulty in answering the question tends to manifest itself in one of two ways: a high nonresponse rate and a large number of implausibly high or low answers. Psychologically, people do not usually consider the question "What is my reservation price?" because few real markets operate in this manner. In the typical hypothetical market (and to a lesser degree in a simulated market), a respondent does not have very strong incentives to devote a lot of effort to formulating the correct response to this question, but many people will give an answer, nonetheless. This may result in a larger number of extreme responses, that is, zeros and very large numbers. These problems have spawned the search for a better elicitation method. The direct question method is now most commonly used to value multiple public goods that do not have a natural relationship to each other in terms of WTP.

The second most obvious elicitation method is to start with some WTP amount and in response to "yes" replies, increase that amount progressively

ANNUAL HOUSEHOLD: INCOME BEFORE TAXES
$20,000 - $29,999
(Average annual amount in 1982 taxes and prices paid for some public programs)

$ 0	$190 — Police	$ 620	$1140
10	210 — and Fire	650	1180
20	230 — Protection	680	1220
30	250	710	1260
40	270	740	1300
50 — Space	290	770	1340
60 — Program	310	800	1380
70	330	830	1420
80	350 — Roads and	860	1460
90	380 — Highways	890 — Public	1500
100	410	920 — Education	1540
110	440	950	1580
120	470	980	1620
130	500	1010	1660
140	530	1040	1700
150	560	1070	1740 — Defense
170	590	1100	1780 — Program

FIGURE 5.2
Payment card.

until the respondents reply "no." Conversely, one should decrease the amount until a yes response is obtained if the respondent says no to the initial amount. This approach is known as the *bidding game* and was proposed by Davis (1963) and developed to its present form in the classic Randall, Ives, and Eastman (1974) paper. The problem with the bidding game is a phenomena called *starting point bias*. Starting point bias arises from two separate sources. First, the starting point is likely to convey some information about what the value of the good should be, and hence the starting point is likely to influence the magnitude of the respondent's final willingness to pay for the good. The second source, which is the process of getting from the starting point to the respondent's final answer, may influence that answer. If the starting point is far away from the respondent's true value, the respondent may be tempted to prematurely say yes or no to end the bidding, or the respondent may engage in yea saying, or to put it more simply, may agree with the interviewer.

A third method, known as the *payment card* (Mitchell and Carson 1981), gives respondents a card with an array of dollar numbers starting at zero (see figure 5.2). A respondent is asked what number on that card (or a number in between) represents his or her maximum willingness to pay for the good in question. The objective of the payment card is to avoid the awkwardness (that is, high nonresponse rate) of the direct question and the starting point bias problem of the bidding game.[13] The origin of the payment card lies with Hanemann's (1978) checklist and more generally with multiple choice survey

[13] It should be noted that the payment card can subtly introduce its own implied value cue through the range of numbers on the card.

questions. Cameron and Huppert (1987) have raised the issue of whether payment card responses are really people's maximum willingness to pay or whether the amount given by a respondent simply indicates the interval in which his or her maximum willingness to pay lies. Certainly, a checklist or a payment card used in a mail survey has this property and, econometrically, this raises some interesting issues. The appropriate estimator in such a case involves interval censoring and requires one to make some fairly strong assumptions about the distribution of responses within each interval.

The payment card can be used to succinctly inform the respondent about how much they are paying for various other goods. Mitchell and Carson did so in their 1981 study, which first put forth the payment card. Essentially, the choice is one of a classic bias-variance trade-off. Telling respondents what they are paying for some other goods stands a chance of biasing the results. Giving them this information also tends to reduce unexplained variance.

The fourth elicitation method is to obtain a single *discrete response* to a take-it-or-leave-it type of question. In environmental economics, this method stems from the seminal 1979 Bishop and Heberlein paper. Bishop and Heberlein advocated this method because it was easier for respondents to answer and, in particular, easy to implement in a mail survey. To those in the field of public choice, it looked like a referendum.[14] The binary choice format has the advantages of being incentive compatible if two other conditions are met. The first condition is that the participant believes some type of plurality decision rule is being used to make the decision and everyone will have to abide by it. The second condition is that the price is set exogenously and the participant does not perceive his or her answer as influencing the conditions of future choice situations he or she may face.

To implement the simple binary discrete choice approach participants are asked whether they would prefer to have the good at a specified price or do without it. If the participants are individually and randomly assigned to a set of prechosen prices, then it is possible to trace out the percentage of respondents who are willing to pay as a function of price. This approach has two related disadvantages. First, a discrete indicator of the participant's actual willingness to pay is necessary to specify either a utility function, or equivalently, a willingness-to-pay function. Second, a discrete indicator conveys substantially less information than knowing the participant's actual maximum willingness to pay.

There are two major debates over the use of the binary discrete choice elicitation method. The first is over whether one is estimating a random utility

[14] It is necessary, however, to distinguish between a political goods (e.g., referendum) market and a binary discrete choice question because it is always possible to phrase the referendum question in such a manner as to say, What is the most that this referendum could cost you in increased taxes and still have you vote for it?

[15] "Incentive compatible" in this usage means that it is in the participant's selfish interest to say yes if he or she prefers to have the good at the stated price and to say no otherwise. Strategic behavior and truth telling coincide for the rational individual in this case.

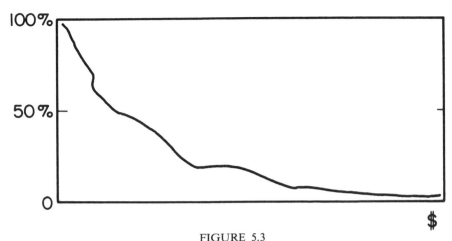

FIGURE 5.3
Percent willingness to pay as a function of required payment.

model (Hanemann 1984b) or a willingness-to-pay function (Cameron and James 1987). The second is over whether it is possible to accurately estimate the mean of the willingness-to-pay distribution from discrete choice data. Both debates revolve around the estimation of the model

$$\Theta(p_i) = f(X,t_i) + \epsilon, \tag{5.3}$$

where p_i is the percentage of respondents willing to pay tax price t_i, X represents respondent characteristics, and Θ is a transformation, possibly linear, of p_i.

The trick, as Bishop and Heberlein (1979) showed, is to estimate the area under the curve defined by equation (5.3) that traces out the percentage of the public that is willing to pay each possible tax price. The vertical axis (figure 5.3) gives the percentage while the horizontal axis depicts the dollars. One of the problems with the discrete choice becomes apparent immediately: the definite integral of the curve defines the mean WTP, but what should the limits of integration be? Setting the lower limit to zero rules out someone having a negative WTP, but most of the time this situation is plausible. Setting the upper limit is more troublesome. In their original study, Bishop and Heberlein set the upper limit equal to $200, the largest dollar amount they asked about in their study.

Let us examine the issue of the upper limit. To make things simple, assume that $\Theta(\cdot)$ is a probit function Φ, $f(X,t_i)$ is linear, and X consists of only a constant term. Equation (5.3) can then be written as

$$\Phi(p_i) = \alpha + \beta t_i + \epsilon. \tag{5.4}$$

In this case, Cameron and James (1987) have shown that WTP $= -\alpha/\beta$. Their approach allows the incorporation of individual characteristics and

Cameron (1988) has extended the approach to cover logit formulations of equation (5.3) as well. Cameron and James seem to avoid the issue of where to truncate the integral, a fact that bothered Bishop and Heberlein. But do Cameron and James really avoid the issue? The answer is no. Cameron and James' major insight is that if t_i is the stimulus variable and t_i is measured in the same unit as WTP, then the estimated coefficient on t_i can be used to recover the scale parameter of the underlying model — a property that is not true in the ordinary probit case. What is less apparent in their paper is that the normal distributional assumption is being heavily exploited in arriving at a closed-form solution for WTP and that this solution implicitly assumes that the upper limit of integration is infinity. Cameron and James have thus provided a very easy-to-use method of estimating WTP if researchers are prepared to make a strong distributional assumption about the shape of the largely unknown tail region. What becomes evident quite quickly in the binary discrete choice models is that the estimate of the median WTP is quite robust to the distributional assumption made and to the transformation of t_i as long as it is restricted to be monotonic.[16]

The other half of the debate revolves around what (5.3) is estimating. Cameron and James see the function $\text{WTP}_i = X_i\beta + u_i$, where X is a vector of respondent characteristics, and they assume that the respondent compares t_i with WTP_i and says yes or no depending on whether WTP_i is greater than t_i or less than t_i. Hanemann (1984b) sees the yes or no response as the result of comparing two indirect utility functions and that estimating (5.3) is justified on the basis of a random utility model. All of this might simply be semantics, but Hanemann shows that the most popular — that is, typically best fitting — form of (5.3), $\Phi(\cdot) = \alpha + \log(t_i) + \epsilon$, is inconsistent with utility theory. This conflict may be resolved in two ways. One is to assume that $\log(t_i)$ is only an approximation to a valid utility function. The other is to assume that every person has a utility function with different parameters and that an equation like (5.4) then, is only a statistical method of describing the population distribution of WTP. Finally, it should be noted that because participants are randomly assigned to a t_i, in large samples t_i will be orthogonal to all individual characteristics so that estimation of the parameter or transformation of t_i is not influenced by the inclusion or exclusion of the participant's characteristics from the estimated equation.

The next issue to be examined is the amount of efficiency that is lost when a discrete choice estimator is used. Alberini and Carson (1990) have recently addressed this issue. They showed that for the simple model given in (5.4), the maximum (Pitman) asymptotic efficiency relative to the discrete choice estimator for the mean WTP relative to any technique that yields observations on actual willingness to pay is approximately $2/\pi$, a little over 60 percent.

---

[16] Monoticity is probably the weakest restriction imposed by economic theory if equivalent subsamples of participants are assigned to each t_i. All this says is that people prefer low prices to higher prices for the same good.

This means researchers will need at least 66 percent more observations with the simple discrete choice estimator.

This maximum relative efficiency is achieved using Finney's (1971) method of picking the t_i's to minimize the *fudicial confidence interval* — an approximation to a standard confidence interval — around the particular point of interest when the mean and variance of the underlying process are assumed to be known exactly, a priori; in this case, the mean which is estimated by $-\alpha\beta$. Finney's method is fairly robust to a bad guess about the variance; however, relative efficiency falls off dramatically as the guess about the mean deviates from the actual population mean. The other drawback of the Finney approach is that it is highly optimized for estimating a single quantile in the distribution and can do poorly for estimating other quantiles far from the design emphasis.

A second method for determining the location of the t_i's is based on the criteria of D-Optimality (Silvey 1980). The *D-Optimality criteria* is based on picking the t_i's to maximize the Fisher information matrix with respect to the parameters, α and β. The D-Optimality approach has two advantages and two drawbacks relative to the Finney approach. It is fairly robust to bad guesses about the mean but not the variance. It estimates the mean much less efficiently than does Finney's method, but on the other hand, it does much better for estimating quantiles far from the mean.

A third method for choosing the t_i's, given initial guesses for the mean and the variance, is to place the t's at *equal distant quantiles*. The researcher determines how many equivalent subsamples will make up the sample as a whole and assigns a different t_i to each subsample. This method has properties that fall between that of Finney's and the D-Optimal methods and is perhaps the one most natural to standard survey administration procedures.

It is important to note that more subsamples, or equivalently more t_i's, is not preferable to fewer. Finney's method and D-Optimality methods will never yield more than three distinct t_i's. The smaller the subsample, the less precisely estimated is the percentage who will pay the subsample's t_i's. The gain is that the more the t_i's are spread out, the less the risk of a bad guess on the mean. The typical two-point, D-optimal design, under the assumption of normality, places one t_i at approximately $m - 1.14s$ and the other at approximately $m + 1.14s$ (where m is the estimate of the mean of willingness to pay and s is the estimate of the variance of the WTP distribution). Finney's method places them at $m + 0.37s$. With Finney's method, a bad guess on the mean can easily place all of the observations on one side of the t_i's; whereas with D-optimal design, a bad guess on the variance can easily place all of the observations in the center of the two t_i's. For these reasons, the equal distant quantile design seems to be a good compromise for contingent valuation studies. However, even with this latter method, bad guesses for the mean and variance can still dramatically reduce the asymptotic relative efficiency of the discrete choice method to close to zero. This should emphasize the strong need for pretests to ensure good estimates for the mean and variance. These pretests, at least the initial ones, should probably use an open-ended response format.

Recognition of the inefficiency of the single binary discrete choice question has led researchers to other discrete choice formats. The first of these is best represented by the Bergstrom, Rubinfeld, and Shapiro (1982) paper. They essentially asked respondents a "more, less, or about right" question. And the respondents appeared to be quite able to answer this question. The drawback of the approach is that the statistical model is fairly complex to estimate, and much more specific assumptions have to be made about the form of the utility function.

More in keeping with the simple binary discrete choice question is to repeat it once. Carson, Hanemann, and Mitchell (1986) showed that a Neyman double-sampling scheme could be used to achieve a very large increase in the efficiency of the estimate. If a respondent answered yes to a question, he or she was randomly assigned a higher number and asked again; if a respondent answered no, he or she was randomly assigned a lower number. If repeated often enough, this scheme turns into the bidding game, and thus the source of the inefficiency of the single discrete choice response is made clearer. The single repeat, with a random assignment exploiting the previously revealed preference, seeks to exploit the gain of the bidding game without setting up the yea-saying syndrome or losing the incentive compatibility property.

Seeing that the trick was to narrow the interval where the participant's maximum willingness to pay lay, Carson (1988) and Carson and Steinberg (1989) showed that the appropriate statistical technique was *interval data survival analysis.*[17] Here, price rather than time is the stimulus variable. The variance of the estimates can be shown to be closely related to the width of the intervals and survival analysis easily handles intervals with zero as the left endpoint and right censored endpoints, thus naturally resolving the infinite willingness-to-pay situation that had bothered Bishop and Heberlein. The estimated survival function is simply the estimated demand curve, and the estimated hazard function is closely related to the elasticity of demand. Survival analysis is a well-developed statistical technique. There are survival distributions that force a constant elasticity, such as the exponential; others allow increasing, decreasing, or constant elasticities with respect to price while maintaining monotonicity, such as the Weibull; and still others make it possible to go the complete nonparametric route forcing no restrictions on the shape of the demand curve. Survival analysis can handle covariates and very complicated assignment schemes.[18]

---

[17] The binary logit and probit models can be shown to be the simplest type of survival model.

[18] Carson (1988) showed that utility theory can be further exploited in double-sampling schemes with certain survival analysis estimators if different amenities asked about have known preference relationships.

5.5 Market Administration

Market instruments may be read to the participants in person or over the telephone, or they may be sent in the mail with a request to complete and return.[19] In recent years, the high costs of in-person surveys and methodological developments in telephone survey technology have led the major academic survey research centers to experiment successfully with telephone interviews, a methodology which commercial polling houses have used for many years (Groves and Kahn 1979). The sampling problems presented by unlisted telephone numbers have been overcome by the use of computer-based random digit dialing techniques.[20] An even less expensive survey method is the mail survey, which unlike telephone interviews, permits the use of visual aids. Here, too, methodological advances have improved the technique. It was once thought that low response rates of 20 to 30 percent were inevitable in mail surveys, but techniques are now available that can result in more respectable 50 to 70 percent response rates. These techniques, it should be noted, require considerably more effort and expense.

Which characteristics of constructed market questions should influence the choice of method? First, constructed markets often involve complex scenarios that require careful explanation and that benefit from the use of visual aids and close control over the pace and sequence of the interview. Second, the need to obtain dollar values requires a method that motivates respondents to exert a greater-than-usual effort. Third, the need to extrapolate data from the sample to estimate benefits for various populations requires that researchers use survey methods that support techniques that compensate for missing data — a topic to be considered in the next section.

For most situations, the method that meets all of these criteria is the in-person survey conducted in the respondent's dwelling place. For example, the physical presence of the interviewer offers the greatest opportunity to motivate the respondent to cooperate fully with a complex or extended interview, and the interviewer has the opportunity to probe unclear responses and to provide observational data (Schuman and Kalton 1985). In-person interviews also lend themselves to the use of various types of visual aids, or "display cards," which help to convey complex ideas or bodies of information. Furthermore, they support missing data techniques.

The large potential cost savings in using telephone and mail surveys has not gone unnoticed by constructed market researchers, however. Several have used mail surveys (Bishop and Heberlein 1979; Schulze, Brookshire, et al.

---

[19] The discussion here refers primarily to contingent valuation surveys. Simulated markets may also be implemented in person, over the telephone, or through mail surveys. Simulated markets sometimes use a variant of personal interviews where individuals are invited into the researcher's lab and a variant of mail surveys. For experimental purposes, a variant of the mail survey is used where students are asked to fill out an in-class questionnaire.

[20] See Frey (1983) and Dillman (1978, 1983) for a discussion of random digit dialing and other aspects of telephone survey methodology.

1983; Walsh, Loomis, and Gillman 1984; Bishop, Heberlein, Welsh, and Baumgartner 1984; Bishop and Boyle 1985) and others have conducted surveys by telephone (Oster 1977; Roberts, Thompson, and Pawlyk 1985; Carson, Hanemann, and Mitchell 1986; Sorg et al. 1985; Mitchell and Carson 1986b; Sorg and Nelson 1986). Randall et al. (1985) compared all three methods in their study of the national aggregate benefits of air and water pollution control.[21] Excluding costs, what are the trade-offs between these methods and the more expensive in-person technique?

First, the more impersonal nature of the telephone survey compared with the in-person interview reduces the ability of the interviewer to motivate the respondent. Second, the absence of visual cues during the telephone interview makes it more difficult for the interviewer to adjust the interview to the respondent's circumstances. In addition, the interviewer cannot use visual aids to help communicate the scenario. The result is that respondents' attention spans for descriptive material are much lower in telephone surveys than in surveys where the interviewer is present. This makes it difficult, if not impossible, to maintain respondent interest and attention while communicating even moderately lengthy constructed market scenarios. It may sometimes be possible to mail materials to households before conducting the telephone interviews. Sorg et al. (1985) provide an example of this.

Although mail surveys have the advantage over telephone interviews of being able to use visual aids, and an advantage over both in-person and telephone interviews in avoiding the possibility of interviewer bias, they suffer from several important shortcomings when applied to constructed markets. One shortcoming is they require the respondent to read and understand the description given in the scenario. Unfortunately, the reading level of a surprising number of Americans is quite low. According to the National Assessment of Educational Progress, which conducted a study of literacy among a national sample of 3,600 young adults between the ages of 21 and 25, 6 percent were unable to read a short sports story in a newspaper, 20 percent could not read as well as the average eighth-grade student, 37 percent could not present the main argument in a newspaper column, and only 43 percent could use a street map (Kirsch and Jungeblut 1986). These data understate reading comprehension problems because the young adult sample has a higher level of education than that of comparable cohorts of older people. Unless the scenario in a mail questionnaire is very short and simple, or the respondent is reasonably well educated and also highly motivated, there is an unacceptably large chance that the respondent may miss important details or misinterpret

---

[21] On the basis of their study, which obtained relatively similar findings for mail and in-person interviews, Randall et al. (1985) concluded that the in-person interviews were not superior to their mail questionnaires. Unfortunately the response rates they achieved for each methodology were too low (44 percent for in-person and 36 percent for mail) to make a definitive judgment on this issue. Nor did they address the important sample nonresponse problem to which mail surveys are particularly vulnerable.

one or more aspects of the scenario. Another set of problems results from the self-administered character of mail surveys. This causes difficulties in using skip patterns, where the choice of follow-up questions depends on the respondent's answer to previous questions, or in tailoring the interview to the individual respondent's needs. A well-trained interviewer can pace the interview according to the circumstances of the interview and can (within the limits imposed by the interview protocol) answer respondent's questions.[22]

The self-administered character of mail surveys provides no way of keeping the respondents from browsing through the questionnaire before they start to fill it out. This precludes the use of multiple scenarios where it is desired to have the respondents answer the questions in a fixed sequence without knowledge of the following scenarios. Mail surveys can also distort the sample because those who fail to fill out and return the questionnaire are typically those who have the least degree of interest in the amenity being valued.

While in-person interviews are clearly the technique of choice for constructed markets, experience with telephone and mail surveys suggest, except for the sample nonresponse bias problem that is discussed later, their shortcomings may be largely overcome provided the respondents are very familiar with the amenity[23] or the scenario is relatively simple.[24] For example, when Bishop and Heberlein (1979) sent a mail questionnaire to goose hunters, those receiving the questionnaire were well acquainted with the hunting opportunity they were asked about, and the nonresponse rate was extremely low for a mail survey. The off-shore recreational divers interviewed by Roberts, Thompson, and Pawlyk (1985) over the telephone were also familiar with the type of diving amenity they were valuing, and consequently, were willing to answer the questions.

As the material becomes more complex and less familiar to the respondents, however, the results are less satisfactory. Mitchell and Carson (1986b) used a relatively simple referendum format in a telephone survey of people's values for reduced risks of contracting giardiasis from San Francisco's water supply. In this case, the use of the telephone method involved a clear trade-off between cost and precision. Even though the survey was developed by an academic survey research organization experienced in conducting difficult telephone interviews, during the interview the researchers had to omit from the scenario

---

[22] It must be emphasized that standard survey practice forbids interviewers from providing ad hoc explanations when respondents look puzzled or improvising answers to respondent questions. They are instructed to read *only* the material provided to them which may, however, include set answers, previously prepared by the researcher, to questions which the pretesting showed might pose difficulties for some respondents. This additional material is only used if the respondent specifically raises the issue.

[23] This is why mail and telephone interview techniques are likely to work best for recreational users.

[24] Discrete choice formats (where a respondent is offered a single price on a take-it-or-leave-it basis) are usually required under these circumstances with some loss of information and additional complexity in statistical analysis over the continuous choice format.

a number of important aspects of the hypothetical situation, aspects which could have been easily incorporated into a personal interview.

Irrespective of how it is administered, a major requirement of a survey is to ensure that the data it obtains are comparable — that is, the information is gathered in a standardized fashion so that one person's answer can be compared with the answer given by another. To this end, survey organizations devote considerable care and resources to pretesting questionnaires and training interviewers. Pretesting is the survey equivalent of the test flight. Just as no plane manufacturer would go into production without rigorously testing its latest design, so too, no survey writer would assume that a questionnaire on a new topic, especially if the questionnaire were complex, could be sent directly into the field without careful tryouts under field-like conditions. Even experienced survey practitioners are often surprised when certain questions obtain better results than they had anticipated while others that they thought were winners turn out to be fatally ambiguous. Pretests normally consist of an extended period of trial and error with draft versions of the questionnaire. If the topic is novel, the pretest process may include preliminary in-depth research, perhaps using focus groups (Desvousges, Smith, Brown, Pate 1984; Randall et al. 1985; Mitchell and Carson 1986b; Krueger 1988) to learn how people conceptualize and talk about the topic.

Comparability also imposes demands on how interviewers conduct themselves in surveys. As David Riesman (1958) once observed, the basic task of the interviewer is to "adapt the standardized questionnaire to the unstandardized respondents." Except for mail surveys, questioning is a social process. Each interaction between an interviewer and a respondent is unique owing to the particular circumstances in which the interview occurs and the personal characteristics of the two participants. In order to "adapt the questionnaire" without distorting or changing it, the interviewer must motivate the respondent to enter into a special kind of relationship. Sudman and Bradburn (1982) describe how interviews differ from ordinary conversations.

> The survey interview... is a transaction between two people who are bound by special norms; the interviewer offers no judgment of the respondent's replies and must keep them in strict confidence; respondents have an equivalent obligation to answer each question truthfully and thoughtfully. In ordinary conversation we can ignore inconvenient questions, or give noncommittal or irrelevant answers, or respond by asking our own question. In the survey interview, however, such evasions are more difficult. The well-trained interviewer will repeat the question or probe the ambiguous or irrelevant response to obtain a proper answer to the question as worded.

It is precisely at the point of probing and handling respondent queries that comparability can be lost unless the interviewer rigorously follows instructions not to offer any information or explanations other than those described in the handbook for the study.[25]

5.6 Sample Design

Probability sampling procedures provide survey researchers with a straight-forward way to generalize from the responses of a relatively small number of respondents to much larger populations. These procedures are based on the principle that each economic agent, such as an individual or a household, in the population of interest has a known probability of being selected. Sampling issues had not received much attention in the constructed market literature until recently, even though they represent a substantial threat to the accuracy of aggregate WTP estimates.[26] Deciding who to interview for a constructed market study and how to locate and interview these people involves a series of decisions. First, the researcher must decide how to define the population of economic agents who are likely to be influenced by the change in the level of the public good. Do they include the residents of a particular town or other geographic areas? And does this group include those who use the amenity? Among the other choices the researcher makes is whether the agents are to be individuals or households. Next, the researcher must decide how to actually identify, or list, this population. This list or method of generating such a list is known as a *sampling frame*. It is from this list that the actual sample is drawn. The third step is to attempt to obtain valid WTP responses from each of the economic agents chosen to be in the sample frame. Unfortunately, there will be a sizable number of respondents who fail, for some reason, to give valid WTP amounts. These nonresponses can lead to nonresponse and sample selection biases unless corrective steps are taken. The eventual benefit estimates can become biased as a result of the sampling decisions and procedures at any or all of these stages. Four types of potential sampling design and execution bias can be identified. They are summarized in table 5.4.

Population choice bias occurs when the researcher misidentifies the population whose values the study intended to obtain. Populations may be defined in terms of the element, sampling unit,[27] extent, and time. For example, the element could be an individual recreator; the sampling unit, the number of cars entering recreation areas; the extent, two counties in northern California;

---

[25] The Research Triangle Institute's 1979 publication *Field Interviewers General Manual* offers an informative overview of the interviewer's role and training.

[26] See Desvousges, Smith, and McGivney (1983), Mitchell and Carson (1989), Bishop and Boyle (1985), Moser and Dunning (1986), Edwards and Anderson (1987).

[27] "Unit" is often used although "element" is technically the correct term in what follows because households were frequently defined as the relevant definition of an economic agent. In this and many other instances, the population unit and the population element will be equivalent.

TABLE 5.4
Potential sampling and inference biases in CV surveys.

Sample design and execution biases

Population choice bias	Where the population chosen does not adequately correspond to the population to whom the benefits and/or costs the provision of the public good will accrue.
Sampling frame bias	Where the sampling frame used does not give every member of the population chosen a known and positive probability of being included in the sample.
Sample nonresponse bias	Where the sample statistics calculated by using those elements from which a valid WTP response was obtained differ significantly from the population parameters on any observed characteristic related to willingness to pay; this may be due to unit or item nonresponse.
Sample selection bias	Where the probability of obtaining a valid WTP response from a sample element having a particular set of observed characteristics is related to their value for the good.

Inference biases

Temporal selection bias	Where preferences elicited in a survey taken at an earlier time do not accurately represent preferences for the current time.
Sequence aggregation bias Geographical sequence aggregation bias	Where the WTP amounts for geographically separate amenities that are substitutes or complements are added together to value a policy package containing those amenities, despite the fact that the amenities were valued in an order (for example, independently) different from the appropriate sequence.
Multiple public goods sequence aggregation bias	Where the WTP amounts for public goods that are substitutes or complements are added together to value a policy package containing those amenities, despite the fact that the amenities were valued in an order (for example, independently) different from the appropriate sequence.

and the time, July 1988. Choosing the correct population is simplest when the population who will pay for the good, or who is presumed to pay according to a given payment vehicle such as a local tax, coincides with the population who will benefit. The greater the divergence between those who pay and those who benefit, the more problematic it becomes to choose the correct population. Consider the case of the huge Four Corners power plant at Fruitland, New Mexico, (Randall et al. 1974). Residents of the area and visitors who come to enjoy the scenery use the public good of air visibility without paying the cost of maintaining it. This payment obligation is (would be) borne by those in Los Angeles (and elsewhere) who purchase their electricity from the utility that owns the plant. Nevertheless, area residents and visitors may be the crucial population for a WTP study of the aesthetic benefits of local air visibility because they experience the benefits directly.

After the population of interest has been identified, the sampling frame must be defined. The frame may be an existing list of the sample units of interest, or more commonly, a method of generating a list. If the population and the sampling frame diverge, *sampling frame bias* can occur. This type of bias makes it difficult, if not impossible, to accurately generalize the results of the study to the population initially defined by the researcher, even if there are no other problems in conducting the survey.

The procedures for defining the sampling frame vary according to the type of survey method used — personal, phone, or mail.[28] The sampling frame for in-person surveys of people who live in a given area are normally based on a physical enumeration of geographically-defined occupied dwellings. When the area is large, various types of area stratification and clustering techniques have been developed that make the enumeration costs manageable (Cochran 1977). Nongeographically-based populations often pose more difficult problems for in-person surveys. Suppose those who use a beach or visit a park comprise the population of interest. A valid sampling frame should make it possible for the sample to represent the visitors according to the time of day they visit, the day of the week, the season of the year, and possibly, by how they use the facility. The sampling frame for telephone surveys can either be chosen from the numbers listed in phone books, with the very real problem of unlisted numbers (both voluntary and involuntary),[29] or more preferably, from random digit dialing. The latter method, which selects numbers at random from the universe of usable numbers for the population of interest (Frey 1983), ensures that unlisted as well as listed numbers are included in the sample. Mail surveys' sample frames are based on lists of potential sampling units. With this method, researchers face the problem of obtaining lists of up-to-date addresses for every economic agent in the population of interest. This is often difficult for surveys of the general public because people in our society frequently change their residence.[30]

The remaining types of bias — *sample nonresponse bias* and *sample selection bias* — occur because of nonresponse. No matter what sampling plan and survey method is used in a CV survey, some level of nonresponse to the WTP questions is virtually inevitable with the consequence that the number of those who give valid WTP amounts will be smaller than the number of originally chosen sample elements. There are two distinct ways in which a member of the sample can fail to respond to a WTP question. In the first, unit nonresponse (Kalton 1983), the person or household fails to answer the

[28] For nontechnical descriptions of sampling frame development procedures see Sudman (1976) or Tull and Hawkins (1984).

[29] Approximately 95 to 96 percent of American households have telephones. Rich (1977) reports that the rate of unlisted numbers in urban areas soared 70 percent between 1964 and 1977. Groves and Kahn (1979) report an unlisted rate of 27 percent for their latest national sample. According to Frey (1983), "when you add new, but unpublished, listings to this figure, it is possible that at any one time nearly 40 percent of all telephone subscribers could be omitted from the telephone directory."

[30] There are likely to be fewer problems of this type where the appropriate sample frame consists of a current list of addresses held by a government agency as the holders of fishing or hunting licenses.

entire questionnaire. This occurs when people cannot be reached at home either by phone or in-person, when they refuse to be interviewed, or when those sampled in a mail survey fail to return the questionnaire.

The second way, item nonresponse, occurs when a respondent answers some or most of the questionnaire but fails to answer a particular question of interest, such as the WTP question.[31] With the exception of questions that ask for the respondent's income, item nonresponse rates exceeding 5 to 7 percent are rare in ordinary surveys (Craig and McCann 1978). In CV surveys, however, nonresponse rates of 20 to 30 percent for the WTP elicitation questions are not uncommon when: (1) the sample is random and therefore includes people of all educational and age levels; (2) the scenario is complex; and (3) the object of valuation is an amenity, such as air visibility, which people are not accustomed to valuing in dollars. Up to a certain point, these higher levels of nonresponse to the WTP questions are acceptable or even desirable. It is unrealistic to expect that 95 percent of a sample will be able and willing to expend the effort necessary to arrive at a well-considered WTP amount for certain types of amenities. Given the choice between having someone offer an unconsidered guess at an amount or having him say he does not know how much it is worth to him, the latter behavior is preferable, provided appropriate procedures to compensate for the resulting item nonresponse are used.

Both unit and item nonresponse result in the loss of valid WTP amounts from those originally chosen for the sample, and both can contribute to sample nonresponse and sample selection bias. For example, if 1,000 households are selected by probability-based methods for a CV sample, and valid WTP amounts are obtained for only 800 of these households, the researcher has to determine what effect the missing 200 households have on the WTP estimate. Put another way, can the values for the 800 people in the realized sample (those for whom valid WTP amounts are available) accurately represent the values for the amenity held by the population from which the original 1000 household sample was selected? If nonresponse in a CV survey was not associated with the WTP values held by the original sample, the failure to interview some respondents from the original sample would not cause bias (provided the sample size was reasonably large),[32] although it would affect the reliability of the estimates. A lack of association cannot be assumed, however. In the first place, researchers have found that a respondent's refusal is often associated with a lack of interest in the topic of the survey (Stephens and

[31] Item nonresponses on WTP questions fall into four general categories: (1) don't knows, (2) refusals, (3) protest zeros, and (4) responses which fail to meet an edit for minimal consistency.

[32] Many CV surveys in the literature use relatively small sample sizes (less than 500, often much less). The loss in statistical power may severely limit the ability of such surveys to conduct methodological experiments or to estimate population statistics within a meaningfully narrow confidence interval. These matters are discussed in detail in Mitchell and Carson (1989: Appendix C).

Hall 1983). Therefore, it seems reasonable to assume that people who are less interested in the amenity will value it differently than will their more interested counterparts. Second, response rates typically vary across population subgroups, such as lower income people, and there is ample evidence that WTP amounts are often associated with the characteristics of these subgroups.[33]

To determine whether observed nonresponse results in bias for a given study, two questions need to be addressed. One question is whether there are differential response rates across identifiable categories or groups of households — for example, users versus nonusers, different educational levels, and so forth — and the other is whether there are systematic differences between those within a particular group who responded and those who did not. Bias will occur to the extent that these between- and within-group differential response rates exist and are related to the value for the good. A given CV study may suffer from a between-group sample nonresponse bias, a within-group sample selection bias, or both.[34] Sample nonresponse bias will occur if, for example, the sample underrepresents the proportion of low-income households in the population, and these households hold different WTP amounts for the amenity than do households of other income levels. Even if the proportion of low-income households in a study's sample were representative, the study could still suffer from sample selection bias if somehow — either by differential selection or by a higher rate of item nonresponse once interviewed — the low-income people who gave usable WTP amounts differed in their preferences for the good from those low-income people who did not.[35]

The in-person, telephone, and mail survey methods have different vulnerabilities to the sample nonresponse and selection biases. But mail surveys are particularly prone to errors from these sources, especially the latter. This occurs because the unit response rates for mail surveys are lower than those for phone or in-person surveys. Also, the potential for sample selection bias is higher because the questionnaires are self-administered. In this situation, researchers lack control over the process of receiving the respondent's cooperation and eliciting his or her answers.

With telephone and in-person surveys, it is normally possible to assume that the nonresponses are not related to the subject matter of the survey. In the first place, the failure to interview people who are not found at home or

---

[33] As are other types of survey variables (Kalton 1983).

[34] The term "nonresponse bias" as used in the survey research literature often refers to both the between and within-group biases.

[35] It should also be clear that the failure to observe a characteristic related to WTP (e.g., income) can change a sample nonresponse bias into a sample selection bias and that obtaining a previously unobserved characteristic can change a sample selection bias into a nonresponse bias. To be more explicit, let $WTP = f(X, \beta) + U$ where $f(X\beta)$ is a regression function based on X, a matrix of predictor variables, and U is a vector of error terms. Sample nonresponse bias occurs when the sample distribution of X's differs significantly from the joint population distribution of X's and sample selection bias occurs when the sample distribution of U differs significantly from the population distribution of U.

who are too incompetent to be interviewed has nothing to do with their personal reaction to the survey's topic. Second, those who refuse to be interviewed in these types of surveys usually do so before the specific topic of the survey is made known to them.[36] Third, studies of people who refuse personal or telephone interviews suggest that refusals occur because of general rather than survey-specific reasons (Stinchcombe, Jones, and Sheatsley 1981; T. W. Smith 1983).

These assumptions cannot be made for those who receive a mail survey and fail to return it. Unless the recipient throws the package out without opening it, his or her decision whether or not to respond, including the decision to lay it aside, is likely to be influenced by his or her examination of the cover letter and the questionnaire. Research has shown that the less salient a mail questionnaire is to a potential respondent, the less likely the respondent is to fill it out and send it back (Heberlein and Baumgartner 1978; Tull and Hawkins 1984).[37] Because in the case of public goods the respondent's interest in the subject matter is likely to correlate with the value the good has to the respondent, there is a likelihood that nonrespondents will hold lower or even $0 values for the good compared with respondents of equivalent demographic categories. In short, mail surveys have a strong potential for sample selection bias, which suggests that information from those who happen to give valid WTP answers cannot be used to infer or to impute WTP values for the nonrespondents.[38] This is one of the reasons market research texts (e.g., Tull and Hawkins 1984) do not recommend their use for general populations.[39]

Richard Bishop has suggested putting in zeros for nonresponses to mail

[36] This presumes, as is the case with many surveys, that the interview topic is described in general terms when the respondents' cooperation is first requested to avoid this type of bias. For example, the interviewer would say they are conducting a study of "people's views about certain kinds of environmental issues" instead of the more specific "how much people are willing to pay to reduce the risk of cancer from trihalomethane contamination in their drinking water."

[37] Undoubtedly some of those who neglect to respond to mail surveys do so for reasons unrelated to the topic. The nature of mail surveys is such, however, that no interviewer is present to record that a potential respondent is sick or has traveled abroad for a month and these nonresponses cannot be distinguished from those who refuse to answer the surveys.

[38] For a discussion of the techniques available to compensate for bias due to nonresponse see Mitchell and Carson (1989).

[39] Some CV researchers have argued that nonresponse bias is not likely to be significant on the basis of the findings of a study conducted by Wellman et al. (1980). The Wellman et al. study compared early and late respondents with a mail non-CV outdoor recreation survey that achieved a 70 percent response rate. The authors argued, on the basis of apparent similarities between these groups on a number of characteristics, that "time, effort, and dollars spent in intensive follow-ups to increase recreation survey response rates might better be expended on other phases of the research process." This finding is an insufficient basis to assume random nonresponse as Wellman et al. did not study the 30 percent of their sample who failed to respond to their survey. There are no grounds for believing that late respondents to mail surveys such as theirs are a valid surrogate for the nonrespondents; there is *a priori* and empirical (Anderson, Basilevsky, and Hum 1983) evidence to the contrary.

surveys as a conservative assumption that also encourages agencies to fund extensive efforts to get high response rates. Almost no completed survey will represent a simple random sample of the population of interest. When using the results of the survey to make estimates, the effects of stratification and cluster, which appear in the best full probability samples, should be taken into account. Weighting to correct for sample nonresponse should also be taken into account. Imputation should be done for item nonresponse and corrections should be made for sample selection bias. This attention to sampling and response issues is extremely important and often strongly influences results.

5.7 Family of α-Trimmed Means

The family of α-trimmed mean estimators drops the α largest and α smallest observations and then calculates the mean value of the remaining observations. The mean is the extreme case where α equals zero and the median is the other extreme where α is 50 percent. For a large class of symmetric distributions, the maximum likelihood estimator can be written in terms of α going from zero to 0.5 as the tails of the distribution become "fatter." Constructed market data tends to be characterized by thick-tailed distributions. These distributions appear to become increasingly asymmetric as the mean willingness to pay becomes larger and more closely tied to the participant's income level — a finding which should not be too surprising. It is a finding, though, which forces the researcher or the policy maker to choose an α. A good way to display the implications is to display a table of the α-trimmed means for different α and to do the benefit-cost analysis using each of these values. Note that because for all individuals i, WTP_i must be nonnegative, the left-hand side outliers are constrained to be zero so that the use of any positive α will typically reduce the estimate of mean WTP.

Even if mean WTP is the desired statistic, using a small nonzero α value may be appropriate, particularly if a hypothetical rather than simulated market is being used. For mean WTP, the main difference between behavior in a hypothetical versus a simulated market appears to be that participants in a hypothetical market take the exercise less seriously than those in a simulated market; however, this does not appear to be the case for WTA markets. For WTA markets, Bishop and Heberlein (1979; 1986) found large differences between hypothetical and simulated markets. This usually manifests itself in mean WTP having a large standard error. Examination of the data usually exhibits a number of implausible large outliers. Use of an α of 0.05 or 0.1 will eliminate the dominant influence of these observations.

5.8 Experimental Design

As researchers have gained more experience with constructed markets, experimental design has taken on a more important role. This is due, in part, to the increasing recognition that many of the early experiments had low power and, in part, to the increasing cost of doing experiments, particularly experiments involving simulated markets in which a considerable amount of money is at stake. The experiments being performed are also taking on new complexity as the hypotheses being tested become more complex.

In designing an experiment involving any type of constructed market, the researcher first needs a clear null hypothesis to be tested and alternatives. Designing an experiment with a clean test between two well-defined specific alternatives is very difficult. Drawing conclusions from rejecting the null hypothesis if more than one possible specific alternative exists, is always dangerous. Bishop and Heberlein's (1979) goose hunting experiment is one of the best known examples. In their study, WTP from a contingent valuation experiment was much smaller than WTA from a simulated market experiment. Bishop and Heberlien concluded, and this was accepted by most contingent valuation researchers, that CV WTP underestimated true WTP since true WTP and true WTA were, according to Willig's results, supposed to be close, and the simulated WTA was accepted as a good estimate of true WTA. Bishop and Heberlein had been unable, for legal reasons at the time, to conduct the simulated WTP experiment. Their later results (1986) showed the simulated WTP and CV WTP were close but quite different from both simulated and CV WTA.

Two other typical problems with constructed market experiments exist. The first is the lack of random assignment of participants to treatments. This usually occurs when two populations are presumed to be similar so that the treatment effect is confounded with the two populations. The second, already alluded to, is the lack of statistical power. By this I mean that in many CV experiments the treatment effect would have to be so large for the null hypothesis to be rejected that for all practical purposes the test is meaningless. Then worse, the failure to reject the null hypothesis may lead the researchers to conclude that the effect is not present. Constructed market experiments are particularly prone to a lack of statistical power due to the large coefficients of variation typical of this type of data and due to the presence of a significant number of outliers.[40] Mitchell and Carson (1989) provided a lengthy appendix on designing experiments that recommends, among other things, a test on medians instead of means (due to the much smaller coefficients of variation and hence smaller sample size needed for a given level of power) and the use of nonparametric statistical tests that are less sensitive to outliers.

---

[40] The coefficient of variation is the standard deviation divided by the mean. For constructed market data, the coefficient of variation is typically greater than one, which is quite large by experimental standards but is reflective of the degree of income variation in the United States.

5.9 Estimation of Valuation Functions

For many environmental amenities, such as air quality and water quality, the economic question the policy analyst is often asked is: What are the benefits of improving the quality level from A to B when level B is assumed to be preferred to A? There are two ways this question can be answered in a constructed market framework. The first is simply to ask a respondent what he or she is willing to pay to have the quality level rise from A to B. The second is to estimate a valuation function that describes willingness to pay for marginal changes in the quality level. The advantage of the first approach is that the analyst does not have to make assumptions about the form of the utility or the willingness-to-pay function. The first approach's disadvantage, of course, is that it is not very informative on changes other than from A to B, except possibly as an upper or lower bound. The valuation function approach has the opposite advantages and drawbacks. The need to estimate the benefits of a change other than A to B or the desire to trace out a large part of the total or marginal benefits curve leads researchers in the direction of estimating valuation functions.

Estimation of a valuation function raises a number of issues. These issues can be divided into two groups.[41] The first group concerns statistical issues; the second concerns economic issues. The statistical issues revolve around how to optimally estimate the region of the response surface — that is, the benefits curve — in which the researcher is most interested. This problem can be thought of as a special type of experimental design. The economic issues revolve around which, if any, restrictions to impose on the utility or willingness-to-pay function, and which, if any, characteristics of individual respondents to consider. Often, distinctions between the statistical and economic issues become blurred.

Statistically, one wants to estimate the relationship:

$$\text{WTP} = f(\text{environmental quality level}). \tag{5.5}$$

Clearly, the more quality levels that one asks for, the more flexible is the form for $f(\cdot)$ that can be supported by the data. For instance, if only two quality levels are asked about, the researcher can only fit a straight line or a curve with a constant elasticity. Thus, to allow for the possibility of a different curvature, the researcher must either ask individual respondents about more levels of the good or increase the sample size and ask the additional respondents about different levels. The choice of the quality levels will also influence what

---

[41] For simplicity, it was assumed that the elicitation method has already been chosen and hence whether the data will be of the continuous or discrete type. Of course, the requirements of estimating a valuation function may influence the elicitation method chosen. In particular, the amount of information in discrete responses is substantially less than that in continuous responses, thus making the task of estimating a reliable valuation function with discrete data more difficult.

can be estimated. If two quality levels very close together are chosen, then it is likely that the WTP function will appear linear. One of the best guides to choosing optimal levels is to inquire about levels just above and below the range defined by existing levels and likely policy options. Box and Draper (1987) provide a good guide to response surface estimation.

The specification in (5.5) can be enriched by the incorporation of covariates. There are two reasons for doing so. The first is to increase the statistical efficiency by reducing the unexplained variance. The second is to test whether or not WTP appears to be driven by predictable factors, particularly those suggested by economic theory. If researchers randomly assign subsamples to different quality levels or if they ask each individual about each quality level and then stack the observations, the quality level will be orthogonal to the individual's characteristics.[42] Estimation of the model with covariates, of course, raises the issue of consistency with utility theory and the issue of whether utility theory imposes any restrictions on the model which should be tested.

A couple of other issues should be raised when considering the estimation of a valuation function. The first is how to treat protest zero responses. The approach used most often is to discard them. This is clearly wrong from a statistical point of view. A better approach is to explicitly model them using some type of maximum likelihood or nonparametric framework. Another problem with constructed market data is the presence of outliers (usually on the right side). Again the typical course of action has been to discard them. Robust regression techniques that down weight these outliers seems to be a better approach and one much more justifiable on statistical grounds.

5.10 Open Issues

While many of the fundamental issues in constructed markets are now settled, there are, nonetheless, a number of open issues with respect to constructed markets. These fall into four main categories: (1) the use of constructed markets in new application areas; (2) the role of information in constructed markets; (3) the exploration of theoretical issues using constructed markets; and (4) the statistical issues in the design and analysis of constructed markets.

One logical way of depicting the history of constructed markets is in terms of the process by which researchers determined how to use constructed markets to value a particular environmental amenity. Perhaps the best example is the long chain of air quality studies that started with Randall, Ives, and Eastman (1974). The main focuses of these studies was how to portray changes in air

---

[42] If the individual is asked about several levels (and those observations stacked for the purposes of estimation) then it may be reasonably expected that there is a panel data type correlation structure induced.

quality to participants and how to define a market structure for air quality. Each new study produced insights into what participants thought they were buying. Occasionally, there was a major advance or failure in describing air quality or the market in which it was sold. Now, a researcher desiring to do an air quality study in a different location has a firm foundation upon which to start. Each new environmental amenity produces a new challenge to researchers. They must determine how to describe it to participants, why the participants want it, and what reservations the participants may have about a program to supply it. This is a new experience to economists who generally have been able to ignore what actually motivates someone to purchase a good.

One of the most exciting new areas for the use of constructed markets is valuing risk reductions from environmental pollutants (e.g., Smith, Desvousges, and Freeman 1985). Psychologists have long argued that changes in low-level risk are very difficult for people to understand. Researchers have been experimenting with a number of different ways of expressing risks and are enjoying some success. Work is currently being conducted on risk from groundwater contaminants, pesticides, and radon. Another new area receiving considerable attention is natural resource damage assessment.[43] Natural resource damage assessment creates a host of new problems because the damage usually has already occurred so that it is difficult to obtain an *ex ante* welfare measurement, and because there is usually an easily identifiable "guilty" party thus creating the clear opportunity for strategic behavior that is usually lacking in most contingent valuation studies.

If a researcher accepts the argument that the values obtained in a constructed market exercise are contingent on the information available to participants, then a systematic exploration of how information influences values would appear to be necessary. What would be ideal is a quantification of how various types of information influence WTP responses, in particular, an investigation into the role of uncertainty with respect to likelihood of the amenity actually being supplied and into the role of the agent receiving payment for the amenity.

Constructed markets allow researchers to test a number of fundamental issues related to economic theory. This has been long recognized by experimental economists using simulated markets. With the exception of testing the relationship between WTP and WTA, contingent valuation has been less used for this purpose.[44] Other areas in which constructed markets should be useful are in examining how people actually discount future environmental amenities,

---

[43] See, for instance, Carson and Navarro (1988), Mitchell and Carson (1988), and Schulze (1988).

[44] In part this is due to the strong suspicion that economists have with regard to responses to hypothetical survey questions. The large differences between WTP and WTA consistently found in contingent valuation studies was ascribed to the hypothetical nature of the questions until Bishop and Heberlein's (1979, 1986) simulated market studies began to show the same large differences.

such as risk reductions (Horowitz and Carson 1988), and how to transfer the values obtained in one constructed market study to a new situation where a benefit estimate is needed.[45] The issue raised by Hoehn and Randall (1989) of aggregating benefits across geographic areas and across policies is still largely unexplored.

While the success of contingent valuation has largely exceeded the expectations of its early proponents, one of their great hopes for contingent valuation was that it would provide a cheap alternative to the other benefit measurement techniques. Unfortunately, contingent valuation has not proven cheap to implement. In order to minimize cost for a specified level of precision, contingent valuation researchers are starting to examine whether it is possible to use more efficient sampling plans and experimental designs. Contingent valuation data, in large part because it is survey data, is also not as clean as the macro or financial data with which economists typically work. This feature of the data is leading contingent valuation researchers to look at techniques for handling outliers and missing data and the implications of using those techniques. The shift to discrete choice contingent valuation questions has focused attention on discrete choice estimators. The ability to frame questions in particular ways is giving insight into what the discrete choice question is measuring (Cameron and James 1987) and can be exploited to gain more efficient estimates of willingness to pay.

[45] To date there has been little work done on this topic. Smith and Kaoru (1988) have undertaken the first formal study of benefit transfer but have focused on recreational demand travel cost studies rather than contingent valuation studies. Carson and Mitchell (1988) showed how Smith and Desvousges's (1986b) Mongahella River water quality CV estimate could be obtained from their CV study of national water quality benefits.

Part II

Methods for Valuing Classes of Environmental Effects

Measuring the Demand for Environmental Quality
John B. Braden & Charles D. Kolstad (Editors)
© Elsevier Science Publishers B.V. (North-Holland), 1991

Chapter VI

ENVIRONMENTAL HEALTH EFFECTS

MAUREEN L. CROPPER and A. MYRICK FREEMAN III

University of Maryland, Bowdoin College, and Fellows at Resources for the Future

6.1 Introduction

One of the basic services provided by the environment is the support of human life. Changes in the life support capacity of the environment brought about, for example, by reducing the pollution of air or water, can lead to decreases in the incidence of disease, reduced impairment of activities, or perhaps, increased life expectancy. The purpose of this chapter is to describe and evaluate the available methods for assigning monetary values to the improvements in human health that can be attributed to improved environmental quality.

The standard economic theory for measuring changes in individuals' well-being was developed to interpret changes in the prices and quantities of goods purchased in markets. During the past 15 years or so, this theory has been extended and applied to a wide variety of nonmarket or public goods and social programs, including public housing and other transfer programs, public investments in parks, transportation, the development of water resources, and improvements in environmental quality and health (Freeman 1979a). This theory is based on the assumption that individuals' preferences are characterized by substitutability between income and health. The trade-offs that people make as they choose among various combinations of health and other consumption goods reveal the values they place on health.

According to the simplest models of individual choice, researchers can interpret an individual's observed trade-off between income and health as a measure of his willingness to pay (WTP) for improvements in his health. However, there are two qualifications to this statement. First, society has developed several mechanisms for shifting some of the costs of illness away from the individual who is ill and onto society at large. Examples include medical insurance, which spreads the costs of treatment among all policyholders, and sick leave policies, which shift the cost of lost work days onto the employer, and ultimately, onto the consumers of the products. An individual's expressed willingness to pay to avoid illness would not reflect those components of the costs of his illness borne by or shifted to others. But the value to society of avoiding his illness includes these components. Empirical measures of the value of reducing illness must take account of these mechanisms for shifting costs. More will be said about this point in subsequent paragraphs.

The second qualification concerns the emphasis given to the individual's concern for his own illness. This emphasis does not preclude altruism because an individual may have preferences about the health and well-being of others, especially close relatives and his spouse. It does mean, however, that an individual may be willing to pay for and may derive benefits from improvements in the health of others. This form of altruism has been discussed by Needleman (1976) and Bergstrom (1982), among others. Also, since it may not be appropriate to talk of children's willingness to pay for improvements in their own health, these improvements can be valued at the parents' willingness to pay for them.

Environmental pollution that impairs human health can reduce people's well-being through at least the following five channels: (1) medical expenses associated with treating pollution-induced diseases, including the opportunity cost of time spent in obtaining treatment; (2) lost wages; (3) defensive or averting expenditures associated with attempts to prevent pollution-induced disease; (4) disutility associated with the symptoms and lost opportunities for leisure activities; and (5) changes in life expectancy or risk of premature death. The first three of these effects have readily identifiable monetary counterparts. The latter two may not. Since reducing pollution may be beneficial to individuals because it reduces some or all of these adverse effects, a truly comprehensive measure of benefits should capture all of these effects. Measures based solely on decreases in medical costs or lost wages are inadequate because they omit major categories of beneficial effects.

Methods for obtaining monetary values for improvements in health can broadly be categorized as those that rely either on observed behavior and choices (revealed preferences) or on responses to hypothetical situations posed to individuals (contingent valuation or bidding games). The first category includes all of those techniques that rely on demand and cost functions, market prices, and observed behavior and choices. (These methods are reviewed in chapters 3 and 4.) The second category includes asking people directly to state their willingness to pay or accept compensation for a postulated change, how their behavior would change, or how they would rank alternative situations involving different combinations of health and income or consumption. (This approach is reviewed in chapter 5.)

In examining health valuation methods, an important distinction is whether the method separates the valuation of health outputs from the estimation of a dose-response function. For example, one approach uses estimates of dose-response or exposure-response relationships taken from the biomedical literature to predict changes in some measure of health: for example, risk of death, incidence of disease, or number of symptom days. The approach then uses direct or indirect methods to estimate the values to individuals of the change in health. By contrast, the averting behavior approach (see chapter 3) views the relationship between pollution and health outcomes as a behavioral one than cannot be estimated independently of the valuation of the health outcome. This approach yields inferences about values from observations of how people change their behavior in response to environmental changes.

This chapter reviews and evaluates all of the methods that have been developed for estimating the various components of benefits. The evaluations are based, in part, on the degrees of correspondence found between the measurements of individuals' willingness to pay for health that are derived using the empirical method and the individuals' true willingnesses to pay. In other words, the evaluations consider how closely the method comes to capturing the whole range of beneficial effects related to improvements in health and the reduced threat of disease or premature death. Section 6.2 discusses the types of health effects that can be valued with economic methods. Sections 6.3 and 6.4 describe the economic measures of benefits of reduced risk of premature death and reduced morbidity, present the economic methods and models for estimating these economic value measures and discuss key empirical results. The concluding section assesses the state of the art of economic valuations of improvements in health and discusses some of the problems and limitations of the existing methods.

6.2 Defining and Measuring Changes in Health

Health has many dimensions, and environmental changes can affect people's health in a variety of ways, ranging from changes in the frequency of mild illness or irritating symptoms to increases in the risk of contracting a serious or fatal disease. This chapter follows the conventional economic practice in distinguishing between mortality and morbidity effects, where in the former case the primary endpoint of concern is death, while in the latter case, the focus is on nonfatal illness or a set of symptoms. This section describes the major categories of health effects and how they typically are measured in empirical economic research. It does not discuss the various approaches to estimating the effects of environmental change on measures of health, even though that would be valuable, because such a discussion requires a paper unto itself.

6.2.1 Mortality

For mortality, the measure of a change in health is the change in the probability of dying, or more specifically, the change in the conditional probability of dying at each age, for an identified group of individuals at risk.[1] Sometimes risk of death is stated in terms of lifetime risk, which is the probability that an individual will die from the cause in question at some time during his or her life. The problem with this measure is that it does not indicate the extent to which a disease shortens life expectancy.

A number of environmental contaminants ingested through various routes are known to cause or are suspected of causing increases in the incidence of fatal diseases such as cancer. Examples of environmental carcinogens include asbestos, which is found in building materials, trihalomethanes (THM), which are byproducts of drinking water chlorination, and polychlorinated biphenyls (PCB), which are persistent byproducts of manufacturing and industrial processes. One problem in valuing changes in risk of death due to exposure to these substances is that there is typically a lag between exposure to the substance and the production of cancerous cells. Because the individual is safe from cancer during this latency period, the benefits of reduced exposure do not occur until the end of the latency period.

6.2.2 Morbidity

Morbidity is defined by the U.S. Public Health Service as "a departure from a state of physical or mental well-being, resulting from disease or injury, of which the affected individual is aware." The last phrase in this definition is the key to answering an important question in air pollution control policy in the United States: What constitutes an adverse health effect from an economic perspective? This question arises, for example, when clinical studies reveal that exposure to an air pollutant under controlled conditions leads to detectable changes in the structure or function of people's organs without necessarily causing them pain, impeding their activities, or reducing their life expectancy. For example, air chamber experiments with heavily exercising young adults show that exposures to ozone at concentrations near the current primary air quality standard can reduce the subjects' forced expiratory volume (U.S. Environmental Protection Agency 1987). The changes are detectable; but are they adverse?

From an economic perspective, the answer to this question depends on whether the changes are perceived by the individual and whether the individual reveals or expresses a willingness to pay to avoid the effect. For example, the

[1] The conditional probability of dying at age t is the probability that one dies before his $t + 1$st birthday, given he is alive on his tth birthday. Age-specific mortality rates provide empirical estimates of the conditional probability of death.

subjects in the air chamber tests could be asked whether they could tell the difference between the high- and low-exposure tests, and if they could, whether they would require a higher compensation to go through the higher exposure level test again. If the answer to either question were no, and if the effect were reversible, then the benefits of avoiding the higher level of exposure and its attendant decrement in lung function would be zero. From an economic perspective, the effect would not be considered adverse. If the effect were irreversible and cumulative, then there might be an increase in the risk of chronic illness. This would be considered adverse. In this chapter, only health effects that are adverse in this economic sense are considered.

From the viewpoint of valuation, an important distinction to make concerning the health outcome that is affected by pollution is whether it occurs often enough and to a sufficient percentage of the population that it may be viewed as certain from the viewpoint of a single individual, or whether it is rare enough that its occurrence to an individual must be viewed as uncertain. For example, although all persons do not experience even minor respiratory symptoms each year, in practice these symptoms are treated as certain occurrences whose frequency is affected by air pollution. Illnesses, such as a severe asthma attack or cancer, that are severe enough to cause either death or an extended period of disability are relatively rare. Therefore, it is appropriate to treat them as uncertain and to examine the value of changing the risk of contracting these illnesses.

Morbidity can be classified in a variety of ways. One classification is according to duration: chronic versus acute. Acute morbidity refers to illnesses that last no longer than a few days and have well-defined beginnings and ends. Chronic morbidity refers to illnesses that are longer term and last indefinitely.

Another way of classifying morbidity is by the degree of impairment of activity. There are several categories of degree of impairment of activity for which published data are available. *Restricted activity days* are those days on which a person is able to undertake some, but not all, of his normal activities. *Bed disability days* are those on which a person is confined to bed, either at home or in an institution, for all or most of the day. Work loss days are those on which a person is unable to engage in his ordinary gainful employment. These categories of morbidity reflect measurements of responses to ill health rather than to health conditions themselves. Whether a given clinical manifestation of ill health results in any restriction on activity, bed disability, or work loss depends upon a number of socioeconomic variables, such as labor-force status, nonlabor sources of income, whether there are other income earners in the household, and so on. This distinction between illness and behavioral response to illness may be important if, for example, restricted activity days is used in an epidemiological study of the effect of air pollution on health.

A third way of classifying morbidity is by type of symptom or type of illness. Some studies measure morbidity by *symptom days,* that is, by the number of days individuals are afflicted with specific symptoms, such as a cough, headache, or throat irritation. Symptom days are used most often to measure illnesses that are not severe enough to restrict a person's normal activities.

The final way morbidity can be classified is based on the number of reported cases of a disease. These measures can be used for either acute or chronic diseases, but they typically do not distinguish the degree of severity of the disease in the affected individuals.

In addition to the morbidity and mortality effects described here, there is increasing evidence that the accumulation of relatively small quantities of some substances in the body can have subtle effects on neurological and physical development of infants and young people. For example, relatively low levels of lead in blood have been implicated in causing low birth weights, decrements to IQ in children, decreased growth rates, and presumably, shorter heights at maturity (U.S. Environmental Protection Agency 1986a, 1986b). These effects pose two kinds of problems for empirical analysis of the benefits of improved health. One is whether these effects are adverse, applying the definition given previously in this chapter. To the extent that subtle effects, such as low birth weight, are associated with other adverse health effects and illness, they pose no special methodological problems from an economic perspective. However, as with small decrements in forced expiratory volume, it is questionable whether smaller height is really an adverse health effect.

The other problem for empirical analysis involves determining whose willingness to pay for health is relevant for accurate measurements of benefits. For the examples given in the previous paragraph, it may be meaningless to talk of a child's willingness to pay to avoid effects of this sort. In other words, this situation is probably pushing the individualistic basis of welfare economics too far. Parents, however, can be expected to gain utility from the health and well-being of their children. Thus, their willingness to pay to avoid these kinds of effects in their children can be taken as the appropriate measurement of benefits.

6.3 The Benefits of Reducing Mortality Risks

6.3.1 Introduction

Because some forms of pollution may increase mortality or shorten life expectancy, economists have had to confront the question of the economic value of life and of preserving life. This is perhaps the most difficult and controversial aspect of valuing the health benefits associated with controlling pollution. To some people, the idea of putting a price or monetary value on human life is insensitive, crass, or even inhuman. Economists' efforts to deal with this issue have generated a large volume of critical literature. (See, for example, Schelling 1968; Mishan 1971, 1981; Broome 1978, 1982, 1985; Jones-Lee 1979; Williams 1979; Bailey 1980; MacLean 1986; Sagoff 1988, especially pp. 114-116; and Ulph 1982.) Some of these papers raise ethical and philosophical questions that go well beyond the scope of this chapter.

Economists have identified two alternative approaches to defining a measure of the value of lifesaving activities. The first approach is based on measurements of the economic productivity of the individual whose life is at risk. This is often referred to as the *human capital* approach because it uses an individual's discounted lifetime earnings as its measure of value. This approach has a long tradition. In fact, Landefeld and Seskin (1982) trace the idea back almost three hundred years. It has been the basis of some widely cited estimates of the benefits of controlling air pollution (e.g., Lave and Seskin 1971, 1977). The second approach is to use some indicator of the individual's willingness to pay to reduce his risk of death as the measure of value. In this section, these approaches to assigning monetary values to lifesaving activities are reviewed. This section also discusses some of the important economic and ethical issues that have been identified in this literature. This is followed by a more detailed discussion of models of willingness to pay. The section concludes with a review of empirical methods for implementing willingness-to-pay measures of the value of reduced mortality risks.

6.3.2 The Human Capital Measure of Value

The human capital measure is based on two assumptions: that the value of an individual is what he produces and that productivity is accurately measured by earnings. With the death of the individual, that output is lost. The human capital approach calculates the value of preventing the death of an individual who is presently of age j as the discounted present value of that individual's earnings over the remainder of his expected life. Earnings are usually calculated before taxes to reflect the government's, and therefore society's, interest in each individual's total productivity. Formally, the present value of expected lifetime earnings is given by $\Sigma_{t=j}^{T} q_{j,t}(1 + r)^{j-t}y_t$, where $q_{j,t}$ is the probability of the individual surviving from age j to age t, y_t is earnings of the individual at age t, r is discount rate, and T is age at retirement from the labor force.

Several issues must be addressed concerning the implementation of the human capital approach. One question is: Should measurements of an individual's productivity exclude or include his own consumption? Excluding consumption leaves a measure of the individual's worth as a producing asset to the rest of society, but this measure is the antithesis of the individualistic premise of conventional welfare economics.

Another question concerns the role of nonmarket production in the measure of productivity and value. The omission of nonmarket productivity is particularly troublesome in the case of homemakers. Some studies have attempted to correct for this omission by imputing earnings equal to the wages of domestic servants. Others have argued that the average earnings of employed females provides a measure of the opportunity cost of working at home for women. If women are rational, the value of home production must be equal to or greater than this opportunity cost. But even these adjustments do not capture the nonmarket productivity of other members of the household.

A third issue concerns what discount rate to use in calculating present value. The human capital value of children and young adults is particularly sensitive to the choice of a discount rate. Rates of 6 percent (Cooper and Rice 1976) to 10 percent (Landefeld and Seskin 1982) have been used in the literature. But Lind (1982) argued that the social rate of time preference may be much lower, possibly in the range of 2 to 3 percent.

Some of the implications of the human capital approach are unsettling. Because of discounting and the time lag before children become productive participants in the economy, the human capital approach places a much lower value on saving children's lives than on saving the lives of adults in their peak earnings years. And, because of earnings differences by sex and race, the human capital approach places a lower value on saving the lives of women and nonwhites than on saving the lives of adult white males. Furthermore, the human capital approach assigns zero value to persons who are retired, handicapped, or totally disabled.

The most important criticism of the human capital approach, however, is that it is inconsistent with the fundamental premise of welfare economics; namely, that it is each individual's own preferences that should count for establishing the economic values used in benefit-cost analysis. In this respect, the human capital approach is fundamentally at odds with the individualistic perspective of welfare economics and the theory of value that underlie the methods being described in this book.

Furthermore, both theoretical reasoning and empirical evidence suggest that human capital measures are poor approximations of the desired willingness-to-pay measures of value for small changes in the risk of death. One reason is that the human capital method implicitly assumes that utility equals the value of consumption over an individual's lifetime. Although an individual's earnings and the consumption it allows are positively related to the utility he derives from his own life, the strict equality of utility and consumption places unreasonable restrictions on the form of the utility function. Also, by definition, an individual with no financial wealth could pay no more than the present value of his expected earnings stream to avoid certain death; however, his statistical value of life based on willingness to pay for small probability changes could be several times his discounted earnings stream.

6.3.3 The Willingness-to-Pay Approach

In keeping with the assumption that individuals' preferences provide a valid basis for making judgments concerning changes in their economic welfare, increases in longevity or reductions in the probability of death due to accident or illness should be valued according to what an individual is willing to pay to achieve them. This presupposes that individuals treat longevity more or less like any other good rather than as a hierarchical value (in a lexicographic ordering). Evidence suggests this is a reasonable assumption. Individuals in a variety of situations act as if their preference functions include life expectancy or the probability of survival as arguments. In their daily lives, they make a variety of choices that involve trading off changes in the risk of death for other economic goods whose values can be measured in monetary terms. For example, some people drive to work rather than ride the bus or walk because driving is faster and more convenient, even though it increases their risk of dying prematurely. Also, some people accept jobs with known higher risks of accidental death because these jobs pay higher wages.[2] In both instances, people presumably perceive their situations as having been improved by the alternatives they have chosen despite the greater risks of death. When what is being lost or gained can be measured in dollars, then an individual's willingness to pay, or the amount of compensation he requires, is revealed by these choices. These choices are the basis of the willingness-to-pay approach to defining the economic value of reductions in the risk of death.

6.3.4 Ex Ante Versus Ex Post: The Ethical Issue of Perspective

One way of characterizing the economic approach is by saying that it avoids the issue of valuing life, per se, by recognizing that what people actually "buy and sell" through their choices and trade-offs is not life versus death, but small changes in the probability of dying. Another way of characterizing the economic approach is by saying that the economic value is derived by focusing on choices *ex ante;* that is, before the uncertainty about whether or not one will die is resolved. At some point in time, however, the uncertainty is resolved. Each individual will know if he or she will die immediately or will live a while longer. From this *ex post* perspective, those who will die would be willing to pay their total wealth to change the outcome or will require an infinite compensation to accept it. Critics of the economic perspective, such as Broome (1978), have argued that this difference in perspective can have no ethical or moral significance, and therefore, neither willingness to pay nor compensation measures based on the *ex ante* perspective are morally acceptable.

---

[2] Evidence that risks to life do affect decisions in this way comes from a variety of empirical studies of behavior. See Peltzman (1975) for evidence about risk and driving.

One defense of the economic perspective is based on the observation that people appear to be willing to make *ex ante* trade-offs involving risks of death. If people are rational and if their preferences are taken to be the basis of economic value measures, then their willingness to consent to *ex ante* trade-offs must have some ethical significance. Furthermore, for many of the public policy issues in which value of risk reduction information might be used, it will never be known *ex post* whose deaths were caused by failure to adopt a policy or whose lives were prolonged by a policy to reduce risks of death. For example, suppose a proposed air pollution control policy is predicted to reduce cancer mortality in the population by 100. If the policy is adopted, the identities of the 100 people whose lives were prolonged will never be known. If the policy is not adopted, it will never be known which 100 of the 400,000 or so cancer deaths in any one year were "caused" by the failure to adopt the policy. Economists argue that it is consent, and the veil of ignorance over who dies and who is saved, that legitimizes the *ex ante* perspective and its focus on the value of changes in risks rather than on the value of life versus death.[3]

6.3.5 Models of Individual Choice and Willingness to Pay

Economic models of individual choice under uncertainty have been used to define willingness to pay for a change in the conditional probability of dying and to generate predictions about how willingness to pay varies with age, lifetime earnings, and risk preferences. These models also provide a theoretical foundation for empirical studies of willingness to pay.

These models are based on the assumption that individuals make choices among alternatives so as to maximize the mathematical expectation of utility; that is, the sum of the utilities realized in the alternative states of the world each weighted by its probability of occurrence. Because individuals must choose alternatives before the uncertainty is resolved, expected utility is an *ex ante* concept. In all of these models, the willingness to pay for a reduction in the probability of death is the maximum sum of money that can be taken from the individual *ex ante* without leading to a reduction in his or her expected utility. Similarly, the willingness to accept compensation for an increase in the probability of dying is the sum of money that just compensates for the greater risk by increasing consumption sufficiently to equalize the expected utilities of the two alternatives.

Life-Cycle Models of Willingness to Pay. The life-cycle model of willingness to pay clarifies the time dimension of risks to life. Several authors, beginning with Usher (1973) and Conley (1976), have used a life-cycle consumption-saving model with uncertain lifetime to analyze an individual's willingness to

---

[3] For further discussion of the *ex ante* versus *ex post* perspective, see Broome (1978), Ulph (1982), Thaler (1982), and Linnerooth (1982).

pay at age j for a change in his conditional probability of dying at age k, $k > j$. Given this expression one can:

1. Examine how WTP for a change in current probability of death varies with age.
2. Determine under what circumstances the present value of expected lifetime earnings (the human capital measure) constitutes a lower bound to WTP.
3. Examine the effects of a latency period; that is, compare the WTP for a change in current probability of dying (for example, due to safer roads) with WTP for a change in risk of death in the future (for example, due to reduction in exposure to a carcinogen with a 20-year latency period).

The Model. In the life-cycle model, the individual has a probability distribution over the date of his death. Let j denote the individual's current age and $p_{j,t}$ the probability that he dies at age t, just before his $t + 1$st birthday; that is, that he lives exactly $t - j$ more years. Because the $\{p_{j,t}\}$ constitute a probability distribution and assuming a horizon of T years, it follows that $p_{j,t} \geq 0$, $t = j, j + 1, \ldots, T$, and that $\Sigma_{t=j}^{T} p_{j,t} = 1$. The probability that the individual will survive to his tth birthday, given that he is alive at age j, $q_{j,t}$ is the probability that he will die at $t + 1$ or later. Formally, $q_{j,t} = \Sigma_{s=t+1}^{T} p_{j,s}$. The probability that the individual survives to his $t + 1$st birthday, given that he is alive on his tth, is $q_{j,t+1}/q_{j,t}$. Henceforth, this conditional probability will be denoted as $1 - D_t$, where D_t is the conditional probability of dying at age t.

Expected lifetime utility at age j is the sum of the utility of living exactly $t - j$ more years times the probability of doing so. Assuming that the individual has no bequest motive, this may be written as

$$V_j = \sum_{t=j}^{T} p_{j,t} \, u_t(c_j, c_{j+1}, \ldots, c_t), \tag{6.1}$$

where $u_t(\cdot)$ is the utility of consumption in years j through t. In most applications in the area of risk valuation (Arthur 1981; Shepard and Zeckhauser 1982, 1984), $u_t(\cdot)$ is assumed additively separable, implying that (6.1) may be written as

$$V_j = \sum_{t=j}^{T} (1 + \rho)^{j-t} q_{j,t} U(c_t), \tag{6.2}$$

where ρ is the rate of individual time preference and T is the maximum lifespan. The period utility function, $U(c_t)$, obeys three conditions: it is increasing in c_t; it is strictly concave; and it is bounded from below.

Two points about (6.2) should be emphasized. First, the equation assumes that the utility of living depends solely on consumption, without regard for the length of an individual's life. The concavity of the utility function, however, implies that it is always desirable to spread a given amount of consumption over a longer time span. Second, most authors treat survival probabilities as exogenous to the individual. The exceptions are Conley (1976) and Viscusi and Moore (1989).

The individual is assumed to choose his pattern of consumption over time, given initial wealth, W_j, annual earnings, y_t, $t = j, \ldots, T$, and capital market opportunities. Arthur (1981) and Shepard and Zeckhauser (1982, 1984) assumed that the individual could save by purchasing actuarially fair annuities and borrowing via life-insured loans.

If actuarially fair annuities are available, an individual who invests \$1 at the beginning of his jth year will receive at the end of the year \$$(1 + R_j)$ with probability $1 - D_j$ and nothing with probability D_j. For the annuity to be fair — that is, for it to have an expected payout of $1 + r$, where r is the riskless rate of interest — R_j must satisfy $(1 + R_j)(1 - D_j) = 1 + r$. Because $R_j > r$, an individual who can save via fair annuities will clearly do so. To cover the possibility that the person may die before repaying a loan, it is assumed that the individual also borrows funds at the actuarial rate of interest.

To prevent unlimited borrowing, the individual's budget constraint requires that the present value of borrowing, discounted at the actuarial rate of interest, equal the value of initial wealth:

$$\sum_{t=j}^{T} \left[\prod_{i=j}^{t-1} (1 + R_i)^{-1} \right] (c_t - y_t) = W_j.$$

This is equivalent to requiring that the present value of expected consumption equal the present value of lifetime earnings plus initial wealth:

$$\sum_{t=j}^{T} q_{j,t}(1 + r)^{j-t} c_t = \sum_{t=j}^{T} q_{j,t}(1 + r)^{j-t} y_t + W_j. \tag{6.3}$$

The individual's pattern of consumption over his life cycle thus is determined by maximizing (6.2) subject to (6.3).

Willingness to Pay. We now consider in the context of the model how government health and safety regulations affect lifetime utility. Government regulations alter the probability that a person dies in any given year only if the individual is alive at the beginning of the year. Government programs thus alter the conditional probability (D_k) that an individual will die at age k; that is, the probability that the individual dies between his kth and $k + 1$st birthdays, assuming he is alive on his kth birthday. A program to increase the police force in a city in a single year reduces D_k for that year alone. A program that reduces an individual's exposure at age 30 to a carcinogen with a 20-year latency period reduces the conditional probability of dying at all ages after 50 $(D_{50}, D_{51}, D_{52}, \ldots)$.

It should be emphasized that when the conditional probability of death is altered at age k, it affects the probability of surviving to ages $k + 1$ and beyond, $q_{j,k+1}$, $q_{j,k+2}$, ..., $q_{j,T}$ since, by repeated use of the definition of D_t,[4] $q_{j,k} = (1 - D_j)(1 - D_{j+1}) \ldots (1 - D_{k-1})$.

In practice, the changes in $\{D_k\}$ corresponding to some public project are likely to be small. For example, it has been estimated that the risk of dying of cancer due to all environmental causes is only 3.6×10^{-5} (Doll and Peto 1981). For this reason the literature has focused on marginal changes in $\{D_k\}$. If several D_k's change, WTP for the sum of the changes equals the sum of the willingnesses to pay.

Formally, the individual's willingness to pay at age j for a change in D_k, or $\text{WTP}_{j,k}$, is the wealth that must be taken away from him at age j to compensate him for a reduction in D_k while keeping expected utility constant:

$$\text{WTP}_{j,k} = - \frac{dV_j/dD_k}{dV_j/dW_j}.$$

Applying the envelope theorem to the Lagrangian function that corresponds to (6.1) and (6.2), $\text{WTP}_{j,k}$ can be written as

$$\text{WTP}_{j,k} = (1 - D_k)^{-1} \sum_{t=k+1}^{T} q_{j,t}[(1 + \rho)^{j-t}U(c_t)\lambda_j^{-1} \tag{6.4}$$
$$+ (1 + r)^{j-t}(y_t - c_t)].$$

Willingness to pay at age j for a change in the conditional probability of death at age k equals the loss in expected utility from year k onward, which is converted to dollars by dividing by the marginal utility of income in year j, λ_j. Added to this is the effect of a change in D_k on the budget constraint. A reduction in D_k makes an individual wealthier by increasing the present value of his expected lifetime earnings from age $k + 1$ onward. An increase in survival probabilities, however, has the opposite effect: it decreases the consumption that the person can afford in years $k + 1$ through T, and as a result, his WTP is reduced by the present value of this amount.

Two points about expression (6.4) should be noted. First, $\text{WTP}_{j,k}$ is the rate at which the individual is willing to trade wealth for risk. To compute the dollar value of a change in risk, equation (6.4) must be multiplied by the magnitude of the risk change. Thus, if $\text{WTP}_{j,k} = \$2 \times 10^6$, but the change in risk is only 10^{-6}, then willingness to pay for the change in risk is two dollars. Second, equation (6.4) very likely constitutes a lower bound to WTP since, in the life-cycle model, utility is a function solely of consumption. As Bergstrom (1982) has pointed out, the intertemporal objective function, if derived from preferences on lotteries, should include a term that values survival per se. If this term is an increasing function of the $\{q_{j,t}\}$, any willingness-to-pay measure derived from (6.2) must be regarded as a lower bound to true willingness to pay.

---

[4] A change in D_k also affects life expectancy, which is the sum of survival probabilities from the current age onward, $\Sigma_{t=j}^{T} q_{j,t}$. Note, however, that there are many changes in $\{D_k\}$ that result in equivalent changes in life expectancy.

Equation (6.4) yields several useful insights, including the conditions under which the human capital measure of the value of life, which is the present value of expected lifetime earnings, is a lower bound to WTP.[5] By use of the first-order conditions for utility maximization, the term in brackets in (6.4) can be written as

$$(1 + r)^{j-t} \left[\frac{U(c_t)}{U'(c_t)} - c_t + y_t \right].$$

This implies that if $U/U' - c > 0$ for all t, then willingness to pay exceeds the present discounted value of lifetime earnings. As noted by Conley (1976) and Cook (1978), $U/U' - c > 0$ implies that the average utility of consumption exceeds the marginal utility — a condition that holds for all increasing, concave utility functions as long as an individual's consumption exceeds subsistence.

Equation (6.4) also implies that as k, the age at which risk of death advances, $WTP_{j,k}$ must decline.[6] This implies that the value of reducing a person's current probability of dying, D_j, must always be greater than the value of reducing exposure at age j to a carcinogen that has a latency period of $k - j$ years. In the second case, fewer expected life years are saved (Cropper and Sussman 1988b).

Shepard and Zeckhauser (1982, 1984) have used (6.4) to examine the behavior of WTP for a change in current risk of death $WTP_{j,j}$ over the life cycle. They determined that if an individual's consumption is constant for all t, as happens when the riskless rate of interest is equal to the subjective rate of time preference, $WTP_{j,j}$ will decline monotonically with age: that is, younger people will always have a higher WTP to reduce current risk of death than will older people. If, however, consumption increases during some portion of an individual's life cycle, $WTP_{j,j}$ may also increase, but only to a point, and then it will decline. If, for example, an individual cannot be a net borrower, but can be a lender at the riskless rate of interest, his consumption is likely to be constrained by income at the beginning of his life. This will cause the present value of the utility of consumption, and hence $WTP_{j,j}$, to increase up to a point and then decline.[7]

In the special case in which the utility function exhibits constant elasticity of marginal utility — that is, $U(c) = c^\beta$, $0 < \beta < 1$ — one can examine the effect of changes in risk preferences on WTP. In this case, (6.4) becomes

$$WTP_{j,k} = (1 - D_k)^{-1} \sum_{t=k+1}^{T} q_{j,t}(1 + r)^{j-t}[y_t + c_t(1 - \beta)/\beta], \tag{6.4'}$$

which is a decreasing function of β. This can be given two interpretations.

---

[5] This condition was first noted in a static context by Conley (1976) and Cook (1978).

[6] Strictly speaking, this holds only if $U/U' - c + y$ is positive; however, this will be the case so long as the individual is above subsistence.

[7] The expression for $WTP_{j,k}$ in the no-net-borrowing case is identical to (6.4) except that the term $(1 + r)^{j-t}(y_t - c_t)$ does not appear inside brackets.

First, since $1 - \beta$ is the relative risk aversion index for the constant elasticity utility function, equation (6.4′) indicates that more risk-averse individuals (those with smaller values of β) have a higher willingness to pay to reduce their risk of death than have less risk-averse persons. Second, the rate at which individuals are willing to substitute consumption at $t + 1$ for consumption at t is an increasing function of β. Small values of b are associated with large values of WTP because individuals cannot easily substitute consumption in one time period for consumption in the next.

Extensions of the Basic Model. Arthur (1981) embedded the life-cycle model for an individual into a one-sector general equilibrium model with population growth. From this, he derived an expression for the value to society of a change in people's conditional probability of death. He stated that society's WTP for a change in D_j differs from the individual's WTP because society gains productivity from an increased population if people's risk of dying during their reproductive years is reduced. Arthur also noted that the consumption cost of the additional years of life that is reflected in equation (6.3) may or may not be borne by the individual, depending on institutional arrangements; however, it will be borne by society.

The recognition that each individual typically is part of a family has led to efforts by economists to model individual choice and willingness to pay within a family context. Cropper and Sussman (1988a) assume that an individual's expected utility depends on the size of the bequest that he would leave heirs in the event of his death and that the utility of the bequest depends on the number of heirs and their degree of economic dependence on his earnings. The individual is also assumed to gain utility while alive from the consumption of his dependents as well as from their existence. A bequest motive, other things equal, reduces willingness to pay for a reduction in risk of death because one can provide for one's heirs by saving or purchasing life insurance. On the other hand, the fact that a married person can get more utility while alive than can a single person should raise the married person's willingness to pay above that of the single person. A study of willingness to pay by Jones-Lee (1986) suggests that the latter effect dominates. (See also Jones-Lee, Hammerton, and Philips 1985.)

Static Models of Willingness to Pay. In addition to the life-cycle models described in the previous section, economists have also used simpler, static models of willingness to pay to reduce risk of death (Jones-Lee 1974; Sussman 1984). Such models are more convenient for obtaining comparative static results, and by making it easier to model risk of death as endogenous, they provide a more convenient foundation for empirical work.

The simplest static counterpart to the life-cycle model assumes that the individual either dies immediately, with probability $\hat{p}$ (analogous to D_j above), or survives to enjoy lifetime consumption of C, with probability $1 - \hat{p}$. Expected utility is given by

$$E(U) = (1 - \hat{p})U(C),$$

assuming that the individual has no bequest motive. The individual's willingness to pay for an exogenous change in p from $\hat{p}$ to p is defined implicitly by

$$(1 - \hat{p})U(C) = (1 - p)U(C - \text{WTP}),$$

implying that marginal WTP is

$$\partial\text{WTP}/\partial p = U(C - \text{WTP})/[(1 - p)\partial U/\partial C]. \tag{6.5}$$

In interpreting (6.5), $U(C - \text{WTP})$ must be viewed as the utility of lifetime consumption, which implicitly depends on an individual's age and future survival probabilities. WTP thus equals lifetime utility divided by the expected marginal utility of consumption. As was shown by Jones-Lee (1974), WTP is an increasing function of current probability of dying $\hat{p}$ and an increasing function of endowment C.[8]

The preceding models, while providing useful insights, do not suggest a method of estimating willingness to pay other than through direct questioning. To use indirect market methods, WTP must be linked to voluntary risk-taking behavior; that is, p must be made endogenous. Suppose now that there are three causes of death: one exogenous, one job-related, and one related to the level of consumption of a private good, X. The corresponding conditional probabilities of death are denoted p_e, p_j, and $p_x(X)$. Assuming these causes are independent, the probability of surviving the current period is the product of the probabilities that the individual does not die from each of the three causes.

$$1 - p = (1 - p_e)(1 - p_j)[1 - p_x(X)].$$

Expected utility is given by

$$E(U) = (1 - p_e)(1 - p_j)[1 - p_x(X)]U(C,X), \tag{6.6}$$

where C is consumption of a numeraire good. The individual picks p_e, X, and C to maximize (6.6) subject to the budget constraint

$$C + q \cdot X = I + w(p_j), \tag{6.7}$$

where I is wealth, $w(p_j)$ is the market locus relating the wage to job risk and q is the price of X.

Applying the envelope theorem to the Lagrangian function that corresponds to equations (6.6) and (6.7) yields an equation for an individual's willingness to pay for a marginal change in exogenous risk of death, dI/dP_e:

---

[8] Jones-Lee's model differs from the one presented here by allowing for a bequest motive. In Cook (1978) the individual can invest his initial endowment W in an actuarially fair annuity paying $1/(1 - p)$ per dollar invested. Consumption is therefore given by $C = W/(1 - p) + Y$, where Y is lifetime earnings. In this case the expression for willingness to pay is given by $U(C)/U'(C) - C + Y$, analogous to (6.4).

$$\text{WTP} = (1 - p_j)(1 - p_x)U(C,X)/[(1 - p)\partial U/\partial C]. \tag{6.8}$$

This is the value of the utility lost if the individual dies, $U(C,X)$, converted to dollars by dividing by the expected marginal utility of income, $(1 - p)\partial U/\partial C$, times the probability that he or she does not die due to other causes, $(1 - p_j)(1 - p_x)$.

An interesting question is whether WTP can be estimated by observing risk-taking behavior in consumption or in the labor market. To answer this, note that the first-order conditions for choice of job risk imply that the individual equates the marginal compensation foregone by moving to a safer job to the marginal benefit of a reduction in job risk,

$$w'(p_j) = (1 - p_e)(1 - p_x)U(C,X)/[(1 - p)\partial U/\partial C] \approx \text{WTP}.$$

The latter is almost identical to the value of an exogenous risk change, (6.8), except that the individual's probability of not dying due to other causes is now $(1 - p_e)(1 - p_x)$. If $1 - p_e \approx 1 - p_j$, then $w'(p_j)$, which is the marginal price of risk, can be used to estimate the individual's willingness to pay for a change in exogenous risk, for example, environmental risk.

Likewise, first-order conditions for choice of X imply, for $U_x = 0$,

$$-q/p'_x(X) = (1 - p_e)(1 - p_j)U(C,X)/[(1 - p)\partial U/\partial C] \approx \text{WTP}. \tag{6.9}$$

Assuming $(1 - p_x) \approx (1 - p_e)$, the price of X divided by the effect of X on probability of dying can be used to approximate WTP for an exogenous risk change.

Conclusions. Perhaps the most important conclusion to be drawn from this review of theoretical models of individual choice and willingness to pay is that the value each person attaches to a small reduction in his probability of dying is likely to differ because of differences in underlying preferences, age, wealth, number of dependents, degree of aversion to risk, and level of risk to which he is currently exposed. This conclusion must be kept in mind when interpreting and using the results of empirical estimates of willingness to pay that are based on averages of groups that, most likely, form a heterogeneous population.

A second conclusion is that in the case of multiple risks of death, where the individual can "purchase" reductions in one component of risk, the price or marginal cost of reducing that component of risk can usually be taken as a close approximation of the individual's willingness to pay for reductions in other components of risk. This theoretical result provides the basis for many of the empirical estimates of the value of risk reduction to be described in the next section.

6.3.6 Empirical Valuation of Risk of Death

The preceding model suggests that willingness to pay for a change in the conditional probability of death can be estimated by examining the trade-offs that individuals make between money and small changes in their risk of death. This is commonly termed the indirect market approach and is treated in chapters 3 and 4. The contingent valuation approach described in chapter 5 elicits this information by asking individuals what they are willing to pay to reduce their risk of death.[9] Indirect market methods can be further subdivided into the compensating wage approach and the averting behavior approach. The latter approach uses data on safety-enhancing consumption, such as purchasing smoke detectors or seat belts, to draw inferences about people's willingness to pay for the chance of a longer life.

Before the empirical difficulties in implementing these approaches are examined, it is necessary to introduce a concept encountered frequently in the empirical literature — the *value of a statistical life*. Since risks of death from environmental contaminants are thought to be small, it is customary to express risks in terms of the number of statistical lives lost due to a contaminant. If exposure to some substance increases the probability of death for person i by ΔD_i and there are N persons in the exposed population, the number of statistical lives lost is $\sum_{i=1}^{N} \Delta D_i$.

One drawback of this concept is that it does not reflect the age of the people at risk, and hence, the number of expected years of life that are lost. The value of a statistical life is the sum of the affected people's willingnesses to pay for these risk changes divided by the number of statistical lives saved, or

$$\sum_{i=1}^{N} \text{WTP}_i \Delta D_i \bigg/ \sum_{i=1}^{N} \Delta D_i. \tag{6.10}$$

If the change in risk is the same for everyone in a population, or $\Delta D_i = \Delta D$ for all i, then (6.10) is simply the average value of WTP_i, which is the rate at which each individual is willing to trade risk for wealth. Estimates of the value of a statistical life are found in the empirical studies described in the following paragraphs.

Averting Behavior Studies. The use of smoke detectors (Dardis 1980) and seatbelts (Blomquist 1979) are both examples of activities that will, at a cost, reduce an individual's risk of death. Under the assumption that these activities are pursued to the point where their marginal cost equals the marginal value of reduced risk of death, they can be used to value an individual's willingness to pay to reduce his risk of death.[10]

If an averting good X is infinitely divisible, then equation (6.9) holds for each individual in the population. Given data for each person on the marginal cost of X (which is likely to vary among individuals, especially if there is a

---

[9] The change in foregone earnings — the human capital measure of the value of a risk change — is sometimes used as a lower bound to willingness to pay.

[10] Empirical studies based on seatbelt use and the purchase of smoke detectors predate mandatory seatbelt laws and discounts on homeowners' insurance for smoke detector purchase. Such regulations make it difficult to use these averting behaviors to infer the value of risks to life.

time cost associated with X) and on the effect of X on survival probability, one can estimate WTP. One problem in doing this is that both q and $p_x'(X)$ are hard to measure. In the case of seatbelts, for example, q involves an unobservable disutility component and a time component that is difficult to value. The relevant measure of the effect of the averting behavior on risk of death is the individual's perception of this risk reduction; however, no published studies of averting behavior have used data on risk perceptions.[11]

A more fundamental problem in applying the averting behavior model is that averting behaviors, such as wearing a seatbelt or purchasing a smoke detector, that have been used in actual studies are 0-1 decisions. These types of decisions are undertaken provided their marginal cost is no greater than their marginal benefit, or

$$-q/p_x'(X) \le \text{WTP}. \tag{6.10'}$$

Equation (6.10') holds as an equality only for the last person to purchase a smoke detector; for all other purchasers, WTP exceeds the marginal cost of a reduction in the conditional probability of dying. When Dardis divided the annual cost of operating a smoke detector by the corresponding increase in survival probability, she thus estimated WTP for the last person to purchase a smoke detector and not for the average purchaser.

To estimate an average value of willingness to pay requires data on the cost of the averting activity and on its effect on reducing the risk of death for a cross section of individuals. If q and $p_x'(X)$ vary among individuals in the sample, it is possible to estimate the average value of WTP using a logit or probit model of averting behavior. Blomquist discussed this procedure in his study of seatbelt use, but was unable to implement it due to a lack of data.[12]

A third problem arises when the averting activity produces joint products; for example, when it reduces the risk of injury or property damage as well as the risk of death. This problem is either handled by treating the value of joint products as zero, and thus obtaining an upper bound to WTP, or by assuming that the value of injury is some multiple of the value of life.

In view of the foregoing problems, especially the discreteness of the averting activity, it is not surprising that estimates of the value of a statistical life obtained from averting behavior studies are lower than estimates obtained from other sources. (See table 6.1.)

Compensating Wage Studies. An alternative approach is to infer the value

---

[11] A possible exception is Ippolito and Ippolito's (1984) use of the change in smoking behavior following the Surgeon General's report to estimate the value of a statistical life.

[12] The probability that person i undertakes the averting activity is $P[q_i - \text{WTP}\Delta p_x(X_i) \le u_i]$, where u_i is an error term that reflects the difference between WTP_i and WTP, the average value of WTP for the population. Under the assumption that u_i is independently and identically normally distributed for all persons in the sample, this may be written using the standard normal distribution function Φ, $\Phi[q_i/\sigma - \Delta p_x(X_i)\text{WTP}/\sigma]$. The coefficients in this equation can be estimated by maximum likelihood methods and WTP estimated by dividing the coefficient of $\Delta p_x(X_i)$ by the coefficient of q_i.

TABLE 6.1
Estimates of the marginal willingness to pay for reductions in risks of accidental death.

Study	Mean risk level for the sample[a]	(millions of 1986 dollars)	
		Range of estimates	Judgmental best estimate
Early low-range wage-risk estimates			
Thaler and Rosen (1976)[b]	11.0	0.44-0.84	0.64
Arnould and Nichols (1983)[b]	11.0	0.72	0.72
Early high-range wage-risk estimates[c]			
R. Smith (1976)	1.0 & 1.5	3.6- 3.9	3.7
V. K. Smith (1983)[d]	3.0	1.9- 5.8	3.9
Viscusi (1978)	1.2	4.1- 5.2	4.3
Olson (1981)	1.0	8.0	8.0
Viscusi (1981)	1.0 - 1.5	8.5-14.9	8.5
without risk interaction terms	1.04	5.4- 7.0	7.0
with risk interaction terms	1.04	4.7-13.4	
New wage-risk studies			
Dillingham (1985)	1.4 - 8.3	2.1- 5.8	2.5
Marin and Psacharopoulos (1982)[e]			
manual workers	2.0	2.7- 3.1	2.9
nonmanual workers	2.0	9.0	
Gegax, Gerking, and Schulze (1985)			
all union workers	8.2	1.9	
union blue-collar workers	10.1	1.6	1.6
Moore and Viscusi (1988)	0.79	5.0- 6.5	5.4
New contingent valuation studies			
Jones-Lee, Hammerton, and Philips (1985)	0.8 - 1.0	1.6- 4.4	3.0
Gerking, DeHaan, and Schulze (1988)	4.2 -10.0	2.4- 3.3	2.6
Averting behavior studies			
Dardis (1980)	0.9	0.36-0.56	0.46
Blomquist (1979)	3.0	0.38-1.4	0.61
Ippolito and Ippolito (1984)	varied	0.24-1.26	0.52

Source: Adapted from Fisher, Violette, and Chestnut (1989).
[a] Approximate annual deaths per 10,000 people.
[b] Based on actuarial risk data.
[c] All based on BLS industry accident rates.
[d] Assuming 0.4 percent of all injuries are fatal, as reported by Viscusi (1978) for the BLS injury statistics, and that the risk premium for fatal injuries is 33 to 100 percent of the premium for all risks.
[e] Their age-adjusted normalized risk variable is not directly comparable with the risk levels used in other studies. However, the average risk of death for the entire samle was 2 in 10,000.

of a statistical life from wage premia that workers receive to compensate them for risk of accidental death. Equation (6.8) suggests that workers equate the marginal cost of working in a less risky job to the marginal benefit — that is, they equate the lower wage they receive to the value in dollars of their remaining years of life. In empirical applications, this equation is usually embedded in an hedonic model of job choice in which each worker selects the collection of job attributes, including risk of death, that equates the marginal benefit of each attribute to its marginal cost. (Hedonic models are discussed in chapter 4.)

To estimate the risk premium, which is the partial derivative of the market wage function with respect to risk of death, requires having data on wages, job attributes, and worker attributes. These data are used to estimate an hedonic wage function, an equilibrium relationship between the wage, job characteristics, and variables affecting worker productivity.[13]

If the hedonic wage function can be estimated satisfactorily, it is then possible to calculate the risk premium (evaluate the partial derivative of the function) at a given risk level and set of job attributes. To see how the risk premium varies with individual characteristics, however, it is necessary to regress the risk premium on the level of risk assumed and on these characteristics.

There are three problems with using the compensating wage approach. One is that compensating wage differentials exist in the marketplace only if workers are informed of job risks. Thus, the absence of compensating differentials need not mean that workers do not value reducing risks of death; instead, they may simply be unaware of them. A second problem is that compensating differentials appear to exist only in unionized industries (Dickens 1984; Gegax, Gerking, and Schulze 1985). This suggests that the wage differential approach may provide estimates of the value of a statistical life only for certain segments of the population. Third, if workers base their decisions on biased estimates of job risks, or if the objective measures of job risk used in wage studies overstate or understate workers' risk perceptions, market wage premia may provide biased estimates of the value of a statistical life.

In practice, the major difficulties in estimating hedonic wage functions have come from obtaining data on job-related risk of death and injury, and from obtaining adequate data on other job characteristics, especially those that may be correlated with risk of accidental death. Because of this, it is useful to categorize hedonic wage studies by data source. This categorization is followed in table 6.1, which summarizes empirical estimates of the value of a statistical life.

An early group of studies (Thaler and Rosen 1976; Brown 1980; Arnould and Nichols 1983), which, in general, found low values of a statistical life, used actuarial data on deaths by occupation due to all causes rather than data on occupationally-related deaths. These studies have generally been discredited for using inappropriate risk data.

A second group of studies has used data on job-related deaths by industry without, however, controlling for occupation. Most of these studies merge data from the Bureau of Labor Statistics (BLS) on fatal accidents by industry,

---

[13] The hedonic wage function is the locus of points at which firms' marginal wage offers (as functions of job characteristics), equal workers' marginal acceptance wages. See Thaler and Rosen (1976) for a description of equilibrium in hedonic labor markets.

with micro data on wages and job and worker characteristics (Viscusi 1978,1981; Olson 1974). Moore and Viscusi (1988) used data from the National Institute for Occupational Safety and Health (NIOSH) on job risks by state for broad industry groups. These studies (see table 6.1) generally found very high values of a statistical life, ranging from $4 million to $8 million (1986 dollars). The problem with these risk measures is that within an industry there is likely to be significant variation in risk across occupations. Furthermore, these risk indices may capture interindustry wage differences that are unrelated to job safety.

A third group of studies, which includes those of Dillingham (1985) and Marin and Psacharopoulos (1982), are designed to remedy the lack of risk data by occupation. Dillingham's study estimated a value of statistical life for U.S. workers. He combined workman's compensation data on job risk with wage data from the 1977 Quality of Employment Survey and 1970 New York State Census. When the same risk-by-occupation variable was used with each of the data sets, and when dummy variables for occupation and industry were included in the equation, the implied value of life estimates (1986 dollars) were $2.077 million and $1.702 million, respectively. When risk data by occupation and industry were used, the risk variable was not significant using the New York census data, but was significant using the Quality of Employment data and implied a value of life of $2.667 million (1986 dollars). Marin and Psacharopoulos, using a measure of deaths by occupation, estimated wage premia for workers in the United Kingdom. Their value of life estimated for manual workers was $2.5 million (1986 dollars).

A recent hedonic wage study by Gegax, Gerking, and Schulze (1985) used individual worker data on risk perceptions and job characteristics to estimate the hedonic wage function. Their survey circumvented the problem of inaccurate worker perceptions and provided a more detailed description of jobs than is available in most labor market data sets. An additional advantage of this study is that it collected data on workers' announced willingnesses to pay to reduce risk of death (responses to contingent valuation questions), so that the latter might be contrasted to estimates of the value of a statistical life based on hedonic wage functions. In their hedonic wage studies, Gegax, Gerking, and Schulze found a significant value of life of $1.6 million (1986 dollars) for blue-collar, unionized workers and a value of $1.9 million for all workers.

Although the last three studies used more appropriate measures of risk of death, it should be noted that none of the studies controlled for risk of injury. The compensating wage differentials reported by the authors may thus reflect compensation for risk of injury as well as risk of death.

Hedonic Property Value Studies. Just as wages are higher in risky occupations to compensate workers for their increased risk of death, property values may be lower in more polluted areas to compensate homeowners for their increased risk of death. It is thus possible, in principle, to infer willingness to pay to reduce risk of death from an hedonic housing price function. The hedonic technique has been applied to value reduced exposure to hazardous waste by Michaels, Smith, and Harrison (1987), Kohlhase (forthcoming), and Schulze et al. (1986), who used distance from a hazardous waste site to proxy health risks. To our knowledge, however, there are no hedonic housing studies in which the risk of death, per se, enters the hedonic function.

Estimation of a Willingness-to-Pay Function. The preceding studies were based on a static model in which willingness to pay varies implicitly, but not explicitly, with age. Because the model estimates an average value of WTP, it values the average number of life years lost in the sample. It would be better to be able to express WTP as a function of age and other relevant variables, such as income, marital status, or number of dependents.

In the context of hedonic wage models, this can be done by regressing the marginal price of risk on the level of risk chosen, age, income, and other relevant variables, provided that the identification conditions are satisfied. Viscusi and Moore (1989) estimated a willingness-to-pay function for risk of death using data from the 1982 wave of the University of Michigan Panel Study of Income Dynamics, which was supplemented by data from the National Traumatic Occupational Fatality Survey published by NIOSH. The study is notable for two reasons. First, it estimated both an hedonic wage equation and a marginal bid function for job risk. The parameters of the marginal bid function were identified by estimating the marginal price of job risk for different regions of the United States, which assured variation in marginal price that was independent of the variables entering the marginal bid function. Second, the marginal bid function was derived from a life-cycle model of job choice, in which the benefits of moving to a safer job reflected discounted life years saved. This allowed Viscusi and Moore to estimate the rate at which workers discount future lifesaving benefits. One drawback of the model, however, is that the discounted life years saved did not vary with current age.

Limitations of Indirect Market Methods. For market prices to convey information about individual preferences for risk reduction, individuals must be informed about the risks being valued. Furthermore, risks, as measured by the researcher, must correspond to individuals' risk perceptions at the time their market decisions were made. As noted earlier, most indirect market studies use objective measures of risk rather than measures of individuals' perceptions of risk. Therefore, it would be interesting to know how closely the two correspond. Information about such a correlation, however, is scant.

The only evidence pertaining directly to risk perceptions in indirect market studies is research that compares workers' risk perceptions with data on frequency of job-related death and injury.[14] This evidence suggests that

[14] In averting behavior studies (seatbelts, smoke detectors) it is assumed that individuals' perceptions of the change in risk caused by the averting activity equal the objective risk change. We know of no studies that compare these quantities.

value causes of death that are specific to environmental hazards. Two contingent valuation studies that attempted to value risk of death in an environmental context are Mitchell and Carson's (1986b) study of WTP to reduce trihalomethane levels, and Smith and Desvousges's (1987) study of WTP to reduce exposure to hazardous waste. Two other recent studies, Gerking, De Haan, and Schulze (1988) (see also Gegax, Gerking, and Schulze 1985) and Jones-Lee, Hammerton, and Philips (1985) valued current changes in risk of accidental death.

To solicit information about workers' willingness to pay for reductions in job-related risks, Gerking, De Haan, and Schulze (1988) mailed questionnaires to a random sample of 3,000 U.S. households and to an additional sample of 3,000 households in 105 counties with large concentrations of high-risk industries. The respondents were asked to place their job on a ladder showing annual job-related risks of death ranging from 1 in 4,000 to 10 in 4,000. Half of the respondents were asked what reduction in annual earnings they would accept to reduce their risk of job-related death by 1 in 4,000 from its current value.[16] The other half were asked how much they would have to be paid to increase their risk of job-related death by 1 in 4,000. Although the presentation of the risk information and the wording of questions were both clear; the researchers attained a response rate of only 40 percent. In this study, the mean willingness-to-pay estimate of the value of a statistical life was $2.71 million (1986 dollars), and the mean willingness to accept estimate was $6.95 million.[17]

Jones-Lee, Hammerton, and Philips (1985) questioned a random sample of 1,150 persons in the United Kingdom about their willingness to pay to avoid risk of death in a highway accident. Respondents were also asked questions about risk perception, although no attempt was made to familiarize the respondents with their current chances of dying or with the additional risks associated with activities other than riding in a motor vehicle.

A typical question asked the respondent to assume that he had been given £200 to take a bus trip in a foreign country. A trip costing exactly £200 would entail a risk of death of 8 in 100,000. The respondent was asked how much he would pay to reduce the risk to 4 in 100,000, or 1 in 100,000. Probabilities were represented graphically by darkening the appropriate number of squares on a page containing 100,000 squares. Unfortunately, 42 percent of all respondents gave the same answer to both questions, suggesting that they did not perceive a difference between the 4 in 100,000 risk reduction

---

[16] Respondents whose risk of job-related death was below 1 in 4,000 were asked to assume that their risk was 1 in 4,000.

[17] A large discrepancy between willingness to accept and willingness to pay has been found in other contingent valuation studies (Cummings, Brookshire, and Schulze 1986). A possible explanation for the discrepancy is that individuals are more familiar with the purchase of commodities than with their sale. Coursey, Hovis, and Schulze (1987) found that, in an experiment where individuals were allowed to submit bids or offers for the same commodity, WTA approached WTP after several rounds of transactions.

and the 7 in 100,000 risk reduction. This suggests that contingent valuation studies of small risk reductions may be unreliable.

In general, Jones-Lee, Hammerton, and Philips's mean value of life estimates, which ranged from $2.3 million to $4.3 million (1986 dollars) depending on the question asked, were higher than those found in recent wage-risk studies, but agreed closely with those obtained by Gerking, De Haan, and Schulze. This is not surprising. Hedonic wage studies reflect the valuation of risks to life by the least risk-averse persons in the population — those who voluntarily accept high-risk jobs. The contingent valuation studies reported here are based on samples that are more representative of the general population.[18]

In spite of this advantage, there are at least three shortcomings in using the Gerking, De Haan, and Schulze and the Jones-Lee, Hammerton, and Philips estimates of willingness to pay to value environmental risks. First, the risks valued in both studies are more voluntary than are many environmental risks. Work by Slovic, Fischhoff, and Lichtenstein (1980, 1982) suggested that estimates of willingness to pay obtained in one context may not be transferrable to another. Second, in the two CVM studies, death was instantaneous; whereas in the case of many environmental contaminants, it is likely to be prolonged. Third, the CVM studies valued risk of death during the current year, whereas many environmental hazards take effect only after a long latency period. The contingent valuation studies of Mitchell and Carson and Smith and Desvousges attempted to remedy this shortcoming.

Smith and Desvousges interviewed 609 households in suburban Boston to obtain these consumers' willingnesses to pay for reductions in probability of exposure to hazardous waste. The risk of exposure (R) and risk of dying given that an individual is exposed (q) were represented graphically using pie charts and placed in the perspective of other risks by means of a risk ladder that used a logarithmic scale. Reference was made to the fact that current exposure would not cause death for 30 years. Individuals were asked how much they would be willing to pay (in terms of higher prices and taxes) to reduce their probability of exposure to hazardous waste during the current year.

Unfortunately, one of the key results of this study — the finding that the amount individuals are willing to pay for a ΔR of different sizes (holding q constant) did not increase with ΔR — suggests that respondents did not correctly perceive different risks of exposure. It is difficult to believe, for example, that individuals' mean willingnesses to pay for a change in risk of exposure are the same whether the change is from 1 in 5 to 1 in 10, or from 1 in 60 to 1 in 150. This similarity may have occurred because of inadequate preliminary discussions of risks during the interview or problems with the visual representation of risks.

[18] Gerking, DeHaan, and Schulze (1988) noted, however, that they probably oversampled high-income, high-human-capital workers.

Mitchell and Carson interviewed 237 households in Herrin, Illinois, to obtain values for small changes in risk of death due to trihalomethanes in drinking water. After acquainting respondents with risk of death by age due to all causes, they explained the extra annual risk of death associated with various occupations and the extra risk of death associated with an appendectomy, an airplane trip, and so on. These risks were explained by using a risk ladder and by translating risks into the annual premium that would have to be paid for $100,000 of life insurance. To communicate very low levels of risk, the bottom of the risk ladder was magnified and risks were described in terms of "cigarette equivalents."

Respondents were asked whether they would vote for an increase in their water bill to cover the cost of reducing THMs from one level to another, and were told the change in probability of death associated with this reduction. Persons who voted "yes" were then asked how much they would pay for the change. For a 1.33 in 100,000 reduction in future risk of death, Mitchell and Carson obtained a corrected value of life of $388,000 in 1986 dollars; for a 4.43 in 100,000 reduction the implied value of life was $181,000. When researchers interpret these value-of-life estimates, it is important they remember that the values represent the willingness of people whose average age is 43, to pay now to reduce a risk that will not occur for another 20 to 30 years. These estimates are lower than the value of a statistical life associated with current risks of death for two reasons: first, the number of expected life years lost is smaller if the risk occurs 20 years hence; and second, the individual may discount the value of future life years lost (Cropper and Sussman 1988b).

The Distribution of Willingness-to-Pay Responses. As noted above, a person's willingness to pay to reduce risk of death should depend on the age at which the risk occurs, his lifetime wealth, and family circumstances. If valuations of risks to life obtained from one group of persons are to be transferred to other groups, it is essential that we know how WTP varies with personal characteristics. Because data on socioeconomic variables can easily be gathered along with contingent valuation responses, it is possible to describe WTP as a function of age, income, family status, and other variables that should affect WTP to reduce current risk of death.

From a policy perspective it is important to know how willingness to pay varies with current age. Life-cycle consumption saving models (Arthur 1981; Shepard and Zeckhauser 1982, 1984), in which consumption is constrained by income early in life, suggest that WTP for a change in current risk of death should increase with age until age 40 to 45, and decline thereafter. Although all contingent valuation studies collect age data, the effect of age on WTP is often not analyzed correctly. In examining the effect of age on willingness to pay, the researcher should hold lifetime wealth constant, rather than current income. Age should enter the equation nonlinearly. Holding income constant, Mitchell and Carson found that being over 55 significantly decreases willingness to pay for a reduction in risk of death at the end of a

TABLE 6.2
Present value of future earnings of males by selected age groups, 1977 dollars.

Age group (years)	Real discount rate		
	2.5%	6%	10%
1 to 4	405,802	109,364	31,918
20 to 24	515,741	285,165	170,707
40 to 44	333,533	242,600	180,352
65 to 69	25,331	21,801	18,825

Source: Landefeld and Seskin (1982).
Note: Dollar figures based on the present value of both expected lifetime earnings and housekeeping services at 1977 price levels and an annual increase in labor productivity of 1 percent.

20-year latency period. Jones-Lee, Hammerton, and Philips reported that WTP as a function of age is hump-shaped, with a peak at age 40, as suggested by life-cycle models. Gerking, DeHaan, and Schulz found no significant relationship between age and WTP; however, age enters their regression equation linearly. All three of these studies found that WTP increases with income and with the size of the risk change being valued.

Human Capital Estimates of the Value of Life. As noted above, one contribution of theoretical models of the value of life is that they can be used to obtain the conditions under which the change in expected lifetime earnings is a lower bound to willingness to pay to reduce risk of death. A sufficient condition for this to hold is that the individual consumes at a level where the average utility of consumption exceeds the marginal utility of consumption; a condition that is likely to be satisfied if consumption is above subsistence.[19] It should be pointed out, however, that this lower bound will be a very small fraction of WTP for children, homemakers, retired persons, and others whose incomes come primarily from nonwage sources. For working men and women, it is worthwhile to examine their present values of expected lifetime earnings and to see how these are related to the values of a statistical life estimated from labor market and contingent valuation studies.

To compute the present value of expected lifetime earnings requires data on earnings by age, survival probabilities, and a rate of discount. In the context of a life-cycle consumption model, the appropriate rate of discount is the real after-tax rate of interest.

Table 6.2 shows the present values of future lifetime earnings for males discounted to various present ages. The calculations are presented for alternative real discount rates. To compare these figures with those in table 6.1, it is necessary to know the average age of the individuals for the studies presented in table 6.1, and to convert the figures in table 6.2 to 1986 dollars.

---

[19] This result is derived in models where actuarially fair insurance and annuities are available; however, the result is likely to hold even in the absence of these financial instruments.

If, for example, the results of Gegax, Gerking, and Schulze (average age of respondents equals 41 years) are compared with the present discounted value of lifetime earnings reported in the first column of table 6.2 (2.5 percent real discount rate), the value of a statistical life appears to be 2.65 times greater than expected lifetime earnings. Using values of a statistical life from contingent valuation studies raises this ratio to approximately five.

These findings confirm the theoretical result that the present discounted value of lifetime earnings constitutes a lower bound to willingness to pay. They also suggest that studies employing the former rather than the latter will seriously understate an individual's willingness to pay to reduce his risk of death.

6.4 Valuing Reduced Morbidity

6.4.1 Introduction

Measures of the value of reduced morbidity associated with environmental improvements must take into account the variety of ways in which people can benefit from reduced incidence or prevalence of disease. Consider an individual who experiences one day fewer of asthma attacks because of an improvement in air quality. He may find that he benefits by avoiding the lost wages associated with being unable to work one day, by reducing his costs for medicine and treatment, and by avoiding the discomfort or pain associated with the attack itself. The first two components of these benefits can be calculated in a straightforward manner, given sufficient information. In fact, many of the estimates of monetary benefits of improved health appearing in the literature are based on just such calculations. But the avoidance of lost wages and medical expenses is typically only a partial measure of the total benefits of improvements in health because it ignores the value of avoiding pain and discomfort.

There is an additional channel through which benefits can be realized independent of any observed change in the incidence of disease. Consider a second individual who does not experience asthma attacks at the present level of air pollution. Suppose, to prevent asthma attacks from exposure to ambient air pollution, this individual spends $200 per month on an air purifier and stays indoors at home on days when the air quality is low. If air quality is improved, this individual benefits from being able to reduce the monetary expenditures, the lost wages, and the opportunities for leisure that are associated with defending against the health impacts of air pollution. The reduction in defensive expenditures or averting behavior is also a component of the health benefits associated with reducing air pollution.

The models used for estimating the benefits of reducing environmental pollutants that pose threats to human health must be capable of reflecting the variety of individuals' responses to pollution and the threat of disease. Simple predictions of reduced cases of illness combined with unit values based on avoided medical expenditures or lost wages may be seriously misleading as estimates of benefits because they fail to reflect the richness of human behavior in response to environmental change. In this section, theoretical models are presented for valuing reduced morbidity that attempt to take account of all of the ways in which people benefit from reduced threat of environmentally mediated disease. Empirical techniques are described for estimating values based on these models.

6.4.2 The Theory of Morbidity Valuation when Health Outcomes Are Certain

Economists have used three techniques for valuing reduced morbidity. The first, *direct questioning* (the contingent valuation method, see chapter 5), involves asking people what they would pay to reduce the number of symptom or restricted-activity days they experience. The second technique, the *averting behavior method* (see chapter 3), infers people's willingnesses to pay to reduce ambient pollution levels from the amounts of money they spend to avoid exposure to air pollution (for example, by installing air filters) or to mitigate its effects (for example, by taking an antihistamine to reduce nasal discharge). The third technique for valuing reduced morbidity is the *cost of illness approach*. It uses data on lost earnings and medical expenditures to infer a lower bound to willingness to pay for reduced air pollution.

The relationships among the three approaches can be examined using a model of health production and consumption.[20] Suppose that the health outcome of interest is the number of hours S during a year or a month that a person spends ill with some respiratory ailment.[21] The health production function relates time spent ill to exposure to pollution, E, and to activities that mitigate the effects of exposure, M. Mitigating activities include taking antihistamines or visiting a doctor, and have a unit cost of p_M, which includes time as well as out-of-pocket costs.[22] Pollution exposure is a function of ambient pollution and activities A termed *averting* or *avoidance activities*,

[20] Grossman (1972) first used the houschold production model to examine health decisions. Cropper (1981a), Harrington and Portney (1987) and Gerking and Stanley (1986) modified the Grossman model to examine the health effects of pollution.

[21] Ideally, the health outcome, S, should reflect severity of illness as well as duration. Dickie et al. (1986) have generalized the model presented here to encompass a vector of health outcomes.

[22] A distinction is sometimes made between preventive and recuperative medical expenditures (Harrington and Portney 1987); however, if S measures severity of illness, the distinction seems irrelevant. Both reduce the level of S experienced.

that affect exposure given ambient pollution levels; that is, $E = E(A,P)$. The variable A might include the percentage of time spent indoors, or the percentage of time spent indoors multiplied by whether an air conditioner or air filter was in operation. Let p_A denote the unit cost of A. The health production function may be written

$$S = S[E(A,P),M]. \tag{6.11}$$

Time spent ill directly affects the individual's utility by producing discomfort; it indirectly affects it by reducing the amount of time (and possibly money) available for leisure activities and consumption. Formally, S enters the utility function, together with all other goods X and leisure time L,

$$U = U(X,L,S). \tag{6.12}$$

S also enters the budget constraint by reducing the amount of time spent at work, and hence, the amount of income earned. The individual's budget constraint says that nonwage income I plus earnings must equal total expenditure. Formally,

$$I + w(T - L - S) = p_X X + p_A A + p_M M, \tag{6.13}$$

where w is the wage rate and $T - L - S$ is the time spent at work (T is total time).

The health production model assumes that the individual allocates his (well) time between work and leisure activities and his income between defensive (averting and mitigating) expenditures and expenditures on other goods to maximize utility. The problem for the individual is to choose the mitigating and averting activities M and A, the expenditures on all other goods X, and the leisure time L that will maximize function (6.12) subject to (6.11) and (6.13).

Valuing a Marginal Change in Pollution. An individual's willingness to pay for a small reduction in ambient pollution P is defined as the largest amount of money that can be taken away from him without reducing his utility. Formally, economists define the pseudoexpenditure function as the minimum value of expenditure minus the wage income necessary to keep utility at U^0, or

$$E = \min [p_X X + p_A A + p_M M - w(T - L - S) \tag{6.14}$$
$$+ m\{U^0 - U[X,L,S(A,P,M)]\}],$$

where m is a Lagrangian multiplier. Applying the envelope theorem to (6.14) and substituting from the first-order conditions for utility maximization, willingness to pay for a marginal change in P, $\partial E/P$, is given by

$$\text{WTP} = -(\partial S/\partial P)p_M/(\partial S/\partial M) = p_M (\partial M/\partial P) \tag{6.15a}$$
$$= -(\partial S/\partial P)p_A/(\partial S/\partial A) = p_A(\partial A/\partial P) \tag{6.15b}$$
$$= (\partial S/\partial P)\text{WTP}_S. \tag{6.15c}$$

Willingness to pay is given by the reduction in sick time associated with the reduction in pollution, $\partial S/\partial P$, times the marginal cost of sick time. The latter is given by the cost of an additional mitigating input p_M divided by the reduction in sick time that input produces $-\partial S/\partial M$, or alternatively, by the cost of averting behavior p_A divided by the reduction in sick time that averting behavior produces $-\partial S/\partial A$. In the health production model, in which pollution affects utility only through health, this amount of money is the reduction in the cost of achieving the optimal level of health made possible by the decrease in pollution. If a reduction in ozone levels from 0.16 to 0.11 ppm reduces the number of days of respiratory symptoms from six to five, and if an expenditure on M of $20 has the same effect, then if all else is equal, the individual should be willing to pay no more than $20 for the reduction in ozone.

Three points about willingness to pay must be noted. First, WTP can be written in terms of the rate of substitution between pollution and any input into the production of health because minimizing the cost of producing health requires that the value of all inputs be equal at the margin. Second, as shown by the last terms in equations (6.15a) and (6.15b), WTP can be calculated from the reductions in expenditures on either mitigating or averting behavior that are required to attain the original health status or number of sick days, holding all else constant. In general, this calculation will not be equal to the *observed* reduction in mitigating or averting expenditures associated with the reduction in pollution because people are likely to respond to the lower cost of attaining any given level of health by "purchasing more health." Third, as (6.15c) shows, the WTP for a marginal change in pollution equals the resulting reduction in sick time, $\partial S/\partial P$, times the value of a marginal reduction in sick time, WTP_S. To a rational individual the latter must, in theory, equal the marginal cost of reducing sick time.

The result in equation (6.15) can be generalized (Dickie et al. 1986) to the case of many symptoms and various forms of averting and mitigating behavior.[23] In the special case in which a single input enters the exposure production function $E(A,P)$, and each symptom depends on exposure, an expression similar to (6.15) holds for each symptom S_j:

$$\text{WTP} = -(\partial S_j/\partial P)p_A/(\partial S_j/\partial A), \tag{6.15'}$$

where A is the input that alters exposure. WTP can therefore be estimated using the production function for a single symptom. If the foregoing conditions do not hold, WTP for a marginal change in pollution can still be estimated from market data provided that the number of averting or mitigating activities is at least as great as the number of symptoms affected by pollution. Estimation

[23] Dickie et al. (1986) also allowed some of the averting goods to enter the utility function as well as symptom production functions; however, estimation of WTP using market data hinges on the existence of "pure" averting goods which do not confer utility directly.

of marginal WTP in this case involves estimating production functions for all symptoms.

Computing (6.15) or (6.15′) requires an estimate of the production function for the health outcome of interest and an evaluation of the numerator and denominator of the equation at current levels of all inputs. In practice, this has proven to be difficult. In addition to evaluating data on the relevant health outcome and ambient pollution levels, the costs of averting and mitigating behaviors must be identified and measured. In practice, the most effective method of reducing exposure, given ambient pollution levels, is for people to spend time indoors. And even if it is possible to measure their time spent indoors, determining its cost is difficult. Devices that reduce indoor pollution concentrations (air conditioners, air filters) have costs that can be measured, but produce other services, such as reducing indoor temperature, so that it is inappropriate to allocate all of these costs to pollution avoidance.

For these reasons, it is useful to consider an alternative expression for willingness to pay. Following the example of Harrington and Portney (1987), WTP can be written as the sum of the value of lost time $w(\mathrm{d}S/\mathrm{d}P)$ and the disutility of the change in illness $(\mathrm{d}S/\mathrm{d}P)(\partial U/\partial S)/\lambda$ plus the observed changes in averting and mitigating expenditures, $p_M(\partial M^*/\partial P)$ and $p_A(\partial A^*/\partial P)$:

$$\mathrm{WTP} = w\,\frac{\mathrm{d}S}{\mathrm{d}P} + p_M\,\frac{\partial M^*}{\partial P} + p_A\,\frac{\partial A^*}{\partial P} - \frac{\partial U/\partial S}{\lambda}\,\frac{\mathrm{d}S}{\mathrm{d}P}, \qquad (6.16)$$

where λ, the marginal utility of income, converts the disutility of illness $\partial U/\partial S$ into dollars, and where $M^* = M^*(I, w, p_X, p_A, p_M, P)$ and $A^* = A^*(I, w, p_X, p_A, p_M, P)$ are the demand functions for M and A. The partial derivatives $\partial M^*/\partial P$ and $\partial A^*/\partial P$ give the optimal adjustments of M and A to a change in pollution. An important difference between equations (6.16) and (6.15) is that in (6.16) WTP is a function of the total derivative of illness with respect to pollution, $\mathrm{d}S/\mathrm{d}P$, which incorporates the effect of pollution on averting behavior and averting behavior on illness.[24] To compute $\mathrm{d}S/\mathrm{d}P$, it is not necessary to estimate a health production function; rather it is possible to estimate a *dose-response function*, which is a reduced-form relationship between illness, ambient pollution, and variables that affect averting and mitigating behavior. In the health production framework, a dose-response function is obtained by substituting the demand functions for M and A into the health production function. Full implementation of (6.16) as a measure of value, therefore, requires the estimation of these demand functions.

As a practical matter, the first three terms in (6.16) can be approximated after the fact by using the observed changes in illness and averting and mitigating expenditures. In this way, equation (6.16) can be used to derive a lower bound to individual WTP. Because the last term in the equation is negative ($\partial U/\partial S < 0$), the first two terms — the value of lost time plus the

[24] Formally, $\mathrm{d}S/\mathrm{d}P = (\partial S/\partial M)(\partial M^*/\partial P) + (\partial S/\partial A)(\partial A^*/\partial P) + \partial S/\partial P$.

change in averting and mitigating expenditures — give a lower bound to WTP. These terms are referred to as the *private cost of illness,* or the cost borne by an individual of mitigating and averting expenditures and lost time. In practice, the cost of these items to an individual may differ from their cost to society due to medical insurance and paid sick leave. Therefore, the social cost of mitigating and averting expenditures plus lost time will be referred to as the *social cost of illness.*

In the health literature, the expression *cost of illness* typically refers only to the social cost of lost earnings plus the recuperative (mitigating) medical expenditures associated with illness. This expression therefore ignores two components of our social cost of illness — the social value of averting expenditures and the cost of lost leisure time that results from illness.

Because the private cost of illness represents only a lower bound to WTP, a researcher may choose to question people directly about the disutility of a change in illness, so as to capture the fourth term in equation (6.16). Alternatively, a researcher can ask people to value an additional day of illness, WTP_S, and use (6.15) to compute WTP. To the extent that the individuals questioned do not bear the full social costs of lost time or medical expenses, these WTP estimates must be adjusted, as must estimates obtained using the averting behavior approach.

Valuing Nonmarginal Changes in Pollution. A measure of value for a small change in environmental pollution was defined in the preceding section. This section will define measures for valuing a larger change. Willingness to pay for a nonmarginal improvement in pollution may be defined, using the expenditure function (6.14), as the additional expenditure necessary to achieve U^0 at the original pollution level P^0 compared with the new (lower) pollution level P^1,

$$\text{WTP} = E(p_X, p_A, p_M, w, P^0, U^0) - E(p_X, p_A, p_M, w, P^1, U^0). \qquad (6.17)$$

Evaluating (6.17) is difficult because expenditures at (P^1, U^0) are not observed. One solution, which is based on the work of Bockstael and McConnell (1983), evaluates (6.17) by using the area under the demand function for a mitigating or averting input. This method, however, requires that the expenditure necessary to achieve U^0 be independent of pollution if use of the input goes to zero. This condition is clearly not satisfied.

An approach that can be used is to compute a lower bound to equation (6.17). As Bartik (1988a) has shown, (6.17) is bounded from below by the difference between the original expenditure level and the expenditure necessary to achieve the original health level at P^1,

$$E(p_X, p_A, p_M, w, P^0, U^0) - E(p_X, p_A, p_M, w, P^1, U^0 \,|\, S = S^0). \qquad (6.18)$$

Equation (6.18) is the difference in the cost of S^0 when pollution is P^0 and pollution is P^1 and can be computed once the cost function for S has been estimated.

6.4.3 Empirical Valuation of Certain Health Outcomes

Data Requirements. To implement the averting behavior approach requires having data in the following five categories for a cross-section of individuals:

1. Frequency, duration, and severity of pollution-related symptoms.
2. Ambient pollution levels to which the individual is exposed.
3. Actions which the individual takes to avoid or mitigate the effects of air pollution.
4. Costs of avoidance and mitigating activities.
5. Other variables affecting health outcomes (age, general health status, presence of chronic conditions, and so on).

Under the averting behavior approach, these data are used to estimate health production and input demand functions, which, in turn, are used to calculate willingness to pay for a marginal change in ambient pollution (6.15). The same data could be used to implement the cost of illness approach. Data from categories 1, 2, 4, and 5 could be used to estimate a dose-response function. The predicted effect of a change in pollution on illness are multiplied by the value of lost time plus the change in medical and avoidance expenditures to calculate the cost of illness.

There is, however, an important difference between the data requirements of the averting behavior versus the cost of illness approach. Ideally, to infer willingness to pay using data on avoidance and mitigating behavior, each individual's beliefs about the efficacy of these behaviors should be known because these beliefs motivate the individual's observed behavior.[25] The spirit of the cost of illness approach is to measure observable costs associated with actual changes in illness that are induced by air pollution. In practice, it is the objective effect of avoidance and mitigating behavior that is measured under the averting behavior approach, with the tacit assumption that individuals correctly perceive the effects of their actions.

The contingent valuation approach involves asking people either what they would be willing to pay to reduce pollution or what value they place on reducing symptoms, and then multiplying this answer by the reduction in symptoms corresponding to a change in pollution. In theory these responses could be collected without obtaining any of the information in the categories above. (If necessary, a dose-response function could be estimated separately from the collection of CVM responses.) Collecting information on the age, health status and income of respondents is, however, necessary if CVM responses from a given study are to be used in other contexts.

The Determinants of WTP Responses of Individuals. Because willingness

---

[25] This implies that individuals must believe that pollution affects illness in order to take actions to reduce their exposure. Rowe and Chestnut (1985) reported that half of the asthmatics they surveyed believed that air pollution aggravated their asthma. They also reported a 20 percent decline in "active outdoor activities" by these persons on high pollution days.

to pay for morbidity benefits should vary with health, age, and income, pollution control policies may have very different benefits depending on the characteristics of the target population. To value morbidity benefits of different policies it is therefore necessary to know (1) the distribution of key variables in the population affected by the policy and (2) how WTP (or the cost of illness) varies with these characteristics.

In principle, each of the three valuation techniques presented in this section is capable of describing how the value of reduced morbidity varies with the characteristics of the respondent. The averting behavior approach predicts that willingness to pay depends on any variables that affect the marginal product of pollution, mitigating activities, or avoidance activities. In practice, these would include health (whether the respondent has a chronic respiratory condition), age, and perhaps, education. Willingness to pay will also vary with factors that affect the cost of averting activities, such as earned income or education. The effect of the aforementioned variables on WTP could be calculated from the health production function or from equations describing the unit cost of averting activities.

When willingness to pay is estimated by the contingent valuation method, data are typically gathered on variables that would enter the health production function or affect the level of averting (mitigating) behavior undertaken. Announced WTP values are then regressed on these variables. A similar approach can be taken with the cost of illness, provided that cost of illness data are gathered as part of a survey that also elicits socioeconomic and demographic data.

Averting Behavior Studies. Gerking and Stanley (1986), Dickie et al. (1986) and Chestnut et al. (1988b) have all used the averting behavior approach to value the morbidity effects of air pollution.

Gerking and Stanley (1986) used data on visits to a doctor, which are a measure of mitigating behavior, to infer willingness to pay for ozone reductions by persons living in St. Louis. They estimated WTP from an input requirement function that relates the mitigating good, whether the respondent usually visits a doctor each year, to air pollution, demographic variables, and two health outputs — whether the respondent has a chronic illness and the number of years he has had it.

The data used in the study, which were not gathered for the purpose of applying the averting behavior method, have clear limitations. It is hard to see how current visits to a doctor could influence whether a person has a chronic disease, although they might influence the severity of symptoms experienced. It is also unlikely that air pollution levels, measured from 1974 to 1977, would have a significant effect on the presence of a chronic disease between 1977 and 1980 (the period during which health data were collected) unless they proxy lifetime exposure for the persons in the sample.

The study by Dickie et al. (1986) represents an improvement over the Gerking and Stanley study in the measurement of both inputs and outputs.

The study was based on interviews with 229 persons living in Glendora and Burbank, California, who had participated in a study of respiratory health conducted by researchers at the University of California at Los Angeles. The health outcomes measured whether the respondent experienced each of nine respiratory symptoms on the days of the survey. Each symptom was a function of the number of visits to the doctor, various forms of averting behavior, and three pollution variables: maximum daily one-hour concentrations of ozone, SO_2, and NO_x. Long-term averting behaviors included cooking with electricity instead of gas, living in an air-conditioned home, and driving in an air-conditioned car. Short-term averting behavior was measured by the number of hours spent indoors, and by variables that indicated whether the individual spent a greater-than-average amount of time in bed or took a greater-than-average amount of medication.

Because air pollution levels were not found to be significant determinants of symptom occurrence, Dickie et al. measured WTP to avoid a symptom day rather than WTP for a change in pollution. Willingness to pay to avoid a marginal symptom day was measured by dividing the marginal cost of various avoidance behaviors by their efficacy in reducing symptoms. Unfortunately, the mere presence of an air conditioner in a home or in a car is an imperfect measure of an individual's reduced exposure to air pollution. What a researcher really needs to know is the proportion of time the respondents spend indoors in an air-conditioned environment on the survey day. Even if such data were available, a problem in using avoidance behavior is that its cost is inherently difficult to measure. What, for example, is the cost of spending leisure time indoors rather than outdoors? Measuring the cost of avoidance behavior is complicated by the fact that many avoidance activities — running an air conditioner while indoors, leaving town on polluted days — may be motivated by considerations other than reducing exposure to air pollution. It is thus inappropriate to attribute all of this expense to reducing exposure. Because Dickie et al. attributed the total cost of operating an electric (versus a gas) stove or an air conditioner to averting behavior, their willingness-to-pay estimates must be regarded as upper bounds.

Chestnut et al. (1988b) used the averting behavior approach to value reduced symptoms (pain) in angina patients. These symptoms may be exacerbated by high levels of carbon monoxide. In this study, willingness to pay for reduced symptoms was computed directly by asking respondents what effect they believed various averting behaviors had on the number of attacks they experienced and dividing this number into the cost of these activities. The drawback of the study is that it used averting behaviors that produce joint products, such as hiring people to do yard work, car repair, and other household tasks. These behaviors do reduce health risks to the angina patient; but since they relieve the individual of the responsibility of performing these chores, they also provide him with leisure time. Thus, it is not appropriate to ascribe the full cost of these services to averting behavior.

An example of an averting behavior study that does not encounter such problems is Harrington, Krupnick, and Spofford (1989). They estimated the value of averting and mitigating expenditures and lost work time [the first three terms in equation (6.16)] corresponding to an outbreak of giardiasis, a waterborne disease characterized by diarrhea, cramps, and weight loss. As the authors noted, the relative magnitudes of the components of WTP changed as a function of information about the disease. When water supplies were initially contaminated but the cause was unknown, the main components of WTP consisted of work loss, mitigating expenditures, and disutility. As information about the cause of the outbreak became available, the main components of WTP became the cost of averting behavior in the form of purchasing bottled water or otherwise avoiding contaminated water.

As the researchers noted, estimates of the first three components of equation (6.16) are sensitive to the value of time. The cost of avoiding contaminated water p_A, and the cost of medical treatment p_M both have significant time components. For persons working outside the home, time was valued at the wage rate, which was computed by dividing earnings by hours worked. The average wage rate of employed persons was used to value the time of retirees and homemakers.

Contingent Valuation Studies. In this section, five contingent valuation studies that have been conducted to value symptoms associated with air pollution are discussed.[26] In all cases, it is the symptoms themselves that have been valued; hence, marginal willingness to pay for symptom reduction must be multiplied by the effect of a change in air pollution in reducing symptoms.

There are several issues that arise in examining the CVM studies. One is how the commodity being valued, such as the reduction of a marginal symptom day, is related to equation (6.16). Is the study worded so that it elicits all components of WTP or just the disutility of illness? For minor illnesses this distinction may not be important as the disutility component is likely to account for most of WTP. Is illness being valued before or after averting behavior has been taken? In (6.16), the disutility of illness is multiplied by the total derivative of illness with respect to pollution, suggesting that a day of headache should be valued after mitigating actions have been taken and should include the cost of the mitigating activity.

A second issue common to all CVM studies is whether the respondents carefully consider the budgetary implications of their responses. A problem with many CVM studies of health symptoms is the respondents' large professed willingness to pay for reductions in minor symptoms, a result which may

---

[26] The studies are: Loehman et al. (1979), Rowe and Chestnut (1985), Tolley, Babcock, et al. (1986) (also reported in Berger et al. 1987), Dickie et al. (1987) and Chestnut et al. (1988b). The 1986 study by Dickie et al. contained contingent valuation questions; however, a more thorough attempt was made to use the contingent valuation method when respondents were reinterviewed in the 1987 study.

reflect the respondents' failure to carefully consider the opportunity cost of their responses.[27]

A third issue concerns the size of the symptom reduction that is valued. For most policy purposes, marginal symptom days should be valued. The researcher must be careful to avoid taking the average WTP implied by a large, perhaps 50 percent, reduction in symptoms and applying it to a marginal symptom day.

Of the five studies reported here, three value minor respiratory symptoms in the general population [Loehman et al. 1979; Tolley, Randall, et al. 1986 (also reported in Berger et al. 1987); Dickie et al. 1987]. Dickie et al. asked the respondents their maximum willingnesses to pay without suggesting a starting bid or a payment vehicle, for example, increased taxes. In Tolley et al. (1986), a starting bid of $100 was used and the bid halved or doubled depending on the subject's response. In Tolley et al., the severity and duration of each symptom was described in detail. Dickie et al. (1987) asked respondents to value a reduction in symptom-days of the type they usually experience. The description of each symptom and its severity was somewhat vague in Loehman et al.

Table 6.3 compares the results of the three studies in valuing a marginal symptom day. In Tolley et al., a marginal increase was valued; in the other studies, a marginal decrease. For each study the table reports mean WTP for each symptom, averaged across all respondents, as well as median WTP. In all cases, median bids are below mean bids, which reflects a number of very large bids in each survey. In Tolley et al., these may be explained by the high starting point ($100) the researchers used in their bidding procedure, which may have biased bids upward; also, the detailed symptom descriptions they provided may have suggested to respondents that mitigating behavior was impossible.

Because high bids may result from people's failure to carefully consider their budget constraints, Dickie et al. informed respondents of the total cost of their bids by multiplying marginal WTP for a one-day reduction in symptoms by the number of symptom-days experienced during the previous month. This procedure may be critized on the grounds that the number obtained overstates total WTP if average WTP for symptom reduction is less than marginal WTP. Mean revised bids were often two orders of magnitude lower than mean original bids.

CVM studies by Rowe and Chestnut (1985) and Chestnut et al. (1988b) valued reductions in symptom frequency for asthmatics and patients with angina, respectively. Rowe and Chestnut asked asthmatics how much they would be willing to pay in increased taxes to reduce by half the number of bad asthma days they experienced each year, with the assumption being that

---

[27] Berger et al. (1987) reported contingent valuation estimates of $76 to $109 to avoid a day of coughing or headache. Dickie et al. (1987) reported initial CVM estimates of $175 and $138, respectively, to avoid a day of coughing (headache).

TABLE 6.3
Willingness to pay for acute symptom reduction, 1984 dollars.

| | WTP for a one-day change in symptom | | | | | |
| | Dickie et al. (1987) | | Tolley and Babcock et al. (1986) | | Loehman et al. (1979) | |
Symptom	Mean WTP	Median WTP	Mean WTP	Median WTP	Mean WTP	Median WTP
Cannot breathe deeply	1140.00	1.00				
Pain on deep inspiration	954.13	3.50				
Shortness of breath	7.88	0.00			78.00 (127.00)[a]	8.00 (18.00)
Wheezing	58.00	2.00				
Chest tightness	813.72	5.00				
Cough	355.10	1.00	25.20	11.00	42.00 (73.00)	4.40 (11.00)
Throat irritation	15.00	3.00	28.97	13.00		
Sinus congestion	239.50	3.50	35.05	14.00	52.00 (85.00)	6.00 (13.00)
Headache	178.39	1.00	40.10	20.00		
Eye irritation			27.73	12.50		
Drowsiness			31.41	15.00		
Nausea			50.28	17.50		

[a] Numbers in parentheses refer to severe symptoms. Numbers above them refer to mild symptoms.

tax dollars could be used to reduce pollen, dust, and other air pollutants. No starting bid was given. Chestnut et al. asked angina patients how much they would pay to reduce by one the number of angina episodes experienced per month. The value of a marginal day of angina symptoms was $41. The mean WTP per asthma day, based on a 50 percent reduction in number of days, was $10.

What are the prospects for using contingent valuation methods to value acute respiratory symptoms? The results from the studies of Tolley et al. and Dickie et al. suggest that responses are sensitive to whether the respondent is encouraged to consider averting behavior and to how much thought is given to the budgetary implications of responses. One way to examine the reliability of CVM responses is to see how announced WTP varies with health, income, and the severity and frequency of symptoms usually experienced. Dickie et al. attempted such an analysis; however, their small sample sizes made it difficult to obtain statistically significant results. Tolley et al. found that, for most symptoms, WTP increased with the number of symptom-days experienced and with overall poor health. Education, but not income, was usually positively related to WTP.

Cost of Illness Studies. As noted above, the value of work and leisure time lost due to illness plus any change in averting and mitigating expenditures constitute a lower bound to willingness to pay for reduced exposure to pollution. If these costs of illness (COI) are to constitute a lower bound to individual WTP, then the relevant prices are those that the individual faces. This measure is referred to as the *private cost of illness.* Since the rest of society's WTP to reduce health risks must be added to the sum of individual WTPs if individuals do not face the full social cost of medical care or lost productivity, it is also of interest to value lost time plus averting and mitigating expenditures at their true social costs. This is termed the *social cost of illness.*

By contrast, the phrase "cost of illness" in the medical economics literature usually refers to the cost to society of the lost earnings (but not lost leisure time) and medical costs (but not averting expenditures) associated with illness. It is this definition that is followed in estimating the costs of illness associated with acute respiratory illness. For some of the respiratory symptoms listed in table 6.3, Berger et al. (1987) reported data on medical expenditures net of insurance payments plus lost earnings (see table 6.4). These ranged from $1.80 per day of drowsiness (1984 dollars) to $14.56 per day of itchy eyes. This does not correspond exactly to the definition of cost of illness used in the medical economics literature because it is net of insurance payments. The latter, however, are likely to be small for the light symptoms valued by Berger et al.

Chestnut et al. (1988b) emphasized the distinction between private and social medical costs when estimating an annual cost of illness for angina sufferers. They reported out-of-pocket medical costs of $256 per person, compared with total medical costs of $4,523 per person. They estimated lost wages to be $9,581 per person, with an annual cost of illness of $14,104. Because the respondents in their study averaged one asthma attack per week, this implies an average cost per attack of $271.

Comparison of Estimates Using the Three Approaches. In theory, contingent valuation questions should produce the same valuation of a change in pollution or in a symptom-day as does the averting behavior approach because both purport to estimate WTP as defined in equation (6.15). However, the relationship between the cost of illness and WTP is unclear unless the private cost of illness, that is, the value of lost time and the change in mitigating and averting expenditures borne by the individual, is used in the calculations.

One goal of the studies of Rowe and Chestnut (1985) and Chestnut et al. (1988b) was to establish a relationship between the cost of illness and WTP. Rowe and Chestnut estimated that WTP for a 50 percent reduction in asthma symptoms was 1.6 times the cost of illness when the latter was defined as the sum of medical costs and lost earnings. Chestnut et al. (1988b) estimated that

TABLE 6.4
Consumer surplus and private cost of illness comparisons.

Symptom	Sample size[a]	Mean daily consumer surplus[b]	Mean daily private costs of illness[c]	No. cases CS > COI	t-value[d]
Coughing spells	25	$ 75.98	$12.17	21	1.44
Stuffed-up sinuses	43	27.32	6.79	37	1.94
Throat congestion	24	43.93	14.27	20	1.60
Itching eyes	16	48.48	14.56	12	0.925
Heavy drowsiness	5	142.00	1.80	5	2.52
Headaches	46	108.71	3.45	39	1.45
Nausea	17	47.88	2.50	14	1.29
All symptoms experiences	44	80.63	3.93	34	2.54

Source: Berger et al. (1987).
[a] Only those experiencing the symptom are included.
[b] Consumer surplus for avoiding one extra day of the sympton.
[c] Calculated as expenditures on doctor visits and medicine net of insurance reimbursements plus lost earnings, expressed on a daily bases.
[d] Test of the null hypothesis that the mean consumer surplus is less than or equal to the mean private costs of illness.

this ratio was between 0.17 and 0.43 for a reduction in one angina episode.[28] Although a desire to establish such a relationship is understandable, given the comparative ease with which illness costs can be measured, the relationship may be of limited value. First, the ratio of WTP to COI will be illness-specific, as a comparison of these two studies indicates. Second, the ratio will depend on the fraction of medical costs plus lost work time that the individual bears himself. These are likely to differ from one population group to another.

The more important question is how contingent valuation estimates of WTP compare with averting behavior estimates of the same quantity. The studies of Dickie et al. (1986, 1987), which were conducted using the same respondents, permit comparisons between the two approaches. The mean revised WTP estimates obtained from Dickie et al.'s CV study ranged from $0.60 to avoid a day of shortness of breath to $5.97 to avoid a day of pain on deep inhalation. These figures were generally lower than Dickie et al.'s mean WTP estimates that were based on the averting behavior approach. These estimates ranged from $0.97 (shortness of breath) to $23.87 (chest

[28] One reason why this ratio is so low is that the individuals in the sample did not bear most of the costs of illness reported above.

tightness). The latter, however, must be regarded as upper bound estimates because the full cost of averting activities that produce joint products were attributed to reducing pollution exposure. Attributing the entire cost of operating a home air conditioner to pollution avoidance is, in particular, likely to produce large overestimates of WTP. If estimates based on home air conditioners are eliminated, the range of mean WTP estimates was reduced to $0.97 (shortness of breath) to $16.15 (pain on deep inhalation). Thus, the two methods produced estimates that were about one order of magnitude apart. These results must be regarded as preliminary, however, due to the small sample sizes involved.

6.4.4 *The Theory of Morbidity Valuation when Health Outcomes Are Uncertain*

Treating environmental health effects as certain is clearly inappropriate when the effect is the risk of cancer or chronic lung disease. In these cases, the appropriate model is one in which pollution increases the probability of entering an undesirable state of health.[29] Berger et al. (1987) formulated a model in which there were two health states: absence or presence of a chronic disease. The probability of having the chronic disease H is assumed to be an increasing function of pollution P and a decreasing function of averting expenditures A. Berger et al. showed that, in this framework, WTP for a marginal change in pollution is, analogous to equation (6.15), the rate of substitution between pollution and defensive expenditures,

$$\text{WTP} = -\frac{\partial H/\partial P}{\partial H/\partial A}.$$

Thus, in theory, expenditures on averting behavior, such as purchasing bottled water to avoid trihalomethanes, could be used together with information on increased risks from ingesting a pollutant to estimate WTP. To our knowledge, this approach has not yet been used to value the risk of chronic illness associated with pollution.

An alternative approach, employed by Viscusi, Magat, and Huber (1989), is to ask individuals to compare alternative risk-income bundles to infer directly the individual's WTP to reduce probability of contracting chronic disease (chronic bronchitis). Because individuals may find it difficult to trade reduced risk for income, Viscusi, Magat, and Huber also asked persons to

---

[29] It is important to distinguish between the willingness to pay to avoid a chronic health state and willingness to pay to avoid symptoms associated with the disease. For example, for individuals with chronic heart disease, their exposure to carbon monoxide can lead to increased incidence of angina. An individual's willingness to pay to reduce the number of angina attacks given that he has the chronic disease can be estimated by using a model of acute morbidity, such as that presented in the preceding section.

compare bundles consisting of risk of chronic disease and risk of death in an auto accident. The individual's rate of substitution between risk of illness and risk of death can be converted to dollars using an estimate of the value of a statistical life.

6.5 Summary and Conclusions

This chapter described the methods and models developed by economists to estimate the values that individuals place on reducing pollution-related illness and risk of death. The benefits of a reduction in the risk of death or incidence of illness have been defined as the sum of what each of the affected people is willing to pay to reduce his own risk of death or illness plus the sum of what everyone in society is willing to pay over and above this amount to reduce the risk of death or illness for any of the other exposed people. The second component of willingness to pay might arise because of altruistic feelings, especially toward one's children or other loved ones, or because members of society are forced to bear the costs of illness and premature death of others, for example, through health insurance and paid sick leave.

To assert that individuals have a WTP for a reduction in risk or illness implies that individuals can perceive and are aware of changes in these determinants of their well-being. It does not require that the individuals know that the reduction is attributable to a decrease in pollution. If the value of reduced risk or symptom days is known, policy analysts can calculate benefits if they can predict the magnitude of the reduction caused by pollution control. Assumptions about individuals' knowledge do play important roles in the empirical estimation of WTP in some circumstances, however. For example, if WTP for risk reductions is to be inferred from wage-risk premia, then it must be assumed that individuals know the risk levels of different jobs. Also, if WTP for reduced illness is to be inferred from an individual's averting or mitigating behavior in response to changes in pollution, then it must be assumed that the individual knows the relationship between pollution and illness.

6.5.1 Mortality

In valuing risks to life, two forms of the indirect approach have been widely used: one is based on compensating wage differentials received by workers in risky occupations; the other is based on the cost of activities that increase safety, such as wearing a seatbelt or operating a smoke detector.

Averting behavior studies based on the cost of safety equipment are 0-1 activities whose benefits exceed their costs for most persons; hence, basing risk valuations on these activities is likely to understate the average value of a risk reduction. The compensating wage approach may also understate the value of a risk reduction because persons who are willing to be paid to accept increased risk (for example, structural iron workers) may undervalue risk compared with the average person.

The averting behavior and contingent valuation studies produce quite different results for the value of reducing mortality risks, with the latter yielding values that are four to six times larger than the averting behavior values. There is also variation of about one order of magnitude in the estimates produced by various compensating wage studies (see table 6.1). However, the newer wage-risk studies described here show a greater tendency to agreement, with the highest and lowest differing only by a factor of two and with the range overlapping that of the contingent valuation studies.

A serious shortcoming of studies of job risk premia and use of safety equipment is that they value only voluntarily assumed risks of accidental death. By contrast, environmental risks are largely involuntary and may lead to a painful illness before death occurs. These considerations suggest that individuals might be willing to pay more to reduce environmental risks than to reduce risks of death from such events as on-the-job or automobile accidents. On the other hand, to the extent that environmental risks may not occur until after a long latency period, fewer years of life will be lost compared with risk of death in an automobile accident during the current year. This suggests that willingness-to-pay estimates for a reduction in current risk of death may overstate willingness to pay to reduce environmental risks. The only study that attempted to deal with this set of issues explicitly (Mitchell and Carson 1986b) suggested that the effect of a latency period dominates: the value of a reduction in risk of death 20 years hence, as a result of reduced exposure to trihalomethanes, is worth about one-tenth of the value of a comparable reduction in current risk of accidental death.

6.5.2 Morbidity

Turning to morbidity, the indirect approach has been used to value changes in respiratory symptoms and reductions in the incidence of waterborne diseases. The value of a reduction in morbidity is inferred either from activities that reduce exposure to pollution, such as running an air purifier or buying bottled water, or that mitigate the strength or duration of symptoms, such as taking medication or visiting the doctor.

The advantage of the indirect approach is that it is based on observed behavior; however, to implement it requires a considerable amount of data. The researcher must be able to measure and estimate the relationships among the health outcome of interest, the amount of mitigating or averting behavior undertaken, and its price. In addition, the researcher must know either how effective the individual thought his averting behavior was or how effective the behavior actually was and assume the individual correctly perceived this.

In addition to data problems, there are at least two conceptual problems with the indirect approach. One problem is that researchers must be able to assume that the averting activity was undertaken to the point where its marginal cost equaled the value of the reduced illness or risk. If the averting activity is a 0-1 activity, such as purchasing an air purifier, the equality of marginal cost and marginal benefit holds only for the last person who buys the purifier. For all other persons, the marginal cost of the activity divided by its effectiveness is less than the value of the health benefit, implying that this approach understates benefits.

The second problem is that averting activity may produce joint products. For example, while an air conditioner filters particles from the air, it cools the house. If the cost of the activity is attributed entirely to producing the health outcome of interest, people will appear to pay more for the health outcome than they are actually willing to pay.

The direct method avoids the above problems by surveying a random sample of the population and asking each person to value a small change in symptom frequency. Although less data must be gathered using this approach, it has the drawback of basing valuations on answers to hypothetical questions. Answers to hypothetical questions may be unreliable for three reasons: (1) individuals are not incurring actual expenses, therefore, they may not carefully consider their budgets in answering questions; (2) individuals may not give reliable answers if the commodity valued is not familiar to them (presumably the commodity being valued is familiar to persons engaging in averting behavior); and (3) individuals may not have been encouraged to think about opportunities for averting behavior.

6.5.3 Directions for Future Research

Is the range of uncertainty in the estimates produced by the methods discussed in this chapter so large as to render them of no use to policy makers? We do not think so. Although it is unrealistic to believe that benefit-cost analysis can ever provide an exact picture of the welfare consequences of pollution control policies, we believe that as long as the nature and magnitude of the uncertainties is understood, properly derived benefit-cost information on the health consequences of proposed pollution control policies can be useful to decision makers.

Having said this, however, there is much that can be done to improve the quality of the estimates of health benefits that economists provide to policy makers. Five areas in which future research could be especially useful are the following.

The first is the development of more refined models for estimating values of mortality risks that attempt to take account of the characterisitics of the risk, such as its timing, its degree of voluntariness, and the cause of death (cancer versus accident).

Second, in applying the averting-mitigating behavior approach, it is usually difficult to measure the cost of activities that reduce exposure to air pollution or that prevent symptoms from occurring altogether. An avenue that has not been explored is using mitigating behavior in the form of medication to measure WTP. Medication has the advantage that it is a pure averting good with a cost that is easily measurable. If a product can be found that is appropriate for each symptom of interest and that affects only that symptom, then the willingness-to-pay estimates produced by equation (6.15) can be summed across symptoms.

Third, for both mortality risk and risk of chronic illness, those of us who undertake an indirect benefit estimation exercise need to know more about the relationship between actual risks and the perceptions of risks that guide individuals' observed choices.

Fourth, theoretical reasoning suggests that willingness to pay will vary not only with the nature of the health effect of concern but also with the socioeconomic characteristics of the individuals. It is important to develop a more systematic body of knowledge of the effects of age, income, education, and family status on WTP.

Finally, it should prove useful to do a series of comparative studies using two (or three) methods to value the same policy output or health effect. This research strategy should improve our understanding of the relative strengths and limitations of each method and help clarify the circumstances in which each method is likely to perform better.

Measuring the Demand for Environmental Quality
John B. Braden & Charles D. Kolstad (Editors)
© Elsevier Science Publishers B.V. (North-Holland), 1991

Chapter VII

AESTHETICS

PHILIP E. GRAVES

University of Colorado

7.1 Introduction

This chapter applies the benefits estimation methods introduced in Part I, particularly chapters 4 and 5, to the valuation of aesthetic benefits. Webster defines aesthetics as "the science or that branch of philosophy which deals with the beautiful; the doctrines of taste." In some instances, such as noninvasive haze or view blockage by skyscrapers, the only environmental impact is aesthetic in nature. The challenge in these cases is to determine how the senses are affected and how economic well-being is changed. More commonly, however, aesthetic benefits from environmental improvements are intertwined with corporeal or material impacts. Cleaner air is probably both healthier and more pleasing. Separating the intertwined effects may be helpful to decision makers, but it presents additional analytical challenges.

An aesthetic good is here taken as a primitive and economically relevant concept. This is despite fuzzy boundaries between aesthetics and other aspects of environmental quality and, more fundamentally, the venerable and continuing philosophical debate about the nature of "beauty" and "doctrines of taste."

Judging by the economic literature on aesthetic values, what is special about this class of environmental commodities is the fact of sensory experience as distinct from material affect on the body or possessions. The absence of physical effect differentiates aesthetic impacts from health, recreation, or materials damages. The requirement of a sensory connection differentiates aesthetic values from "nonuse" values based on knowledge rather than experience. These distinctions define the niche of this chapter.

Aesthetic impacts presume a real, measurable change in the environment. By definition, the change stimulates sensory responses without inducing direct behavioral responses. Valuation requires three steps: first, the identification of the sensory channels affected by the stimuli; second, the selection of indirect or surrogate reactions from which the worth of the unobservable sensory reactions can be inferred; and third, the use of appropriate statistical and theoretical constructs for data analysis. The following discussion focuses on the first two requirements. The theoretical and analytical considerations are well covered in chapters 2 through 5.

From the preceding, it is clear that methods requiring direct consumption responses, such as the travel cost approach to recreation demand, have limited usefulness.[1] Rather, from the earliest efforts, studies evaluating environmental aesthetics have employed indirect methods — contingent valuation (CVM) and hedonic price (HPM). The following sections review these developments and the methodological issues of special importance to aesthetics.

7.2 Contingent Valuation

Aesthetic valuation and contingent valuation have more or less grown up together. One of the earliest uses of CVM was Randall, Ives, and Eastman's 1974 study of haze from power plant development in the southwestern United States, and visibility has been a frequent topic in subsequent developments of contingent valuation.[2] As discussed by Carson (chapter 5), survey design is the central challenge of the CVM. Aesthetic goods pose a number of special challenges.

Creating a "real" familiarity with the object of study is essential to the CVM. This is difficult enough for tangible goods that are exchanged in familiar market settings. For the often intangible aesthetic commodities, it can be a tremendous challenge. Fischhoff and Furby (1988), in an excellent recent contribution, discussed how this problem has been addressed in studies of visibility. Their framework was to treat valuations of policies affecting visibility as being like any other transaction, involving (1) a good, (2) a payment, and (3) a social context within which the transaction is conducted. They argued persuasively that features of each of the three components of the transaction can introduce difficulties in interpreting CVM valuations. Their table 1 provides a number of examples for the case of visibility of how valuations can be affected by definition of the good, how it is paid for, and the social context.

A CVM instrument presents a description and elicits a response. Even elaborate or graphic descriptions are not perfect substitutes for actual experience. If the description does not capture the important attributes, then the

---

[1] See, however, the study by Tolley and Randall et al. (1986) of visibility value. One of their approaches to valuation involved paid trips to skyscraper viewing-decks in Chicago. These data were used to value the effects of information of current visibility conditions.

[2] For example, Rowe, d'Arge, and Brookshire (1980), Brookshire et al. (1982) and Schulze and Brookshire et al. (1983). See also the review of visibility studies by Chestnut, Rowe, and Murdoch (1986).

image it creates in the minds of respondents will be of something other than the good which is being evaluated. The survey instrument may inject features of the good that the respondent would not normally consider nor think important while missing features that the respondent cares about.

Because aesthetic goods are often qualitative in nature, description is particularly difficult. The very effort to capture an aesthetic good in words or pictures may cause a respondent to value it differently than he or she would based solely on his or her prior experience. Respondents who are unfamiliar with the aesthetic goods being studied will be especially swayed by the limited information contained in the definition. Variations in response values may stem from divergent perceptions of the good rather than from differences in taste and income.

These thorny problems must in any case be addressed. Research to date suggests two ways of making more compelling definitions of aesthetic goods. One is to supplement words with aids that are explicitly sensory. The most tried approach — one especially relevant to visibility — is to illustrate the range of outcomes with photographs (e.g., Brookshire et al. 1982; Tolley, Randall, et al. 1986). But, tape recordings of noise levels, odor samples, risk simulations, and other devices could contribute in other contexts. The second approach is to establish a familiar point of reference and describe the aesthetic goods as they relate to that point of reference. As an example, noise pollution might be characterized by indicating that the respondent's usual television volume would have to be increased, say, 30 percent in order for broadcasts to be heard clearly. As discussed in Fischhoff and Furby (1988), finding meaningful ways to specify the reference and target states resulting from a specific policy is a very complicated undertaking. The extent to which the aesthetic good has been (or can be) properly portrayed to the respondent, even for visibility, which has been most studied, is an open question.[3]

The survey design is not the only source of information bias. The same can be said of the respondent's mood and previous experience, and of the demeanor of the interviewer. Anticipating such influences, questions could be asked in different ways by the same people or in the same way by different people to judge statistically how important cue-seeking is in expressed values. The central issue is, of course, replicability — if valuations prove to vary greatly at different times or with small changes in interviewer demeanor, the scientific validity of the CVM approach is called into question. Indeed, one of the most important, though perhaps expensive, contributions to the CVM literature would be to explicitly test how robust valuations are to a variety of "shocks." For example, when valuations were found to vary according to "starting point," experimental redesign (payment cards and the like) helped eliminate this problem. Similar corrections could be made for other biases once their importance was known.

[3] For a critical examination of the Tolley et al. study, see Fischhoff and Furby (1987).

Another important contextual factor is the cause of the aesthetic impact. Haze caused by natural vegetation or humidity may be valued differently from haze caused by an industrial development. Thus, the aesthetic impact must be explicitly tied to its cause. The responses will be specific to the context and this may limit their generalizability. However, this is better than leaving causality ambiguous, and later, without clear guidance, attempting to infer what the respondents believed.

In a vein similar to other robustness tests, it might be possible in a single CV study to address several causal contexts. In one approach, a specific causal setting would be described and evaluated, after which respondents would be asked how their answers would change for other specific causations. Although starting point bias could be present, at least relative magnitudes could be elicited. Another approach would leave causation vague until after the initial value response. At this point the respondent could be asked what causation he or she had imagined in responding and how the value would change under different circumstances. The open-endedness of this approach, however, could render meaningless the initial value responses. As suggested above, a better approach for statistically exploring the role of causation would be to have a larger sample wherein subsamples were given different causations.

Overall, the CVM may best be thought of as a "regrettable necessity" for valuation of aesthetics. It gets around the lack of direct behavioral responses to aesthetic changes, but the results will inevitably be affected by the survey design, the scenario presented to the respondents, and the possibility that CV responses reflect attitudes rather than real economic commitments.

7.3 Hedonics

The main alternative to CVM has been the hedonic method (see chapter 4 and Blomquist, Berger, and Hoehn 1988; Graves et al. 1988; Haurin 1980; Roback 1982; Rosen 1974, 1979). The general premise is that aesthetic values are capitalized into land values or wage rates (or both, as has been convincingly argued is appropriate in some of the studies referenced above) and can be isolated through cross-sectional analysis of the factors that determine land prices or wages.

Unlike CVM, hedonic analysis must rely on available data. Few aesthetic attributes of environmental quality are readily measurable, much less routinely measured. Thus, there are few occasions when hedonics can be used. The use of proxy data, such as particulate count for visibility, can introduce confusion because the proxies may be associated with other impacts; for example, particulates are associated with respiratory problems and soiling.

Thus, hedonics begin with the disadvantage common to all market-based alternatives — data needs. In addition, the hedonic method is increasingly recognized as a very fragile artifice, highly sensitive to the techniques used and to the markets studied. The ambiguities of aesthetic goods may only exacerbate these weaknesses.

In a study of the effects of visibility and suspended particulates on land values, Graves et al. (1988) found that the estimate for visibility was highly sensitive to a number of different model specifications. The issues explored were measurement error, functional form, error distribution, and variable selection and treatment, particularly regarding the size of lots. Regarding variable selection and treatment, the effects of air quality on property values were found to be very different according to the set of "doubtful" variables included. Moreover, a full market equilibrium (costless rebuilding) housing model of the type employed in urban economics would adjust the housing stock in response to an aesthetic change. This implies that the aesthetic value derived is a per-acre value and should be multiplied by lot size to derive a total value. The notion is that a desired trait, such as ocean access or aesthetic views, should raise land prices, which in turn, should cause a substitution of housing capital for land in the production of housing. Alternatively, a partial equilibrium model — the usual treatment — would allow lot size to be entered as merely one more explanatory variable. Graves et al. found that the two models produce very different estimates for the value of visibility.

Also troubling was their finding that the estimates of the effect of visibility (and total suspended particulates) on property values were highly sensitive to potential measurement error. This is particularly unsettling in view of the difficulty of measuring aesthetic attributes. In addition, more general functional forms were seen to perform better than the simpler functional forms commonly employed in the literature. This problem is complicated by the fact that alternative variable selections were seen to alter conclusions regarding the best functional form. Using a minimum absolute deviation estimation technique (instead of least squares) did not appear to greatly affect the cocfficients of the aesthetic variables, however. Overall, Graves et al. concluded that existing single market analyses provide a poor evidential base upon which to base important policy decisions. There is simply not much in the literature that yields convincing values for aesthetic goods in light of the difficulties summarized above. An immediate corollary of this conclusion is that it is questionable whether hedonic methods should be used to bolster CVM results (e.g., Brookshire et al. 1982) since virtually any hedonic result desired can be plausibly obtained.

One reason why single-market hedonic estimates might be so fragile is the assumption that all capitalization occurs in one market — the land market or the labor market. This is unlikely, and recent theoretical and empirical studies suggest the virtue of multimarket approaches (e.g., Rosen 1979; Haurin 1980; Jones 1980; Roback; 1982; Blomquist, Berger, and Hoehn 1988).

Aesthetic amenities would generally be expected to affect both land and labor markets. Desirable locations would be characterized by some mix of higher rents and lower wages due to the influx of people demanding the aesthetic goods. The implication is that valuation of location-fixed aesthetic traits would generally be systematically understated if only one market is considered.

The usual justification for single-market analysis is a sequential model of work and location decisions: a job is first selected in a labor market and only later is a residential location selected. But, such behavior must surely be irrational — one does not select a labor market independently of residence traits and housing costs. Although a model of simultaneous choice creates considerable analytical complexity, the potential biases in the simple sequential model may cause very large errors in the search for subtle aesthetic values. Insofar as most previous hedonic studies of aesthetic values have been single market studies, their conclusions are probably flawed.

It is tempting to apply formulae to relationships between labor and land market capitalization; say, 60 percent in the land market and 40 percent in wages. The researcher could then study a single market and apply the ratio to figure out the remainder. Unfortunately, such an approach has neither theoretical nor empirical justification (Blomquist, Berger, and Hoehn 1988; Graves and Waldman 1989). Consider, by way of illustration, two locations that are assumed to be equally attractive to households, one attractive by virtue of the aesthetic appeal of a harbor, the other having relatively low humidity — a trait exactly offsetting the lack of a harbor. Suppose further that low humidity does not affect a firm's cost functions although access to the harbor will lower its transportation costs. The harbor location will, in equilibrium, have higher rents than the low-humidity location: the increased labor demand (hence residential and producer land demand) in the productive location will cause more of the aesthetic values to be in land values there. Hence there is no unique capitalization share that can be employed to "back-out" a full aesthetic value from a single market study.

With regard to aesthetic values, the best multimarket study to date is that of Blomquist, Berger, and Hoehn (1988) who estimated hedonic wage and rent equations considering climatic, environmental, and urban conditions. Their data on households and individuals came from the 1980 U.S. Census, and these data were merged with county, regional, and industry data from a variety of sources. Their sample was huge by economic standards and its size permitted reliable hedonic inference for each of 253 counties from across the nation.

From the hedonic wage equation, wage compensation for a county was calculated by adding up the products formed by multiplying the estimated slope coefficients times the respective amenity levels, holding constant all relevant human capital variables. A similar technique was employed for the rent equation. The values from the two markets were then combined to determine a total value.

While access to such a rich data set is the exception rather than the rule, and is clearly infeasible for issues of local rather than national scope, it seems clear that the general multimarket capitalization model is more reasonable than a single market approach. Studies that look at only one market underestimate amenity values in general, and the value of aesthetic goods in particular.

But even a multimarket hedonic approach has important limitations as it applies to aesthetics. The aforementioned issue of the degree of equilibrium affects model specification and interpretation. The evaluation of specific aesthetic goods would require an extensive data set to separate out the effects of closely related amenities (visibility and other climate variables). People may not measure up to the immense perceptual abilities and knowledge of market alternatives that the model presumes. Finally, a really credible hedonic model would incorporate variations in the spatial distribution of preferences. To presume that aesthetic or other such amenities could be accurately valued even in state-of-the-art hedonic studies would seem unduly optimistic for the near term, and perhaps for the distant future.

7.4 A Brief Literature Review: CVM, Hedonics, and Comparative Studies

A literature review of methods for valuing aesthetics could be organized in numerous ways. For example, it could be arranged according to how the authors dealt with one or more problems, such as hypothetical bias in the CVM or omitted variables in a property value study. Or, it could be arranged according to the subject of focus, such as visibility versus water clarity. The present review is organized in neither of these ways. Because both of the dominant approaches to valuing aesthetic goods — the CVM and hedonic price analysis — have serious though probably surmountable reservations, this review will be brief, a bit negative, and will contain fewer references to the literature and more recommendations for research than is common. The three approaches examined are the contingent valuation method, the hedonic price method, and finally, comparative studies aimed at gauging accuracy.

7.4.1 The Contingent Valuation Method

As noted earlier, when assessing aesthetic values it is particularly difficult to be certain that the respondents do not include the nonaesthetic dimensions of a hypothetical change, such as health, soiling, or ecological effects, in their valuations. However, there is a positive feature in this difficulty. Suppose a researcher is interested in the aesthetic dimension of a recreational experience, but existing CVM studies have only considered the value of the composite recreation experience. It might be possible, at low marginal cost, to survey recreators as to the percentage of the value of a recreation experience they attribute to the aesthetic dimension. Or, should even this approach be beyond the budget of the decision maker, the researcher could select a plausible percentage. For example, one or more prior CVM studies may have estimated a one-day canoe trip down a river as having a particular value. The researcher could argue that aesthetic dimensions of water quality account for 10 percent of the value of the canoeing experience. In any event, the aesthetic dimension will certainly be more than 1 percent of the experience and less than 100 percent — hence, the researcher will at least have order-of-magnitude estimates readily available. He or she could then argue that a 10 percent improvement in water quality might raise the aesthetic component also by 10 percent, or by 1 percent of the value of the entire recreational experience. Adding up these incremental values over all recreational activities affected by the aesthetic improvement — for example, canoeing, picnicking, bird-watching, hiking, and so on — could result in a plausible overall valuation.

Although such *ad hoc* approaches to valuation are unconventional, in far more cases than is commonly recognized, and they give adequate information to decision makers. The reason is that most policy options are discrete, all-or-nothing decisions, such as having or not having scrubbers or secondary waste treatment; therefore, benefits will commonly exceed or fall short of costs by an order of magnitude, or more. The researcher can often quickly determine whether or not more information is required by merely employing a sensitivity analysis to the CVM valuations when using them in benefit-cost analysis, particularly for the difficult to value aesthetic dimensions of a policy. If, in the example of the preceding paragraph, the researcher needed to have aesthetic benefits represent, say, 75 percent of the values of canoeing and other experiences in order for an aesthetic improvement to have benefits greater than costs, then he or she does not need an elaborate independent assessment of aesthetic valuations because the aesthetic component can clearly not be so high a percentage of the value of such recreational experiences. Given the shortcomings of both the HPM and the CVM, this application of CVM may have considerable appeal to decision makers.

Tables 7.1, 7.2, and 7.3 provide a brief annotated sampling of the CVM literature broken down by media — water, air, and other. These tables allow those with a specific valuation difficulty to see what other researchers have done in that or a similar context.

7.4.2 The Hedonic Price Method

The notion that compensation or desirable and undesirable traits can occur in land or labor markets has a long history with perhaps earliest mention more than two century's ago in Adam Smith's *The Wealth of Nations*. However, the active phase for the development of hedonic literature almost exactly

TABLE 7.1
CVM studies: water.

Study	Context
Daubert and Young (1981)	Recreation benefits from instream river flows on the Cache la Poudre River in northern Colorado.
Walsh et al. (1978)	Option and preservation values of improved water quality in the Colorado Platte River Basin.
Mitchell and Carson (1981)	CVM bids for improvements in national water quality (discussed strategic bias).
Brookshire, Ives, and Schulze (1976)	Recreators interviewed at Lake Powell (discussed strategic bias).
Greenley, Walsh, and Young (1982)	Recreation on South Platte River Basin in Colorado (discussed vehicle bias — sales tax versus sewer fee).
Cronin and Herzeg (1982)	Water quality study conducted on beaches of Potomac River (discussed impact of information on willingness to pay (WTP), finding cost information affects bids as well as who pays the costs).
Desvousges, Smith, and McGivney (1983)	Water quality study employing water quality ladders to yield perceptions of outcomes to more closely approximate actual outcomes.
Seller, Stoll, and Chavas (1985)	Study of boating values on four lakes in Eastern Texas (asked respondents how accurate they thought their valuations were, with most indicating "quite accurate").
Loomis (1983)	WTP for trout fishing in 11 western states employing CVM.
Miller and Hay (1984)	Freshwater fishing days in five states valued by travel cost method (TCM).

parallels that of the CVM. This is not particularly surprising, as the interest in the 1960s and 1970s in valuing environmental goods required looking around for nontraditional approaches. Early hedonic work occurred at about the same time as did early CVM contributions. Also, beginning about the same time (with Rosen's 1974 hedonic paper), was a theoretical concern for what the numbers might actually mean. Because the data necessary to do at least rough hedonic analyses are abundant, there are far more hedonic studies than contingent valuation studies.

However, as indicated earlier, the vast majority of existing studies are seriously flawed from the perspective of obtaining accurate aesthetic valuations. The problem is that the single-market studies, which constitute virtually all hedonic studies, look at either land or labor markets in isolation and ignore amenity compensation that occurs jointly in both land and labor markets.

TABLE 7.2
CVM studies: air.

Study	Context
Randall, Ives, and Eastman (1974)	Aesthetic benefits from reduced air pollution; modern CVM applied to aesthetic interest.
Gallagher and Smith (1985)	Value improved air quality in a national park (uses a change in the entire probability distribution of air quality on a given visitor day).
Rowe, d'Arge, and Brookshire (1980)	WTP and willingness to accept compensation (WTA) measures for preserving alternative air quality levels in the Four Corners region of the Southwest (discussed strategic bias).
Brookshire, et al. (1981)	CVM of air quality in Los Angeles (interesting in that both health and aesthetic values are solicited; found that sequencing of benefits matters, as does vehicle bias, with additional interest in starting point bias).
Schulze, Cummings, et al. (1983)	CVM values for reduced ozone concentrations.
Schulze, Brookshire, et al. (1983)	Study of visibility at the Grand Canyon (commonality of perceptions "assured" by showing all respondents the same set of photographs of known visibility levels).
Tolley and Fabian (1988)	Edited book summarizing research at the University of Chicago on visibility, considering how visibility affects values of outdoor recreation, urban recreation, view-oriented amenities, and aviation and auto-traffic safety.

Moreover, as indicated earlier, the ubiquitous property value study is not robust to functional form, measurement error, and so on.

The approach presented in Blomquist, Berger, and Hoehn's 1988 study is state-of-the-art (and includes 16 amenity variables). But it is not readily apparent how to decompose the total amenity value of clean air or water into its aesthetic dimension, which is the central concern here. Perhaps some progress along those lines could be made by conducting a CVM analysis of the percentage of value of these amenities that is aesthetic, rather than nonaesthetic, in nature.

7.4.3 Comparative Studies

Its early association with CVM made visibility an attractive context for comparing contingent valuation with other methods. There have, however, been fewer comparisons between hedonic methods and CVM than for the travel cost method and CVM (see table 7.4). This is due, in part, to the fact

TABLE 7.3
CVM studies: other.

Study	Context
Chestnut et al. (1988a)	Study of the WTP to avoid angina episodes (looked at consistency with averting behavior and costs as well as out-of-pocket costs; the WTP numbers compared well with averting costs).
Crocker (1984)	Value of reduced acid deposition damage to forest stocks.
Sorg and Brookshire (1984)	Valued elk wildlife encounters.
Brookshire, Randall, and Stoll (1980)	Wildlife study with an interest in vehicle bias, employing hunting license fees and utility bills.
Cummings, Burness, and Norton (1981)	CVM to estimate benefits from reduced household soiling due to TSP reductions.
Knetsch and Davis (1965)	Measured benefits of recreation in the woods of northern Maine by using three different methods: interviewed hunters, fishers, and campers (see also table 7.4).
Bishop and Heberlein (1979)	TCM versus two types of CVM to study WTA and WTP of hunters receiving free early season goose hunting permits (see also table 7.4).
Thayer (1981)	CVM versus TCM and site substitution method (SSM) employed in survey in Jemez Mountains of northern New Mexico on WTP (entrance fees) to prevent geothermal development (described noise, odors, and showed pictures of other geothermal developments, along with a map of proposed areas impacted).
Cummings, et al. (1982)	Used hedonic and contingent valuation methods to value municipal infrastructure in 26 boomtowns in Rocky Mountain area.
Brookshire, et al. (1984)	HPM and CVM to determine WTP and WTA for earthquake risk, the latter assumed present if the house was located in a special studies zone (SSZ).

that the goods being valued (for example, recreational experiences of many kinds) often do not exist in areas with good hedonic data. It is also the case that many of the comparisons, such as Cummings et al.'s 1982 study of boomtown infrastructure and Brookshire et al.'s 1984 study of earthquake risk, do not relate to the aesthetic goods of interest here. There are, however, two studies that deserve particular mention.

Brookshire et al. (1982) examined CVM willingness to pay for an increment to air quality from "fair" to "good" and compared this to hedonic payments. Their study supported their *a priori* conjecture that WTP is greater than zero and that rent changes are greater than CVM willingness to pay. However, if

TABLE 7.4
Comparison studies: TCM, hedonic, and CVM.

Study	Context
Comparisons of CVM versus TCM:	
Knetsch and Davis (1965)	Use three methods (CVM, TCM, and willingness to drive) to measure benefits of recreation in woods of northern Maine; concluded that "results were close." Thought that the comparison revealed promise for the CVM, although some odd assumptions were employed in the TCM work.
Bishop and Heberlein (1979)	Compared the TCM with two versions of CVM in a good study valuing goose hunting permits; although the numbers were quite different, they were "ballpark" similar.
Desvousges, Smith, and McGivney (1983)	TCM compared to CVM and contingent ranking of values of water quality improvements; considered three changes in water quality: decline with complete loss of recreation, increase from boatable to fishable, and increase from boatable to swimmable. Again the results could be interpreted as close, though clearly different, under the three methods.
Thayer (1981)	CVM versus TCM in WTP to avoid geothermal development; concluded that CVM is quite accurate.
Sellar, Stoll, and Chavas (1985)	TCM and CVM, with a novel question about how accurate the respondent felt his/her response to be; again the results were "close."
Fisher (1984)	Used the TCM data of Miller and Hay (1984) and the CVM data of Loomis (1983) in comparisons, for the two states common to both studies, of the value of fishing days; concluded that results are "fairly close," noting that information accurate to a factor of 2 or 3 may be better than no information at all.

air quality is capitalized in wages as well as rents, which is a reasonable supposition, there is no compelling reason to expect that rent change will be greater than CVM willingness to pay. Moreover, Graves et al. employed data from the same primary source in their analysis of the robustness of land value compensation discussed above, finding that estimated rent changes are fragile.

Chestnut, Rowe, and Murdoch (1986) reviewed 6 different urban visibility studies, collectively considering 15 different "scenarios," or changes in visual range. Using a regression analysis, they found that WTP across scenarios is significantly affected in the expected direction by the change in visual range. This finding provides support for the potential of CVM to provide reasonably accurate aesthetic valuations.

TABLE 7.4 *Continued*

Study	Context
Comparisons of CVM versus HPM:	
Brookshire, et al. (1982)	WTP for an increment to air quality (from fair to good) was examined, with their *a priori* conjecture that WTP > 0 and dR > WTP being supported. As noted in the text: if, as seems reasonable in this context, air quality is capitalized in wages as well as rents, there is no compelling reason to expect that dR > WTP.
Cummings, et al. (1982)	Boomtown study of the value of infrastructure, employing 26 Rocky Mountain areas; conclude that estimated values are comparable.
Brookshire, et al. (1984)	Earthquake risk (location in a special studies zone, (SSZ) was examined, with hedonic prices (of living out of a SSZ) being found similar to CVM estimates of WTP.
Consistency Checks:	
Chestnut et al. (1988a)	Found, in a study of heart disease patients, that the averting behavior questions yielded answers ($28 to avoid an angina episode) that were similar to CVM solicitations of WTP to avoid an episode ($38).
Rowe and Chestnut (1985, 1986)	Found similar consistency in a benefit analysis of oxidants and asthmatics in Los Angeles.
Chestnut, Rowe, and Murdoch (1986)	Reviewed 6 different urban visibility studies, collectively considering 15 different "scenarios" (changes in visual range). Found, in a regression analysis, that the WTP across scenarios was significantly affected by the change in visual range, as would be expected.

7.5 Assessment and Conclusions

CVM seems to have more to offer for aesthetic valuation than does the hedonic approach or other alternatives. The advantages of applying CVM include its abilities to: (1) more sharply distinguish the aesthetic dimension of a policy-induced change; (2) generate data rather than relying on remote proxies; (3) impose fewer behavior assumptions, with the important exception of presuming that intentions accurately portray actions; and (4) yield results that are at least plausible when the studies are conducted carefully, particularly in applications to visibility. The most important difficulties in applying CVM to aesthetic goods are attaining a high degree of respondent familiarity with the aesthetic good and taking care to avoid introducing extraneous influences.

The hedonic price method is severely limited by data, particularly in applications involving aesthetic goods. Many important aesthetic goods are located where neither labor nor land markets are well developed. Even where the markets are functioning, few aesthetic goods are well measured. In the few cases of high-quality data, a large number of observations would be needed to conduct a multi-market study to avoid systematic bias in the results.

Regardless of these problems, both CVM and hedonics will continue to be employed and improved. Progress in public decision making about aesthetic goods depends on these developments. The challenges are both conceptual and methodological: better understanding of aesthetic goods and how to characterize them; better measurement of aesthetic goods; better preference revelation methods that bridge the gap between intentions and commitments; and less restrictive econometric models for interpreting preference and market data.

Measuring the Demand for Environmental Quality
John B. Braden & Charles D. Kolstad (Editors)
© Elsevier Science Publishers B.V. (North-Holland), 1991

Chapter VIII

RECREATION

N.E. BOCKSTAEL, K.E. McCONNELL, and I.E. STRAND

University of Maryland

8.1 Introduction

The number and quality of recreational opportunities are measures of how well the U.S. economy is helping to meet one of our explicit goals: "the pursuit of happiness." Recreation is an end that, under plausible assumptions about income elasticities, ought to increase with economic growth. And thus the use of resources for recreational ends may become more important as real incomes rise.

The fact that recreation is an important end in itself does not make it a critical matter of public policy or an obvious topic for discussion in this book. What separates recreation activities from many other end activities is the failure of the market to provide efficient quantities and qualities of natural resources for the satisfaction of these activities.

There are several kinds of market failures that relate to natural resources and recreation. For example, the technical conditions that prevent the market from supplying efficient quantities of recreational services are akin to the forces that create natural monopolies (Knetsch 1974). The marginal social cost of producing another unit of recreational service, say in the form of a visitor day, is typically close to zero. Consequently, competitive pricing cannot support an industry that produces these services because when price is equated with marginal cost, firms will suffer losses. It is difficult to imagine a competitive pricing scheme that would sustain Yosemite, Yellowstone, Glacier, and other large natural resources. In fact, the market's inability to provide recreational services on large, undeveloped natural resources at competitive prices was one of the initial forces that involved the government in preserving national parks for recreational purposes.

Also, there remain questions of acquisition or divestment of public lands and the optimal diversion of lands from their natural state for the production of market goods. For one thing, the diversion or development of natural lands might be irreversible, raising important questions about the long-run and dynamic forces that shape future demand. Even if markets existed here, market forces might well discount the future at a higher rate than is socially optimal. Market forces would also be unable to appropriate the benefits that accrue to people who do not visit the resource. To the extent that a resource is valued simply for its presence or because of its symbolic contribution to society, benefits exceed those that private firms could count when deciding on the most profitable use of the resource.

Market failure influences the quality of recreational services in the form of externalities generated in the production and consumption of goods. This occurs in obvious ways, such as the discharge of visible contaminants into a river used for swimming. But it also occurs in less obvious ways, such as the elimination of breeding habitat for game and fish. This form of market failure — the erosion of the qualities of resources — is more insidious, more varied, and probably more prevalent than the failure of the market to provide the correct quantities of recreational services. The materials balance perspective on the production of output suggests that the unfettered market will steadily erode the quality of natural resources because the weight of residuals to be disposed of is roughly equal to the weight of inputs. Users of natural resources are gradually deprived of the services of uncontaminated resources without being able to voice their preferences. As a consequence, the deterioration will occur without the "due process" usually afforded by a market.

Hence, the role of the economics of recreation: Recreation is obviously important to the well-being of society. The production process for many recreation services utilizes a substantial natural resource base. But the private forces will produce this recreational service at a socially insufficient quantity. Furthermore, industrialization and the concomitant production of externalities will steadily impair the ability of natural resources to produce recreational services. Tied as it is to problems of environmental pollution and land use conflicts, recreation economics has a role to play in a number of critical social questions.

The history of the development of recreation economics reflects economics' role in the policy debate over the allocation of natural resources. There were two basic stimuli for this development, each reflecting some aspect of the federal government's influence in the allocation of natural resources. The first stimulus arose from the federal government's role in allocating land and water in the West. The government is the largest landowner in the West, and governmental subsidy of water resource development has been a critical factor in the growth of agriculture in this arid region. In each of these activities, the need for accountability grew as the government expanded its influence in the region. Because recreation was, in most cases, a secondary output of these activities, it was natural that some intellectual resources be devoted to measuring its output. The first of these was the U.S. Army Corps of Engineers' water resource projects, which attempted to keep track of recreational use.

But keeping track of recreation can mean many things, from counting visitors to observing recreational capacity. The famous Hotelling letter to the Park Service represents a new era by arguing for an empirical method which would allow the calculation of the value of the service flow, rather than simply counting the use. Hotelling was responding to a request from a Department of Interior official who was searching for methods which would value the service flow as economists now do: in terms of surplus. And while Hotelling laid the groundwork for the travel cost method now widely used, it was not for over a decade after Hotelling wrote his letter that the method was attempted, even in crude form.

The initial travel cost models were estimated and used in the decade of the 1960s and they were applied to the problems of valuing access which were raised by the federal government's involvement in water resource development and land ownership. Little attention was given to the micro details of these models. The emphasis reflected the interests of economists at the time. At a general level, economists debated the rationale for government intervention, using both efficiency and equity criteria for judging its merit.

At a project level, the issues of concern were intrinsically temporal. A typical Corps of Engineers dam would have a life of more than 50 years. The development of some projects might even be irreversible: such was thought to be the case for the destruction of the natural environment to provide electricity production and flat-water recreation from dams proposed for Hell's Canyon. Hence recreation demand and valuation models were developed during a period when there was substantial concern for temporal issues. The discount rate for project evaluation (Marglin 1963), preservation versus development in the context of irreversibility (Arrow and Fisher 1974), asymmetric technical change (Smith 1974), and a variety of other public investment questions arose (U.S. Joint Economic Committee 1969). Economists working on recreational demand models and valuation techniques during this period were preoccupied with broad issues of valuing access to natural resources.

The second stimulus for the development of recreation economics stemmed from concern over the quality of the environment. This concern was manifested in research which began in the 1960s but developed in its current form in the late 1970s and early 1980s. This intellectual development is a natural extension to valuing access. If recreational demand models can be used to value the service flow from a resource, then why can they not be used to value the change in a service flow induced by a change in environmental quality? Much of the work in the past 15 years has focused on the problem of valuing changes in environmental quality. Researchers have attempted to value a wide variety of environmental dimensions of recreational resources, from aspects of water quality such as turbidity and Secchi depth to different types of visibility reflecting air pollution to risk of health impairment.

Concurrent with the growth in importance of environmental quality issues was a change in the focus of research. Economists paid increasing attention to the details of the models from which benefit estimates were derived. Papers in mainstream economics justifying the use of estimated demand functions in benefit measurement gave impetus to this work. From these forces, there evolved and continues to evolve, a set of tools for measuring the benefits from natural resource-based recreational service flows (e.g., Willig 1976; Hausman 1981; and Bockstael and McConnell 1983).

This chapter will examine the approaches that attempt to form public policy by valuing aspects of recreational resources, with special attention to valuing environmental amenities in the context of recreation. It will analyze the various approaches that economists have developed, the challenges they have successfully met, and the challenges that remain.

8.2 Resource Issues and the Definition of Analytical Methods

The convergence of recreational economics and environmental policy is most often studied in the context of water resources. The special connection between recreation and water quality is the natural result of the original Federal Water Pollution Control Act which defined water quality standards in such terms as fishable and swimmable. The connection was further strengthened by early suggestions that 80 percent or more of the benefits of cleaner water would accrue to recreationists (Freeman 1982). This does not mean that recreational benefits do not accrue through other environmental policies (for example, air quality control). Nor does it mean that economists have not attempted to measure recreational benefits associated with other types of environmental improvements. But, the emphasis on water quality has shaped the direction of recreational benefit estimation.

Just as there is a predominant object of study in recreational benefit assessment, there is also a predominant method of analysis. The travel cost method was designed for recreation. Simple travel cost is now part of a broader group that is labeled recreation demand models, but they are all based on the same premise that the "price" of a recreational trip can be measured, at least in part, by travel costs to recreational sites. Recreation demand models, having an early start, still represent the most prevalent method for obtaining recreation-related benefits, although nonbehavioral-based studies of recreation, such as contingent valuation, are common and frequently used in conjunction with travel cost.

Outdoor recreation typically requires the participant to go to a site to enjoy its services. This simple fact, when combined with the emphasis on water resources, has caused recreation economics to deal with specific resources at specific places. From the individual's perspective, recreation takes place at specific sites that exhibit observable characteristics and measurable travel costs, and so recreational service flows are described as site specific. In their earliest applications, recreation demand models were used to value the recreational service flows of a particular dam. These models were irrelevant for assessing the general policy of building dams. Likewise, the value of an improvement in water quality has been viewed as a relevant question in the context of specific sites known by both recreationist and researcher, not in the broad context of national clean water policy. The principal exception to this statement is the study by Vaughan and Russell (1982) in which the authors attempted to obtain aggregate freshwater sportfishing benefits attributable to national water pollution policy.

Numerous studies have valued recreation-related benefits associated with specific resources. Examples include Desvousges, Smith, and McGivney's (1983) Monongahela River study; Caulkins, Bishop, and Bouwes's (1986) Wisconsin lakes study; Hanemann's (1978) work on Boston beaches; Bockstael, McConnell, and Strand's (1988) Chesapeake Bay analysis; and Sellar, Stoll, and Chavas's (1985) analysis of Texas lakes recreation. Site-specific analysis is not limited, however, to a narrowly defined geographical resource. It is possible to apply site-specific analysis to a broad geographical area — for example, recreational fishing along the East Coast — as long as the individuals in the sample have clearly defined sets of fishing sites. All individuals in the sample need not visit the same sites, but the researcher must be able to measure characteristics and travel costs of relevant sites for each individual.

This description of the site-specific commodity is most critical to the application of recreation demand models. However, it plays a role in non-behavioral-based studies as well. A large number of contingent valuation method (CVM) recreation studies are also site specific. For those which are not, one is led to compare the unique benefits of CVM with the drawbacks of a vaguely defined valuation good.

The notion of site-specific service flows helps to define the welfare measures whose estimation is a principal objective. Two definitions are needed: one for access to a site and one for the change in the quality of a site. Let the utility function be $U(x,q)$ where x is an m-dimensional vector of service flows and q is an n-dimensional vector of attributes of the service vector. The service flows themselves, x, may be marketed commodities as well as recreational services. The attributes of the service flows, q, are specific. That is, it is assumed, at least for now, that each element q_j modifies only one x_i, though the details of that mapping have not been given here. To complete the definitions of the welfare measures, the consumer must be endowed with income and given a set of prices. Let y be income and p be an n-dimensional vector of prices such that p_i is the price of good x_i. The welfare measures of neoclassical economics are strongly dependent on these prices being exogenous to the consumer.

Welfare measures, as defined in chapter 2, are based on the minimum cost function (expenditure function):

$$e(\boldsymbol{p},\boldsymbol{q},u) = \min[\boldsymbol{x}\boldsymbol{p} \,|\, U(\boldsymbol{x},\boldsymbol{q}) = u]. \tag{8.1}$$

The benefits or losses of a change in the state of the world are represented by the change in the value of this expenditure function (admitting, of course, the lack of consensus over the reference utility level). The expenditure function is a function of parameters exogenous to the individual. To value the elimination of a site, it must be reflected in terms of changes in these parameters. The common convention is to equate elimination of access to a price change for the good sufficiently high to cause the individual to exit the market of his or her own accord. Let x_1 be the service flow to the site which is of interest, and let p_1 be the price of access to the site. The first welfare measure, w_1, is simply the value of access, which is given by:

$$w_1 = e(\tilde{\boldsymbol{p}},\boldsymbol{q},u) - e(\boldsymbol{p}^0,\boldsymbol{q},u) \tag{8.2}$$

where $\tilde{\boldsymbol{p}} = (\tilde{p}_1, p_2^0, \ldots, p_m^0)$ is the vector of prices that includes $\tilde{p}_1$ the choke price of x_1, the price high enough to eliminate the service flow from the site of concern and $p^0 = (p_1^0, \ldots, p_m^0)$ is the vector of current prices. Depending on the choice of utility reference level, w_1 may be an equivalent or compensating measure (see chapter 2 for a discussion).

The second welfare measure, w_2, is the value to the individual of a change in an attribute of the site. Let q_1 be the original level of a particular attribute at the site, and let q_1^* be its new level. The welfare effect of this change is

$$w_2 = e(\boldsymbol{p}^0,\boldsymbol{q},u) - e(\boldsymbol{p}^0,\boldsymbol{q}^*,u) \tag{8.3}$$

where $\boldsymbol{q}^* = (q_1^*, q_2, \ldots, q_n)$. This also may be compensating or equivalent variation, depending on the reference utility chosen. Here, w_2 is defined so that it has the same sign as the welfare change. When q_1 is a "good" attribute (fish density rather than water turbidity) and it increases from q_1 to q_1^*, w_2 is positive.

8.2.1 Preview of Methods of Measuring Recreation Value

There are three basic approaches to obtaining these measures empirically: models of behavior that derive from demand for services of recreational sites; models of the demand for generic attributes; and contingent or hypothetical valuation approaches that value access or quality changes directly from consumers' responses to questions.

In recreation analysis, the most prominent of these methods, and the ones that will require the most attention, are based on the demand for a site's services. They include travel cost models and random utility models (see chapter 3 for a discussion). The fundamental insight that drives these models is that a consumer must visit a site to consume its services. Travel costs serve as surrogate prices, and variation in these prices causes variation in consumption. Observations on the joint variation in prices, consumption, and sometimes quality characteristics are an essential ingredient in the estimation of demand functions and the derivation of welfare measures.

The two classes of models grouped under the term *recreation demand* use observations on individual visits to sites and also exploit travel cost as a surrogate price. One essential difference between the models is the assumption about the recreationist's planning horizon. The travel cost model is a model of demand for the services of a site over a period of time, say a season or year. The random utility model describes how people choose among a group of sites each time a choice is to be made.

Modeling consumer choice for generic characteristics, in contrast to site specific goods, has become standard fare in housing studies but not in recreation economics. The use of hedonic models for housing assumes that consumers purchase bundles of attributes, and that an attribute in one bundle is the same as an attribute in another bundle. For example, one bedroom is one bedroom, regardless of the house. Of course, preferences for bundles depend on the attribute composition.

The hedonic travel cost model (Brown and Mendelsohn 1984) is an attempt to utilize insights from the housing market, where the market insures that the attributes that are valued are priced at the margin. The first stage of the hedonic procedure involves regressing the travel costs individuals in the sample incur in visiting a recreational site on the measures of quality characteristics at the site. The marginal cost of each quality characteristic is computed from the estimated equation and used in estimating the demands for each characteristic.

But there are several difficulties with the hedonic paradigm in the context of the demand for recreation. First, there are conceptual difficulties. No market intervenes between buyers and sellers in the recreation context as does the housing market. As a consequence, the relationships estimated when applying this approach must be considered carefully.

There is also an obvious departure from the housing market hedonic concept. The relationship between travel costs and the level of quality characteristics obtainable in the recreation problem will be determined by nature and not by the market. It is true that including only visited sites will exclude sites that are dominated by others, but that does not guarantee the remaining sites are arrayed so that desirable site characteristics will cost more to attain. Nature's array of sites and quality characteristics, along with traditional population centers, will have much influence on whether positive marginal costs can be obtained for characteristics of interest. Additionally, individuals' relative preferences for characteristics will affect the set of chosen (dominant) sites. As a simple example, consider two desirable characteristics of lakes, scenic beauty, and fish catch. Suppose sites were arranged such that sites progressively farther from the individual's origin exhibited increasing fish catch but declining scenic beauty. Individuals may be observed going to several sites along the continuum, but no positive marginal cost for scenic beauty may be derivable.

Even if an array of available sites were sufficiently rich to preclude the dilemma suggested above, there are additional problems. Smith (chapter 3) argues that the conceptual difficulties of a meaningful hedonic price function are exacerbated because costs should include the opportunity costs of travel time, not just monetary costs, and these time costs will vary among individuals. This suggests that each individual will face a different hedonic price function, but estimating different functions is, of course, precluded by the data.

Another criticism has been lodged by Smith and Kaoru (1987). The demand equation estimated in the second stage of the hedonic travel cost procedure relates the "price" of acquiring the quality characteristic with the level (desirability) of the quality characteristic. For example, individuals may have to pay a higher price for greater scenic beauty or to guarantee a lower level of fecal coliform at a beach. If the good is a one-time purchase, as in housing, this treatment of quality makes some sense. But it does not account for the quantity dimension intrinsic to recreation problems, such as when several trips are chosen. Hence, there can be serious empirical problems when estimating the hedonic travel cost model, such as the negative marginal prices encountered by Bockstael, Hanemann, and Kling (1987) and Smith and Kaoru (1987).

Even if the economist accepts the conceptual validity of the hedonic travel cost model, its relevance for public policy is questionable. Most policy questions are asked with respect to specific resources, not generic characteristics. The hedonic travel cost model cannot, for example, value the introduction or elimination of a site or a resource. It might appear to be an appropriate vehicle for valuing national environmental policies, such as changes in national water quality standards, but even here it is less than useful. First, the hedonic model produces a "price" function specific to one origin only (i.e., one city) and is conditioned on the sociodemographic characteristics and the array of sites and quality characteristics relevant to that population. Second, the values generated can purport to reflect only the marginal values of characteristics. These values do not hold for discrete changes in quality characteristics that will change the configuration of the cost function.

After initial explorations in the early 1980s, the approach has received little further attention in the literature. As a consequence, the hedonic travel cost model will not be pursued in further detail in this chapter.

The contingent, or hypothetical, valuation approach to measuring the value of access or the value of changes in quality, while not as frequently encountered in applications to recreation as in other valuation problems, nonetheless holds an important place in this literature. Unlike recreational demand models, the hypothetical valuation approach does not rely on observations of behavior to make inferences about value. Practitioners of this approach attempt to infer surplus measures directly from consumer responses to hypothetical questions.

As in the travel cost model and to a lesser extent the random utility model, there are many variants of the contingent valuation method. The discussions that follow will describe the range of contingent valuation experiments that have been applied in recreation.

8.3 Using Hypothetical or Contingent Methods in Recreation Valuation

While the bulk of this chapter addresses those methods that exploit observations on individual behavior, there is an alternative method for valuing recreation service flows. In simple terms, this alternative involves asking the individual what he or she would be willing to pay for some hypothetical change. There is a substantial literature on the topic, including the books by Cummings, Brookshire, and Schulze (1986) and by Mitchell and Carson (1989), as well as chapter 5 in this book. This literature deals with the many important issues of conducting a hypothetical study, including sample selection, questionnaire design, and so on. As these issues are all covered in depth in chapter 5, the discussion here is restricted to an overview of the application of contingent valuation to recreation.

8.3.1 Applications in Recreation

Hypothetical valuation was originally employed in a 1963 study of recreation in the Maine woods by Robert Davis. The method was revived by Randall, Ives, and Eastman (1974) in a study of the aesthetics of recreation in the Southwest. These studies were motivated by apparent difficulties in using behavioral methods. Since the Randall, Ives, and Eastman paper, there has been a flood of work applying this method to resource valuation questions. This flood reflects a growing acceptance of contingent valuation as a method of valuing goods and services not provided by the market.

Because recreation is one of the most prominent service flows that provides value to people but is not valued by the market, it has been the subject of many contingent valuation studies. In an appendix summarizing contingent valuation studies, Mitchell and Carson list over 100 completed works. Of those, about half deal with recreation. Some of these studies value the recreational resource directly or access to the recreational resource. Bishop and Heberlein's (1979) study of hunting permits is a well-known example. Others include Majid, Sinden, and Randall's (1983) valuation of public parks, Milon's (1988) study of artificial reefs, and Walsh, Loomis and Gillman's (1984) valuation of wilderness. Most studies value quality dimensions of the recreation resource. Contingent valuation methods are especially appropriate when the impact of the quality characteristic is difficult to capture in a recreation demand model. An example is congestion, an endogenous and typically undesirable characteristic of recreational resources. By its nature, the congestion of recreational facilities varies across locations and within a season. Hence, it is difficult to model econometrically. Contingent valuation studies have estimated welfare losses due to congestion, beginning with congestion in wilderness recreation (Cicchetti and Smith 1973) and beach use (McConnell 1977). More recent applications include congestion in wilderness areas (Walsh and Gilliam 1982) and ski area congestion (Walsh, Miller, and Gilliam 1983).

Because of the obvious flexibility of contingent valuation methods, they have been used to value a remarkable array of goods, services, and characteristics of these goods and services, with emphasis on environmental amenities. Not all of these environmental studies are linked with recreation, but many are. An example is the Sellar, Stoll, and Chavas (1985) study which valued recreational boating at lakes of different water quality. A host of other studies value changes in fish catch or game bagged, changes that could come about from new environmental or management policies. (See Wegge, Hanemann, and Strand 1985; Carson, Hanemann, and Steinberg 1988; Brookshire, Randall, and Stoll 1980; Cameron and James 1987.)

A common characteristic of these studies is their attempt to value a dimension of the quality of recreational services that for some practical reason cannot be attempted by behavioral methods. The problems of obtaining good observations on quality variables will be discussed in the subsequent section. "Expected encounters of elk per trip" may seem like an indicator of the quality of elk hunting that fits into the neoclassical model, but practically speaking, there is sometimes no good way of observing this variable and hence no means of capturing econometrically its effect on behavior.

For some studies, such as Desvousges, Smith, and McGivney (1983), the authors ask hypothetical questions to elicit valuations for improvements in water quality from boatable to swimmable levels. These are subjective water quality criteria that individuals can understand. However, they are not criteria that can be observed and measured by the researcher, and as such they cannot be used in a recreational demand setting. Similarly, Bockstael, McConnell, and Strand (1988) ask willingness to pay for an improvement in the water quality of the Chesapeake Bay to a level that was deemed "acceptable" for swimming and other water-related activities by the respondent.

8.3.2 The Appeal of Referendum Models

One difficulty that plagues nonmarket valuation in general is that the desired welfare value, whether consumer's surplus, compensating variation, or other measure, is never observed. This makes comparisons between contingent valuation and other methods of limited use, as neither is being compared with a correct standard. Behavioral methods partly compensate for this difficulty by building value estimates on observations of behavior. But for contingent valuation, neither the true welfare values nor any corroborating behavior can be observed to support the estimated values.

There are recent developments in contingent valuation that allow a closer connection to behavioral models. The referendum model (Hanemann 1984b) utilizes yes or no responses to hypothetical choices between two situations. For example, one alternative might be swimming in water that is of poor quality, while the other would be swimming in cleaner water but at an annual fee. As Hanemann (1984b) and Cameron and James (1987) showed, these responses can be used to estimate parameters of the utility or expenditure functions. The referendum model has some attractive features; for example, it does not require the respondent know how much he or she would pay, only whether one situation is preferable to another. By carrying the results one step further to the utility or expenditure function stage, this approach has the added appeal of allowing predictions of behavior that would be implied by the parameters of these functions. This provides a door to observability that has been lacking in contingent valuation.

8.3.3 Some Critical Observations

To many skeptics it is remarkable that the contingent valuation method has become so widely accepted. After all, the general line of criticism might be: ask a hypothetical question and you get a hypothetical answer. The acceptance is due to a variety of reasons. The most compelling reason for using contingent valuation methods is that in some settings no behavioral methods can be used to value the service flows of interest. The best example is existence value, which by definition does not involve behavior. When people value the presence of a resource, they need not visit the resource, and so there is no way to infer the value of the resource from their behavior. This is a relevant motive for recreational resources when people who use the resource also value it for its presence. In such a case, behavioral methods would capture only part of the value.

The contingent valuation method can also be used as a substitute for travel cost or random utility models. Why is it that, even when behavioral methods are available, contingent valuation methods are used? First, there is no evidence that contingent valuation gives answers that are dramatically different from or more random than behavioral methods, such as the travel cost models. This reflects as much the arbitrary assumptions required in using behavioral models as it does the accuracy of contingent valuation. Second, there is no evidence that strategic considerations are as important as early critics of contingent valuation expected. This fear was fueled in part by Samuelson's famous indictment of free riders: people will not reveal their true willingness to pay for public goods so as to avoid paying for them.

The strongest arguments in support of contingent valuation are due not to its own strengths but to the weaknesses inherent in recreation demand modeling. The difficulties encountered in deriving welfare measures from recreation demand models are outlined in the next several sections. Because modeling based on observable behavior is fraught with specification problems and econometric pitfalls, direct methods such as contingent valuation are often argued to be preferable. Presumably "we know what we are getting." But to some extent direct questioning merely obscures the same problems that recreation demand modeling highlights. Take, for example, the substitute site issue that will be discussed in the next section. In the recreation demand context, inclusion of substitute site data produces multicollinearity and omission biases coefficients. Valuing a change at one site must be accomplished with all other sites' parameters held constant. Additionally, valuing changes at many sites introduces complicated aggregation issues. Direct questioning may appear to circumvent this problem, but it may only obscure it. When researchers ask the value of a change in the water quality, for example, do they know what the individual is holding constant in his or her mind? Has the respondent's hypothetical behavior been adjusted to the new circumstances? Has he or she allowed for substituting to or away from alternatives? There are so many influences that need to be held constant that each cannot be mentioned explicitly.

For good reason, there is still much skepticism about the contingent valuation method. Cummings, Brookshire, and Schulze (1986) described the circumstances under which the method can be expected to work reasonably well and the circumstances in which it may not, calling these "reference operating conditions." Among other things, they suggested that better results will be obtained when the respondent is familiar with the resource being valued and has had experience making choices about the resource. Unfortunately it is precisely those situations, where the respondent has had little experience, that the contingent valuation method is most needed and could be substituted for recreation demand models.

8.4 Issues in Obtaining Recreation Benefit Estimates from Continuous Demand Models: The Travel Cost Model

To anyone who has not attempted benefit measurement in the context of a recreation demand model, the task may appear straightforward. In fact, a whole host of issues arises in the specification and estimation of the model and subsequent calculation of consumer surplus, all of which have enormous bearing on the final benefit estimates. Choice among alternative procedures is made difficult by our inability to observe benefits, even after the fact. Unlike demand analysis where model predictions can be compared with behavior, welfare analysis produces estimates that are never verifiable. However, unlike hypothetical valuation, benefit estimates derived from recreational demand models are only one step away from observable measures.

8.4.1 Valuing Access

The Welfare Measure. This type of valuation question arises when a proposed public action would eliminate a recreational opportunity (site), an outcome that might result from industrial or residential development. Valuation of access to a resource is also relevant in determining the losses that would accrue from pollution or other hazards severe enough to mandate public closure of a recreational site. Suppose the researcher chooses the recreation demand model approach and attempts to estimate a single equation demand function for trips to the site in question. What is known from welfare theory that can be brought to bear on the problem? Recall, w_1, the first welfare measure defined in (8.2). The compensating variation associated with elimination of access is given by

$$w_1 = e(\tilde{p},u) - e(p^0,u) \tag{8.4}$$

where p^0 is the current level of prices and $\tilde{p} = (\tilde{p}_1,p_2, \ldots, p_m)$ the vector that includes the choke price for x_1 (that is, the price that drives the individual out of the market for x_1); u is the reference utility level.

The measure in (8.4) can be expressed in terms of compensated demands very simply. The compensating variation measure, w_1, equals

$$w_1 = e(\tilde{p},u) - e(p^0,u) = \int_{p_1^0}^{p_1} \frac{\partial e}{\partial p_1} \, dp_1$$
$$= \int_{p_1^0}^{\tilde{p}_1} x_1 dp_1 \tag{8.5}$$

where x_1 is compensated demand. This is the well-known result that compensating variation equals the area under the Hicksian demand.

For this scheme to work, and more fundamentally, for the question to make sense, x_1 must be a nonessential good. A good is nonessential if there are combinations of other goods $(x_2', \ldots, x_n')$ that will compensate the individual for its absence:

$$u(x_1^0,x_2^0, \ldots, x_n^0) = u(0,x_2', \ldots, x_n'), \tag{8.6}$$

where the x_i^0's are the utility maximizing demands given the set of prices and income. The equivalent condition for the expenditure function is that the limit of the expenditure function as $p_1 \to \infty$ be finite.

The notion of nonessentiality turns on the ability to compensate an individual completely for the loss of access to x_1. If the good is nonessential and noninferior, not only will its compensating variation be finite (by definition) but its Marshallian consumer surplus will be as well, since the Marshallian demand for noninferior goods lies everywhere inside the Hicksian demand as price is increased.

Nonessentiality is in some sense a trivial matter. One is hard-pressed to think of any specific recreational resource that fails the nonessentiality criteria. It may be difficult to compensate some individuals for the loss of broadly defined, generic recreational opportunities, but the usual site specific resource of policy analysis is rarely essential. The nonessentiality condition is simply a reminder that functional forms that imply essentiality, and therefore infinite compensation, are inappropriate in recreational analysis.

The welfare derivation in equation (8.5) results in the Hicksian demand, but observations are made on behavior implied by Marshallian demand. It is necessary to know how accurate an approximation will be provided by the area under the Marshallian function. One is tempted to appeal to Willig's (1976) arguments here, but these are not strictly applicable in the valuation of access problem. The Willig results draw a comparison between the areas to the left of the Hicksian and the Marshallian functions for a price change. The original and new price are used for both the Hicksian and Marshallian measures. For most functional forms, however, the Hicksian and Marshallian choke prices differ. To see this, look at figure 8.1 where x_1^H stands for the Hicksian demand curve and x_1^M stands for the Marshallian demand curve. Comparing the areas under the Hicksian demand curves for the price change $p_1^0 \rightarrow \hat{p}_1$ would permit the use of the Willig bounds. But that would omit the shaded area under the Hicksian demand curve above the Marshallian choke price but below the Hicksian choke price. The relevant areas under the Marshallian and the Hicksian are better viewed as areas associated with the same quantity change from $x = x^0$ to $x = 0$. Randall and Stoll (1983) translated the Willig results into quantity terms with intuitively similar results. Small income effects mean small differences in Hicksian and Marshallian measures.

Whether small income effects are likely in recreation depends very much on the specific recreational resource in question. The failure of many recent studies to estimate significant coefficients on income variables is suggestive. Income levels are more likely to distinguish participants in a recreational activity from nonparticipants than they are to affect the number of recreational trips a participant takes in a season. Except for relatively unique recreational resources, such as the large national parks of the West, day trips to recreational sites are common and the variable costs associated with these day trips are small. In fact, time costs are often considered more of a constraint to increasing trip frequency than are money costs (Bockstael, McConnell, and Strand 1988). In contrast, there may be obstacles to entry (for example, ownership of automobile) or fixed costs for equipment (for example, boat, recreational

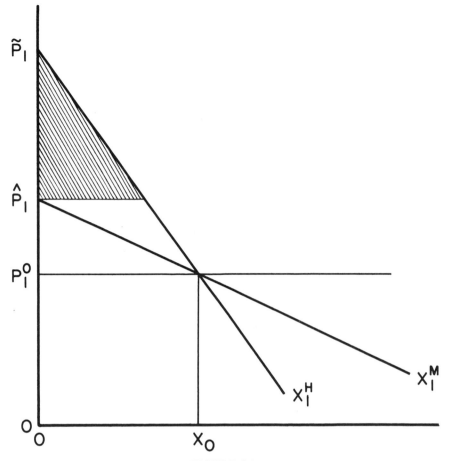

FIGURE 8.1
Marshallian and Hicksian demands and their choke prices.

vehicle, fishing equipment) that cause income to affect participation. The participation decision is explored later in this chapter.

Looking carefully at the well-known relationship between the definition of the welfare effects of a price change and the area under the Hicksian demand is instructive. Recall that

$$\int_{p_1^0}^{\tilde{p}_1} \frac{\partial e}{\partial p_1}\, dp_1 = \int_{p_1^0}^{\tilde{p}_1} \left(p_1 \frac{\partial x_1}{\partial p_1} + x_1 - \mu u_1 \frac{\partial x_1}{\partial p_1} + p_2 \frac{\partial x_2}{\partial p_1} \right.$$
$$\left. - \mu u_2 \frac{\partial x_2}{\partial p_1} + \ldots + p_n \frac{\partial x_n}{\partial p_1} - \mu u_n \frac{\partial x_n}{\partial p_1} \right) dp_1,$$

(8.7)

which simplifies to

$$\int_{p_1^0}^{\tilde{p}_1} x_1 \mathrm{d}p_1 \tag{8.8}$$

because $p_j(\partial x_j/\partial p_1) - \mu u_j(\partial x_j/\partial p_1) = 0$ since $p_j - \mu u_j = 0$ by first-order conditions from the problem $\min\{\boldsymbol{p} \cdot \boldsymbol{x} - [u(\boldsymbol{x}) - u]\}$. To obtain an estimate of the potential lost welfare, it is only necessary to estimate the demand for good x_1 — that is, trips to the site being considered for elimination — and then calculate the area under it. It is not necessary to calculate changes in surpluses under substitute site demands.

As a basic premise, if there is one exogenous change, the welfare measure can always be found in one market. The recreational, or nonmarket goods, case turns out to be simple because even an infinite change in the "price" of one site will not alter the "prices" of other recreational sites because prices are not market determined. Even if demands to other sites shift, prices, which are travel cost related, will not change. The derivation above is a reminder that the economist need look no further than the area under the demand curve for x to obtain a welfare measure of the change in x's price. (See Just, Hueth, and Schmitz 1982, for a complete discussion of welfare measures in multiple markets).

Specification Considerations. This analysis begs the prior question of estimating the demand function. The welfare measure is captured by integrating the *estimated* demand function over the price change where the coefficient on price reflects (at least approximately) the effect of a change in price holding all else constant. Many kinds of misspecifications can cause a violation of the above, only some of which involve substitute sites. If the demand function is specified with substitute prices included, then all goes well and the simple area under the demand curve for x_1 will be appropriate. If substitute prices are omitted from the demand function and they are positively (negatively) correlated with own price, then the "welfare triangle" will tend to be an under- (over-) estimate of the welfare measure we seek. But correlation among prices is inherent in cross-sectional observations on recreational activity. Individuals in a sample who live far from a coastal recreational resource, for example, will live far from all shore sites, leading to positive correlation. In other applications where sites and people are distributed more evenly, people who live far from the site of concern may live close to a substitute site and vice versa, causing negative correlation. Multicollinearity problems seem the rule rather than the exception in recreational demand modeling.

An additional source of bias is the omission or misspecification of the opportunity cost of time. The predominant component in the monetary price, or cost, of taking a recreational trip is the cost of travel. Frequently, though, the opportunity cost of the time spent traveling is the greatest real cost to the individual. The value of a unit of time will vary across individuals depending on the returns from their alternative uses of time. However, the amount of time and the amount of money spent traveling will tend to be correlated because both are correlated with the distance traveled. The omission of time costs in the demand function will bias the coefficient on money costs upward in absolute value terms (make it more negative), flattening the demand curve and biasing the consumer surplus measure downward.

This suggests that careful treatment of the value-of-time issue is central to recreational benefit estimation. An extensive literature has emerged on this topic (e.g., McConnell and Strand 1981; Bockstael, Strand, and Hanemann 1987; Smith, Desvousges, and McGivney 1983). Theoretically consistent and intuitively meaningful measures for the value of time require very specific data, including information about the individual and the family's labor market constraints, and such data are difficult to obtain in a survey. As a result, researchers often resort to using the percentage of an individual's wage rate measures, admitting the sensitivity of benefit measures to what is essentially an arbitrary formulation.

Sample Selection Issues. In addition to the choice of explanatory variables, there is another issue of specification that characterizes recreational demand problems and that has significant implications for estimation and benefit calculation. A sample of the population in a given geographical area will generally exhibit a low participation rate in the recreational activity of interest. If activity at a given site is the focus of attention, the participation rate will be even lower still. To ignore the nonusers is to discard useful information about what determines whether people participate in the activity or at the site. Deleting nonusers from the sample has the additional consequence of selecting the sample on the value of the dependent variable and thus altering the distribution of the error term. Inclusion of nonparticipants adds information but complicates the statistical model; the dependent variable is no longer truncated but it is censored.

A sample selection problem exists when there are systematic influences on the decision to participate in the consumption of a given good. (For the purposes here, the good will be defined as trips to a given site for a given recreational activity.) These influences ought to be accounted for in the estimation process. But the way they are accounted for should depend on how the researcher views individuals' decision making. There are many choices. He or she may consider site choice conditional on water recreation choice conditional on leisure choice. Furthermore, the quantity choice may be modeled as an integer count variable. This modeling has been undertaken by Smith (1988) and by Shaw (1988).

Here the sample selection problem is illustrated in a simple way. Suppose that both the participation decision and the quantity decision are driven by a function of exogenous variables plus a normal error, and that assumptions about the errors determine the estimation techniques. Several simple variants of models designed to handle the problem are described in the following paragraphs.

The sample selection problem is inherently a site choice problem. If individuals always chose positive quantities, there would be no problem. The relationship that described the demand for trips by individual i, for example, might be given by

$$x_i = f(z_i;\beta) + u_i \tag{8.9}$$

where x_i is the number of trips to a site by the ith individual, $f(\cdot)$ is the functional form of demand, z_i is a vector of explanatory variables associated with individual i and site z, β is a vector of unknown parameters, and u_i is a stochastic element with mean zero. The expected value of x would be given by

$$E(x_i) = f(z_i;\beta) \tag{8.10}$$

when there are corner solutions for some individuals ($x_i = 0$), the expected value of x_i will equal

$$E(x_i) = \text{prob}(x_i > 0) \cdot E(x_i | x_i > 0) + \text{prob}(x_i \le 0) \cdot 0. \tag{8.11}$$

This expression shows the expected value of x_i is much affected by the sample selection rule, the assumption that is used to estimate whether the individual is in or out of the market (a participant or not).

The Tobit (Tobin 1958) is perhaps the oldest and best known of the econometric models used to estimate relationships with "censored" data. The model can be defined simply as

$$\begin{aligned} x_i &= f(z_i;\beta) + u_i && \text{if } f(z_i;\beta) + u_i > 0 \\ x_i &= 0 && \text{otherwise,} \end{aligned} \tag{8.12}$$

for all individuals $i = 1, \ldots, n$, where u_i is $N(0,\sigma_u^2)$.

In this case the "sample selection rule" is simply $u_i > -f(z_i;\beta)$. That is, the probability that the individual participates equals $\text{prob}[u_i > -f(z_i;\beta)]$. Given this "sample selection rule," expression (8.11) can be rewritten as

$$E(x_i) = \Phi(t_i) \cdot E[x_i | u_i > -f(z_i;\beta)] + [1 - \Phi(t_i)] \cdot 0 \tag{8.13}$$

where

$$\begin{aligned} E[x_i | u_i > -f(z_i;\beta)] &= f(z_i;\beta) + E[u_i | u_i > -f(z_i;\beta)] \\ &= f(z_i;\beta) + \sigma_u h(t_i) \end{aligned}$$

where $t_i \equiv f(z_i;\beta)/\sigma_u$. Φ and ϕ denote the distribution and density functions of the standard normal and $h(t) = \phi(t)/\Phi(t)$.

A key behavioral aspect of the Tobit is that the same relationship, including the error structure, is assumed to determine both an individual's quantity and whether he or she participates. If the value of $f(z_i;\beta)$ becomes small enough [that is, if $f(z_i;\beta) + u_i \le 0$], then the individual leaves the market. In the Tobit model, the same characteristics that cause the person to visit the site influence the frequency of his or her visits, and the coefficients of these characteristics have the same sign and magnitude.

A different way of addressing the problem is to partition the z vector into two sets of variables (z_{1i}, z_{2i}) with z_{2i} affecting the participation and z_{1i} affecting the frequency decision. These need not be mutually exclusive variables. Now let

$$x_i = f_1(z_{1i}; \beta_1) + u_i \qquad \text{if } x_i > 0 \tag{8.14}$$
$$x_i = 0 \qquad \text{otherwise}$$

as before, but also allow a second function to determine whether x_i is positive. Thus

$$x_i > 0 \qquad \text{if } w_i^* > 0 \tag{8.15}$$
$$x_i = 0 \qquad \text{if } w_i^* \leq 0$$

where $w_i^* = f_2(z_{2i}; \beta_2) + v_i$. The stochastic variable, v_i, is assumed here to be normally distributed with mean 0 and variance σ_v^2. Covariance between u and v is denoted σ_{uv}, which equals $\sigma_u \sigma_v \rho$, where ρ is the correlation between errors.

The sample selection rule is now $v_i > -f_2(z_{2i}; \beta_2)$, leading to an expression for the expected value of x_i equal to

$$E(x_i) = \Phi(t_{2i}) \cdot E[x_i \mid v_i > -f_2(z_{2i}; \beta)] + [1 - \Phi(t_{2i})] \cdot 0 \tag{8.16}$$

where $t_{2i} = f_2(z_{2i}; \beta)/\sigma_2$, and

$$E[x_i \mid v_i > -f_2(z_{2i}; \beta)] = f_1(z_{1i}; \beta_1) + E[u_i \mid v_i > -f_2(z_{2i}; \beta)] \tag{8.17}$$
$$= f_1(z_{1i}; \beta) + \sigma_{uv} \cdot h(t_{2i})/\sigma_v.$$

This model was devised by Heckman (1976); the Tobit is a special case of it. When $\sigma_u = \sigma_v$, $\sigma = 1$, $\beta_1 = \beta_2$, and $z_1 = z_2$, the model underlying the Heckman is identical to the Tobit. The participation-triggering equation and the quantity equations are identical.

Allowing different factors and different error structures to affect the participation and use decisions is an appealing feature of the Heckman procedure, not present in the Tobit. But the Heckman introduces a problem of interpretation when z_{1i} and z_{2i} have no variables in common. The sample selection story must be told like this: factors in z_{2i} determine whether individual i participates in this activity. However, if z_{2i} is such that the individual does participate, then factors in z_{1i} determine how much he or she participates. But if the individual is in the market given the vector of z_{2i}, no amount of increasing or decreasing of z_{1i} elements will cause him to exit the market. (For example, if price determines demand for trips, there is no price sufficiently high to drive the individual out of the market.) The Heckman procedure introduces a different problem if z_{1i} and z_{2i} do have elements in common; problems of multicollinearity tend to arise in the second stage of the estimations. The scalar $h[f_2(z_{2i};\beta_2)/\sigma_v]$ tends to be highly correlated with elements of z_{1i}.

The conceptual difficulty embodied in the Heckman stems from the fact that this model describes a joint probability between w_i^* and x_i. Thus there exists a nonzero probability that x_i is positive even for $w_i^* < 0$ and a nonzero probability that x_i is nonpositive even for $w_i^* > 0$.

There is a third model of behavior that in some respects is more intuitively appealing than the Heckman and perhaps more appropriate for recreational behavior applications. The Cragg (1971) model is superficially similar to the Heckman, but the demand equation is conditional on a positive response in the participation model. The Cragg model takes the form

$$x_i = 0 \qquad \text{if } w_i^* \le 0 \tag{8.18}$$

where

$$w_i^* = f_2(z_{2i};\beta_2) + v_i.$$

Conditional on $w_i^* > 0$, then

$$\begin{aligned} x_i &= f_1(z_{1i};\beta_1) + u_i & \text{if } f_1(z_{1i};\beta_1) + u_i > 0 \\ x_i &= 0 & \text{if } f_1(z_{1i};\beta_1) + u_i \le 0. \end{aligned} \tag{8.19}$$

This model differs from the more general version of the Heckman in that there are two effective switches. If the individual is observed not to participate in the market, it may be for one of two reasons; he or she may have been eliminated because of factors in either the z_1 or z_2 vector.

The first stage of the Cragg procedure is estimated using a probit. The second stage is estimated using only nonzero observations, similar to the Heckman. However, the implicit sample selection rule of this stage is different from the Heckman. In the Cragg model,

$$E(x_i | x_i > 0) = E[x_i | u_i > -f_1(z_{1i};\beta_1)] = f_1(z_{1i};\beta_1) + \sigma_u h(t_1). \tag{8.20}$$

The sample selection rule for this truncated model is based on quantity demanded decision variables (the z_{1i}'s).

Calculating Welfare Measures. Neither conceptual nor technical difficulties end when the demand function is estimated. Welfare measures must now be calculated on the basis of the estimated function, introducing a new set of problems.

The most common procedure is to calculate relevant consumer surplus measures from the estimated Marshallian demand. For a simple linear demand function of the form $x(p) = \beta_0 + \beta_1 p$ where β_0 and β_1 are now scalars, the consumer surplus (CS) formula would be of the form

$$\text{CS} = \int_{p^0}^{\tilde{p}} x(p) \mathrm{d}p = x^2/\beta_1. \tag{8.21}$$

Even with this simple formulation, two difficulties arise. The true value of β_1 is unknown; only the estimated value $\hat{\beta}_1$ is known. Since $\hat{\beta}_1$ is a random variable, $\hat{CS} = -x^2/\hat{\beta}_1$ will also be random. Unbiasedness would be a desirable property of the $\hat{CS}$. However, Bockstael and Strand (1987) have shown that when $\hat{CS}$ is nonlinear in $\hat{\beta}_1$, an unbiased $\hat{\beta}_1$ is not sufficient to ensure an unbiased $\hat{CS}$. Approximate correction factors can sometimes be defined, based on the standard error of $\hat{\beta}_1$. Truncated and censored models add new complications that have not been thoroughly explored.

The second problem that arises in the calculation of $\hat{CS}$ according to a formula such as (8.21) is the interpretation of "x," the quantity demanded. There are two possibilities. The function can be calculated using the observed value of x, where $x^0 = \hat{\beta}_1\mathbf{z} + \hat{u}$, or using the predicted value of x, where $x^0 = \hat{\beta}_1\mathbf{z}$. As has been discussed elsewhere (Bockstael and Strand 1987), good reasons can be given for choosing either value of x. If most of the error implicit in u is believed to be due to omitted variables, then the observed x might give a better estimate of consumer surplus. However, in recreational surveys where recall error abounds, u may reflect measurement error in x as well. If this predominates, the predicted x may be more reliable than the observed value. It is likely that the error due to recall is reduced considerably when nonparticipants are included. People may not remember accurately the number of trips they took, but they generally remember whether they participated in the activity. This argues for choosing the actual value of reported trips rather than the estimated value. Using reported trips produces a sample of CS estimates with far greater variation than when the predicted value is used.

Sample selection rules introduce complexities in the prediction of x. For example, the predicted value (or expected value) of x for the Tobit estimation is given by

$$E(x_i) = \Phi(\hat{t}_{1i})[f(z_i;\hat{\beta}_1) + \hat{\sigma}_u h(\hat{t}_1)] \tag{8.22}$$

where now we return to the broad definition of $\hat{\beta}_1$ as a vector of demand parameters. Expressions (8.16) and (8.17) define the predicted value of x_i for the Heckman model:

$$E(x_i) = \Phi(\hat{t}_{2i})[f_1(z_{1i};\hat{\beta}_1) + (\hat{\sigma}_{uv}/\hat{\sigma}_v)h(\hat{t}_{2i})]. \tag{8.23}$$

Finally, for the Cragg model, the predicted value is

$$E(x_i) = \Phi(\hat{t}_{2i})[f_1(z_{1i};\hat{\beta}_1) + (\hat{\sigma}_u h(\hat{t}_{1i})] \tag{8.24}$$

where $\hat{t}_{ki} \equiv f_1(z_{ki};\hat{\beta}_k)/\sigma_{(k)}$ for $k = 1, 2$; $\sigma_{(1)} = \sigma_u$, $\sigma_{(2)} = \sigma_v$.

As an alternative to using actual or predicted trips with an estimated Marshallian demand curve, the estimated Marshallian function could be integrated to obtain the expenditure function and from this calculate compensating variation. This procedure has the advantage of producing values which are, at least in theory, "exact measures" of welfare changes.

Integrating from the Marshallian demand curve to the expenditure function has been popularized by Hausman (1981). The method follows from the fact that Marshallian demand curves are derived from the indirect utility function:

$$x_1(\boldsymbol{p},y) = \frac{-\partial v(\boldsymbol{p},y)/\partial p_1}{\partial v(\boldsymbol{p},y)/\partial y} \tag{8.25}$$

This expression is useful because along a given indifference curve,

$$dv(\boldsymbol{p},y) = 0, \tag{8.26}$$

and if $\boldsymbol{p}$ and y are adjusted to ensure this equality, it must be that

$$v_{p1}dp_1 = v_y dy$$
$$\frac{dy}{dp_1} = -v_{p1}/v_y \tag{8.27}$$
$$= x_1(\boldsymbol{p},y).$$

This is an ordinary differential equation whose solution is the expenditure function $e(\boldsymbol{p},u)$. In solving the differential equation (8.27), it is customary to equate the constant of integration with the unknown utility level. For example, suppose the Marshallian demand function is semilog:

$$x_1(\boldsymbol{p},y) = dy/dp_1 \tag{8.28}$$
$$= \exp(\beta p_1 + \gamma y).$$

The solution to this differential equation is the semilog:

$$y = e(\boldsymbol{p},u) = -\gamma^{-1}[\ln(-\gamma/\alpha)\exp(\alpha p_1) - \gamma u]. \tag{8.29}$$

Once obtained, $e(\boldsymbol{p},u)$ can be used directly to value access by calculating how it changes with a price change from p_0 to the Hicksian choke price. Analogous solutions can be found for other popular functional forms. And Vartia's (1983) method of numerical integration can be employed with good accuracy when exact solutions cannot be obtained.

If it is so straightforward to obtain the expenditure function, why would anyone want to calculate consumer surplus from the Marshallian demand function? There are several practical reasons for sticking with the Marshallian. To begin with, many recreational problems can be expected to have small income effects, and indeed income effects estimated for recreational demand models typically are small or insignificant. The difference between consumer surplus and equivalent or compensating variation is, therefore, unlikely to be very large, not warranting further investigation.

A second reason has to do with the errors that enter the estimation. Perhaps no variable is measured with as much error as the income variable. It is typically recorded as a value falling within a broad range. Rarely are federal, state, and local income taxes accounted for, nor is the distinction between household and individual income or wealth made clear. The method of integrating back places more weight on the income coefficient than is warranted by the accuracy of income measures.

The rationale for integrating back to produce an "exact" measure rather than an approximation is valid only if the precise Marshallian function, that is, the precise functional form and parameter values were known. Many functional forms can fit approximately the same shape relationship between quantities and prices in "demand space" and the data will not permit distinction among them. All will yield approximately the same consumer surplus because it is measured in this same space. But these different functional forms with different parameter values can integrate back to very different expenditure functions introducing additional error into the measurement process. Though the welfare measure implied by the expenditure function is called "exact," its "exactness" depends on the inferential process that led to the choice of functional form as well as the accuracy of parameter estimates.

8.4.2 Valuing Quality Changes

The second type of valuation question emerges when quality dimensions of recreational sites are altered. Environmental policies that affect the air or water quality at recreational sites provide some of the most important examples. For example, acidification of lakes influences the quality of fishing at lakes. Air pollution affects the visibility and hence scenic beauty of hiking in the Southwest. Policy actions associated with wildlife management fall in this category when the outcome affects the hunter's bag, the fisherman's catch, or the nature-lover's wildlife sitings. Welfare evaluation of this sort requires an extension of the applied welfare theory presented up to this point. The compensating (or equivalent) variation is defined as in equation (8.3), but its realization in terms of observable behavior must be explored.

Assume for a moment that parameters that reflect the outcomes of policies exist and are measurable. Such parameters might be indices of wildlife abundance, water quality, fish catch, and so on. Assuming the existence of these indices of quality, how does one empirically measure the welfare effects of a quality change?

The Welfare Measure. Recreational valuation has been motivated as a site-specific problem. Questions about quality changes have been formulated in the same context. Consider the case of a policy that alters the quality at one site: x_1 denotes the site, q_1 denotes a quality dimension of this site, p_1 is site 1's access price (including, for simplicity, the opportunity cost of time).

The fundamental welfare question concerns the relation between observable behavior and the definition of the welfare effects of a change in quality as defined earlier is

$$w_2 = e(p^0,q,u) - e(p^0,q^*,u) \tag{8.30}$$

where $q = (q_1, \ldots, q_n)$ and $q^* = (q_1^*,q_2,\ldots, q_n)$. The change in the area under the Marshallian demand curve, which is generally taken as an approximation of w_2, is:

$$A = \int_{p_1^0}^{\tilde{p}_1} [x_1^M(p,q^*,y) - x_1^M(p,q,y)]\mathrm{d}p_1 \tag{8.31}$$

Under many circumstances, it is reasonable to believe that A is an approximation of the exact welfare measure, w_2. As with the welfare effects of access, the relation between w_2 and A can be established via the change in the area under the Hicksian demand curves; this connection is more complicated compared with the access case.

Suppose the Hicksian demand function is known and the welfare effect is measured as the change in the area under it. Then

$$\int_{p_1^0}^{\tilde{p}_1} [x_1^H(p,q^*,u) - x_1^H(p,q,u)]\mathrm{d}p_1 = e(\tilde{p},q^*,u) - e(p^0,q^*,u)$$
$$- e(\tilde{p},q,u) + e(p^0,q,u) \tag{8.32}$$

where x^H denotes the Hicksian function and we have exploited Shephard's lemma $[x_1^H(p,q,u) = \partial e(p,q,u)/\partial p_1]$ in the integration. Compare this expression with (8.30); the change in the area under the Hicksian demand curve equals the exact welfare measure when $e(\tilde{p},q^*,u) = e(\tilde{p},q,u)$. This states that the minimum expenditure does not change with quality changes when the price of access to the site is so high that access is precluded. This is of course the familiar condition of weak complementarity (Mäler 1974; Willig 1978; Freeman 1979a). It can be stated in a variety of ways, for example, $\partial u(0,x_2, \ldots, x_n,q)/\partial q_1 = 0$; however the essence is that the individual values the quality of service flows only when he or she uses these service flows. For example, a recreationist values the water quality of a lake only if he or she makes trips to the lake.

This first step in showing the relationship between the change in the Marshallian area and the exact welfare measure is easy because duality and weak complementarity give us the exact relation between the Hicksian demand and the expenditure function. Showing the relation between the change in the areas under the Hicksian and Marshallian is less straightforward. To get a feel for the difficulty, economists can exploit their intuition about price changes. Define E as the error that results from using Marshallian demands rather than Hicksian demands. Calculate E as

$$E = \int_{p_1^0}^{\bar{p}_1} [x^M(\boldsymbol{p},\boldsymbol{q}^*,y) - x^M(\boldsymbol{p},\boldsymbol{q},y)]\mathrm{d}p_1$$

$$- \int_{p_1^0}^{\bar{p}_1} [x^H(\boldsymbol{p},\boldsymbol{q}^*,u) - x^H(\boldsymbol{p},\boldsymbol{q},u)]\mathrm{d}p_1$$

(8.33)

where the first integral equals the consumer surplus measure because x^M denotes Marshallian demand, and the second integral equals the compensating variation measure. Rearranging terms gives

$$E = \int_{p_1^0}^{\bar{p}_1} [x^M(\boldsymbol{p},\boldsymbol{q}^*,y) - x^H(\boldsymbol{p},\boldsymbol{q}^*,u)\mathrm{d}p_1$$

$$- \int_{p_1^0}^{\bar{p}_1} [x^M(\boldsymbol{p},\boldsymbol{q},y) - x^H(\boldsymbol{p},\boldsymbol{q},u)\mathrm{d}p_1$$

(8.34)

Each integral in this expression appears to evaluate the difference between the compensating variation and the consumer surplus of a price change, suggesting that each reflects a small error if income effects are small. This interpretation is correct, however, only for the second integral in (8.34). Only in this case are they starting at a point where compensating and ordinary demands are equal. At q, it is true that $x^M(\boldsymbol{p}^0,\boldsymbol{q},y) = x^H(\boldsymbol{p}^0,\boldsymbol{q},u)$. However, at the new quality level and the existing price, compensated and ordinary demands are not equal, $x^M(\boldsymbol{p}^0,\boldsymbol{q}^*,y) \geq x^H(\boldsymbol{p}^0,\boldsymbol{q}^*,u)$.

To see this, begin at the initial point of equality $x^M[\boldsymbol{p},\boldsymbol{q},\mathrm{e}(\boldsymbol{p},\boldsymbol{q},u)] = x^H(\boldsymbol{p}^0,\boldsymbol{q},u)$, and consider a change in q_1:

$$\partial x^M/\partial q_1 + e_q \partial x^M/\partial y = \partial x^H/\partial q_1,$$

(8.35)

which implies for normal goods that $\partial x^M/\partial q_1 > \partial x^H/\partial q_1$ because $e_q < 0$. In figure 8.2, the two new demand functions $x^M(\boldsymbol{p},\boldsymbol{q}^*,y)$ and $x^H(\boldsymbol{p},\boldsymbol{q}^*,u)$ will not, except under very unusual circumstances, cross at $\boldsymbol{p}^0$. If they cross in the first quadrant, and there is no guarantee they will, it will be at some $\boldsymbol{p} > \boldsymbol{p}^0$. Consequently, the first term in expression (8.34) does not capture the usual difference between the compensating variation and consumer surplus of a price change. The error E in (8.41) can be depicted as area A + C − B in figure 8.2.

The two new demand functions $[x^M(\boldsymbol{q}^*)$ and $x^H(\boldsymbol{q}^*)]$ can cross at $\boldsymbol{p}^0$ only if either $\partial x^M/\partial y = 0$ (zero income effect) or $e_q = 0$ (a quality change does not matter to the individual). In the case of zero income effects, compensated and ordinary demands coincide so that $x^H(\boldsymbol{q})$ and $x^M(\boldsymbol{q})$ are identical and $x^H(\boldsymbol{q}^*)$ and $x^M(\boldsymbol{q}^*)$ are identical. In the case in which quality has no effect, $x^H(\boldsymbol{q}^*)$ and $x^H(\boldsymbol{q})$ coincide as do $x^M(\boldsymbol{q}^*)$ and $x^M(\boldsymbol{q})$. In both cases the compensating variation of a quality change equals the equivalent variation and the

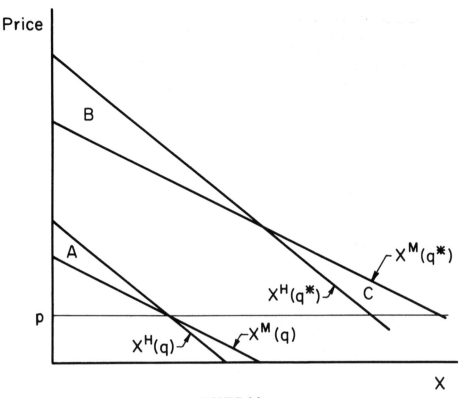

FIGURE 8.2
Quality-induced shifts in Marshallian and Hicksian demands.

consumer surplus of the change. In the latter case this is true because all are zero.

So, in trivial cases consumer surplus and compensating variation are identical. But in the usual case the error in approximation, given by (8.34), will not simply be the sum of two Willig-like errors in approximation. The fact that $x^H(p,q^*,u)$ and $x^M(p,q^*,y)$ do not cross at p^0 makes the error in approximation (area A + C − B) difficult to assess. One approach is to call upon arguments of Hanemann (1980b) to show that when v_q/v_p is independent of income, consumer's surplus for a quality change is *bounded* by equivalent and compensating variation. This condition on the ratio of the partial derivatives of the indirect expenditure function is a condition set out by Willig (1978) and is equivalent to the requirement that incremental consumer's surplus per unit of x be independent of income. This is a useful result to help order our thinking, but there is no particular guarantee that the condition will hold, and even if it does, it does not prevent the difference between compensating variation and consumer surplus from being large.

A second means of bridging the gap between the Marshallian measure and compensating variation involves integrating back and using the weak complementarity assumption to help determine the boundary conditions (see Larson 1988). Integrating back in the context of quality changes suffers from the practical difficulties that plague the price case, but also runs the risk of missing terms in the expenditure function. As an illustration, suppose the indirect utility function is:

$$v(\boldsymbol{p}, \boldsymbol{q}, y) = v^1(\boldsymbol{p}, \boldsymbol{q}, y) + v^2(\boldsymbol{q})\delta(\tilde{p}_1, p_1) \tag{8.36}$$

where $\delta(\cdot)$ is an indicator function which equals one for $p_1 < \tilde{p}_1$ and zero for $p_1 > \tilde{p}_1$. Then Roy's Identity implies

$$x_1(\boldsymbol{p}, \boldsymbol{q}, y) = -v_p^1(\boldsymbol{p}, \boldsymbol{q}, y)/v_y^1(\boldsymbol{p}, \boldsymbol{q}, y), \tag{8.37}$$

which leaves $v^2(\boldsymbol{q})$ out of the Marshallian demand curve. Granted, this preference function implies odd tastes — something like existence value for a site emerges only if a person uses the site — but it illustrates technical difficulties that can arise in integrating back in the quality case. Methods for exact welfare measurement of quality changes remain incomplete.

While the welfare theory of quality change is less complete than the empiricist might wish, a bigger problem arises in trying to measure relevant quality characteristics and incorporate them into demand models. Economists are frequently asked to value changes in water quality or wildlife abundance. Each poses special problems.

What is the value of a water pollution abatement program to an individual? Behavioral methods of welfare measurement attribute no value to the individual from an abatement program unless a link can be established to some quantifiable dimension of water quality. What is perceptible about water quality? Those aspects that can be directly discerned by the senses — turbidity, odor, and flotsam — comprise one type of quality. Another type would be hazardous heavy metals, bacteria, and so on, the presence of which can cause health problems but may not be detectable by the senses. People learn about these aspects through news media and by deducing the sorts of contamination that surrounding industrial and recreational development will create. Other aspects of quality are typically learned through experience. For example, density and variety of species of fish in a lake and congestion along a backpacking trail are dimensions of quality of a recreational experience that can only be learned by spending time at the site.

The importance of perceptions in determining behavior casts some doubt on the role of objective measures of quality in demand functions. Smith and Desvousges (1985) used dissolved oxygen as one measure of site quality. Bockstael, McConnell, and Strand (1988) used the product of nitrogen and phosphorus. It is easy to see how these measures relate to objective assessments of the quality of water, but the link between objective measures and human perceptions is tenuous.

The connection between policy and behavior is more transparent when quality takes on a catch rate or bag rate connotation. It has been difficult for natural scientists to link management or pollution abatement policies accurately to fish or wildlife abundance indicators, but functional relationships between abundance and catch or bag are common. The economist's task is then to relate catch or bag rates to behavior, and it is relatively easy to argue that catch rates are observable quality dimensions of recreational trips. This is not to say that practical problems do not arise. There are, for example, obvious difficulties in choosing species groups to include in the specification. Additionally, catch rates vary over trips and individuals, producing uncertainty for both recreationist and researcher.

Uncertainty pervades all quality augmented demand models. Weather variation and random human activities may cause the water quality at any one site to vary considerably from day to day. But in fishing (or hunting) the variability in catch (bag) is even more dramatic. The individual bases recreation decisions on what he or she expects quality to be at a given site on a given day. Expectations, however, may not be realized. The discrepancy might arise from the individual's limited knowledge of the site, but even the most experienced will witness differences between expectations and reality.

As the extensive literature on option value suggests, uncertainty complicates the use of valuation models. There are no simple recipes for dealing with uncertainty. To the extent possible, economic welfare should be based on the perspective of the planning agent's behavior. In most cases, this implies expectations about quality, but when there are substantial differences between expected and realized quality, they need to be taken into account. It may be appropriate to model individuals as recognizing their uncertainty about outcomes, by including a probability distribution or at least some measure of variance rather than simply a mean expectation. Additionally, expectations might be modeled as revised by realizations, especially if behavior is modeled over time.

Because of the difficulty of getting any data on quality, the distinction between these concepts has not received the attention it deserves. Some argue that if the recreational experience is repeated frequently enough, perfect information is approximated; at the very least, the means of expectation distributions will correspond to realizations. Recent work by Larson (1988) has motivated the introduction of moments of catch per trip as arguments in the demand function. Larson showed that mean and standard deviation of catch influence the demand for striped bass fishing. Furthermore, he showed how the conditional distribution of expected catch varies across individuals.

Specifications When Valuing Quality: Multiple Site Models. In some studies, perceptions of water quality or expectations of fish catch have been elicited from individuals as part of a survey and incorporated as the quality dimension in a recreational demand model. (See, for example, Bockstael, Strand, and Graefe 1986, when individuals were asked both expected catch and the expected chances of coming home empty handed from the fishing trip.) Far more often, average catch rates in a region or gross indicators of quality at a site serve this role. Examples of gross indicators are nitrogen or phosphorous. From an econometrician's standpoint, there is a distinct difficulty with the latter approach. No quality variation is observed across individuals at any one site. This factor, together with the recognition that substitution among sites is often important in recreation, has contributed to the recent development of multiple-site models.

The obvious prototype for a multiple-site model is a neoclassical system of demand equations. If there are n quality differentiated sites, the system would ideally be estimated:

$$x_{1j} = f_1(p_{1j}, \ldots, p_{nj}, q_1, \ldots, q_n, y_j) \qquad \text{for all individuals } j$$

$$\cdot$$
$$\cdot \tag{8.38}$$
$$\cdot$$

$$x_{nj} = f_n(p_{1j}, \ldots, p_{nj}, q_1, \ldots, q_n, y_j)$$

with appropriate restrictions across equations. However, the lack of variation in any q_i over the cross-sectional sample together with the considerable multicollinearity among prices makes estimation of n demand equations impossible.

There are ways around the problem, but all require restrictions on the form of the equations in (8.38). One alternative is to pool the information over sites, so that the equations in (8.38) all have identical structure; that is, $f_1 = f_i = f_n$ and own price and own quality each have a single coefficient irrespective of site. One equation is estimated over the entire set of observations irrespective of site. It is, however, difficult to incorporate substitute prices and qualities in a meaningful way in this structure. One cannot just add $n - 1$ more variables and coefficients for the prices of other sites because the estimation procedure would force the same relationship between the trips to any site x_i and the jth substitute price, while the ordering of substitute prices would be arbitrary. A modification of the model in (8.38) could be estimated by including prices of substitute sites in a stacked regression form, so that own price and quality coefficients were the same across sites but a different set of substitute price coefficients would be obtained for each site. This method could not be applied to substitute qualities because there would be no variation in the quality variables over observations. In this "pooled model" approach, the most common solution is either to ignore substitutes altogether, or to include the price (and sometimes quality) of the "next best alternative," usually defined arbitrarily as the site within the set of nonchosen sites with the smallest access cost.

An alternative means of accomplishing the necessary variation in quality in the system of demands in (8.38) is to introduce quality in cross-product terms with prices and income. This is, in essence, what the varying parameter model achieves (Vaughan and Russell 1982; Smith and Desvousges 1985; Bockstael, McConnell, and Strand 1988.) The general form of the model expresses trips to different sites as different functions of prices and income, as in (8.38) but with qualities omitted. Then the estimated parameters from this model are regressed on quality characteristics of the sites. Consider the linear example

$$x_1 = \beta_{01} + \Sigma\beta_{i1}p_i + \beta_{n+1,1}y + \varepsilon_1$$

$$.$$
$$.$$ \qquad\qquad\qquad\qquad\qquad\qquad\qquad\qquad\qquad (8.39)
$$.$$

$$x_n = \beta_{0n} + \Sigma\beta_{in}p_i + \beta_{n+1,n}y + \varepsilon_n$$

where $\beta_{ij} = \delta_{i1} + \delta_{i2}q_j$ for $I = 0, 1, \ldots, n, n + 1$.

This model is estimated in two steps. First, the n site demand models are estimated separately, accounting for the censoring or truncation appropriately. This stage gives n sets of estimated parameters. In the second stage, estimated parameters become dependent variables and are regressed on site qualities. Because of the limited quality information, the number of estimated parameters treated in the second stage may be restricted. For example, let β_{ii} be the estimated own price coefficient. Then the second step would estimate $\hat{\delta}_{i1}$ and $\hat{\delta}_{i2}$ from the model

$$\hat{\beta}_{ii} = \delta_{i1} + \delta_{i2}q_i + \eta_i. \qquad\qquad\qquad\qquad\qquad\qquad (8.40)$$

The second stage must account for the heteroskedasticity in η_i that emerges because the variance of $\hat{\beta}_{ii}$ depends on the site. This version of the varying parameters model was developed by Smith and Desvousges.

8.5 Measuring Recreation Benefits with Random Utility Models

Several features of the recreation demand problem make the continuous models described in the previous section difficult to specify and estimate. To deduce anything about recreationists' valuation of quality characteristics, it is necessary to observe behavior over a set of quality differentiated sites. As was discussed earlier, practical and econometric problems arise when estimating a system of site demands. These are truncated or censored samples because individuals do not visit all sites. Additionally, it is difficult to incorporate substitute prices and virtually impossible to incorporate substitute qualities in a meaningful way.

8.5.1 The Random Utility Model of Choice

The discrete choice or random utility model (RUM) provides quite a different structure in which to model recreational demand, a structure which focuses attention on the choice among substitute sites for any given recreational trip. As a consequence, the RUM is especially suitable when substitution among quality differentiated sites characterizes the problem. It has been used chiefly to value changes in the specific characteristics of a site, such as catch rates or water quality, because those site characteristics included in the estimation are instrumental in explaining how individuals allocate their trips across sites. However, it can also be used not only to value the losses from eliminating a site but also to value the benefits of introducing a new site, something quite beyond the scope of the continuous model.

Alas, the RUM has a major shortcoming. By itself, it cannot explain (as the continuous model does so directly) the total number of trips an individual takes to a given site in a season. Often a *single* continuous demand function for trips to all sites is appended in a RUM application and used in conjunction with the RUM, which explains the proportional allocation of the total number of trips over sites.

As the same framework is equally useful for valuing access as for valuing quality changes, a discussion of the basic RUM will lead to both types of benefit assessment. To motivate the model, begin with a utility maximizing framework, but with a very restrictive definition of the time period of reference. Normally, when economists formulate a constrained utility maximizing problem and the resultant indirect utility function, demand functions, and so on, these functions are pertinent to a time period long enough to witness repeated purchases of the goods in question. Now consider an indirect utility function relevant to a very short duration, based on a single recreational choice occasion.

On any given choice occasion, decisions to visit different recreational sites are mutually exclusive. Thus, the recreationist chooses the site from a finite set of alternatives that gives the maximum utility. Define the indirect utility function for choice occasion r and site choice i as

$$v_{ir}(q_{ir}, y_r - p_{ir}) \tag{8.41}$$

where q_{ir} is a vector of characteristics of site i in time period r, y_r is the income available to the individual in time period r, and p_{ir} is the price of accessing site i in time period r. Few applications have explicitly allowed variables to change with choice occasion although there is nothing in the model that precludes it. For simplicity of notation the subscript r is suppressed. Given this notation, the individual can be expected to choose alternative site i if $v_i(q_i, y - p_i)$ exceeds all $v_j(q_j, y - p_j)$ $j \neq i$.

For estimation purposes, it is necessary to introduce a stochastic element into this choice model. This stochastic term can have either an omitted variable or random utility interpretation, though they are functionally similar. With additive errors, the individual chooses site i if

$$v_i(q_i, y - p_i) + \varepsilon_i \geq v_j(q_j, y - p_j) + \varepsilon_j \qquad \forall \text{ sites } j \neq i. \tag{8.42}$$

The simplest and most common assumption is that the errors are independently and identically distributed as type I extreme value (Weibull) variates, resulting in the multinomial logit model

$$\pi_i = \exp(v_i) \Big/ \sum_{j=1}^{N} \exp(v_j) \tag{8.43}$$

where π_i is the probability that the individual chooses site i. The associated likelihood function for a sample of M individuals and N alternative sites is

$$L = \prod_{m=1}^{M} \prod_{n=1}^{N} \left[\exp(v_{mn}) \Big/ \sum_{j=1}^{N} \exp(v_{mj}) \right]^{t^{mn}} \tag{8.44}$$

where $t_{mn} = 1$ if individual m chooses site n and $t_{mn} = 0$ otherwise. For simplicity, v is usually specified as a linear function of the attributes of the site and the site price with one set of parameters over all sites and individuals. In the multinomial logit framework (where there are more than two alternatives) it is possible to introduce individuals' characteristics only as interaction terms with site characteristics.

A fundamental difficulty of the multinomial logit model is its implicit restriction of independence of irrelevant alternatives; that is, the relative odds of choosing any pair of alternatives remains constant irrespective of what happens in the remainder of the choice set. When this assumption is clearly violated — that is, when there are obvious patterns of substitution and complementarity among alternatives — then a nested model may be appropriate. The more general nested model avoids the independence of irrelevant alternative assumptions.

Examples of subgroupings of alternatives are prevalent in the recreational demand literature. Bockstael, Strand, and Hanemann (1987) used a nested model of beach use when the relevant beaches included freshwater and saltwater. The individual was viewed as first choosing between freshwater and saltwater and then among the specific beaches in the chosen subcategory. Other obvious applications arise in sportfishing decisions; an individual often has more than one level of choice, including fishing mode, fishing site, and target species.

When there is a general pattern of dependence among the choices, such as would arise if the alternatives included shore fishing in region A or region B, and boat fishing in region C, D or E, then a nested multinomial logit model can be used to avoid the unfortunate but obvious violation of the independence of irrelevant alternatives. McFadden (1974, 1978) derived this model, referred to as the Generalized Extreme Value (GEV) model, as well as the simpler multinomial logit in the context of stochastic utility maximization. If the errors are assumed drawn from a generalized extreme value distribution defined by

$$F(\varepsilon_1,\varepsilon_2, \ldots, \varepsilon_N) = \exp\{-G[\exp(-\varepsilon_1), \exp(-\varepsilon_2), \ldots, \exp(-\varepsilon_N)]\} \quad (8.45)$$

then the choice probabilities take the form

$$P_i = e^{v_i} G_i(e^{v_1}, e^{v_2}, \ldots, e^{v_N})/G(e^{v_1}, e^{v_2}, \ldots, e^{v_N}) \quad (8.46)$$

where $G_i = \partial G/\partial e^{v_i}$. This derivation is somewhat complicated by giving a general form for G. The specific form will vary with the number of subgroups and the number of alternatives in each subgroup. The general form of G is

$$G = \sum_{s=1}^{S} a_s \left\{ \sum_{j=1}^{N_s} \exp[v_{sj}/(1 - \sigma_s)] \right\}^{1-\sigma_s} \quad (8.47)$$

where S equals the number of subgroups of alternatives, N_s equals the number of alternatives in subgroup s (for example, two for shore fishing or three for boat fishing) and σ_s is a parameter in the unit interval related to the level of correlation among alternatives within groups. For the special case in which no pattern of substitutability exists among the alternatives, G takes the form $G = \Sigma e^{v_j}$ giving the simple multinomial logit model of (8.43). Should the alternative set include two subgroups as in the boat and shore fishing example, G equals

$$
\begin{aligned}
G = a_1 &\left\{ \exp[v_{11}/(1 - \sigma_1)] + \exp[v_{12}/(1 - \sigma_1)] \right\}^{1-\sigma_1} \\
+ a_2 &\left\{ \exp[v_{21}/(1 - \sigma_2)] + \exp[v_{22}/(1 - \sigma_2)] \right. \\
&\left. + \exp[v_{23}/(1 - \sigma_2)] \right\}^{1-\sigma_2}.
\end{aligned}
\quad (8.48)
$$

The resulting conditional choice probabilities are

$$P_j|s = \exp[v_{sj}/(1 - \sigma_s)] \bigg/ \sum_{k=1}^{N_s} \exp[v_{sk}/(1 - \sigma_s)] \quad (8.49)$$

where s is boat or shore and j is an alternative in the boat or shore subgroup. The probability of choosing subgroup s is

$$P_s = \frac{a_s \left\{ \sum_{j=1}^{N_s} \exp[v_{sj}/(1 - \sigma_s)] \right\}^{1-\sigma_s}}{\sum_{k=1}^{S} a_k \left\{ \sum_{j=1}^{N_s} \exp[v_{kj}/(1 - \sigma_k)] \right\}^{1-\sigma_k}} \quad (8.50)$$

The usual procedure is to assume that the a_s's and the σ_s's take identical values over subgroups. Considering that v_{sj} can be partitioned into explanatory variables that vary only between groups and thus explain the between group choice (call them Y's) and explanatory variables that vary over all alternatives (call them X's), v_{sj} can be written as $\theta X_{sj} + \beta Y_s$ and the following sequential estimation procedure can be employed:

$$P_j \mid s = \exp[\theta X_{sj}/(1 - \sigma)] \Big/ \sum_{k=1}^{N_s} \exp[\theta X_{sk}/(1 - \sigma)] \tag{8.51}$$

and

$$P_s = \exp[\beta Y_s + (1 - \sigma)I_s] \Big/ \sum_{l=1}^{S} \exp[\beta Y_l + (1 - \sigma)I_l] \tag{8.52}$$

where

$$I_s = \log\left\{ \sum_{k=1}^{N_s} \exp[\theta X_{sk}/(1 - \sigma)] \right\} \tag{8.53}$$

I_s is called an inclusive value and reflects weighted information about the alternatives within the s subgroup. Relating this to the earlier example, P_{js} might be the probability of fishing in region C given that the individual chooses to fish from a boat. P_s may be the probability that he chooses to fish from a boat given the value of the available alternatives for boat and shore fishing.

Any number of subgroups can appear at any level of the decision process and additional levels may be added with additional correlation terms. Thus the economist might model: (1) whether an individual does or does not go fishing; (2) if he or she goes fishing, what fishing mode is chosen; (3) given the mode, what site does he or she choose; and (4) given mode and site, what species is targeted. For an impressive example of this free structuring of decisions, see Wegge, Carson, and Hanemann's (1988) application to Alaska sportfishing. The estimation procedure begins at the end, so to speak, by estimating species choice, and works backward so that choices among subgroups will reflect information about the alternatives within subgroups. The technical aspects of these models are discussed in Maddala (1983) and Domencich and McFadden (1975).

8.5.2 Welfare Measurement

RUMs are appealing because of their attention to site characteristics and to the substitutability among sites, features that are central to most recreation valuation questions. Yet in this discrete world, the usual empirical measures of welfare are absent: there are no estimated demand functions whose areas can be calculated.

Measurement of welfare in the context of discrete choice models is in some sense more direct. If the v_i can be interpreted as an indirect utility function (as did McFadden), then this function can be used to obtain the compensating variation of a change in any explanatory variable (although the procedure is complicated somewhat by the stochastic nature of the problem). Given that special properties are not usually imposed on v, and more important that utility is not an observable entity, the interpretation of v as related to indirect utility may seem a stretch of faith. This contrasts with deriving welfare from Marshallian demands. There is difficulty moving from areas behind Marshallian demand functions to compensating variation measures, but at least economists know something about demand functions and can observe not only the explanatory variables but also the quantity of the good demanded.

The procedures most often used for obtaining welfare measures in the context of discrete choice models are attributable to Small and Rosen (1981) and to Hanemann (1982). First, consider the valuation of a change in the quality of site i from q_i to q_i^*. Additionally, suppose for a moment that site i is the chosen site before the quality change and after it. Then the compensating variation of the quality change should be able to be expressed as C in the following:

$$v_i(p_i, q_i^*, y - C) + \varepsilon_i = v_i(p_i, q_i, y) + \varepsilon_i. \tag{8.54}$$

To calculate C, v must be estimated as a function of price, quality and income, among other variables. However, because income does not change over alternatives, no coefficient on income can be recovered in the estimation. Hanemann points out that income and price should enter the indirect utility function as $y - p_i$ so that the estimated coefficient on price may be assumed to be the coefficient on income, with a sign change.

There is a basic problem with (8.54). The researcher does not know *a priori* which alternative will be chosen. A change in a characteristic of site i when site j is ultimately chosen yields no gain to the individual. *A priori* the model can predict only the probability that each outcome has of being chosen. Compensating variation ought to account for the researcher's uncertainty about which site the individual would choose. One definition of compensating variation, and the one that has been used most frequently because it is tractable, is the measure which equates the *expected value* of the maximum of the indirect utility functions. Thus, C will be defined by

$$E[v(p, q^*, y - C)] = E[v(p, q, y)] \tag{8.55}$$

where

$$v(p, q, y) = E \max[v_1(p, q, y) + \varepsilon_1, v_2(p, q, y) + \varepsilon_2, \ldots, v_N(p, q, y) + \varepsilon_N] \tag{8.56}$$

If the errors are generalized extreme value, then $E[v]$ equals

$\ln G[\exp(v_1), \ldots, \exp(v_N)] + k$. In the simple discrete choice problem where $G = \Sigma_j\exp(v_j)$, this would require solving the following for C:

$$
\begin{aligned}
&\exp[v(p_1,q_1^1,y - C)] + \exp[v(p_2,q_2,y - C)] + \ldots \\
&+ \exp[v(p_N,q_N,y - C)] \\
&= \exp[v(p_1,q_1^0,y)] + \exp[v(p_2,q_2,y)] + \ldots \\
&+ \exp[v(p_N,q_N,y)]
\end{aligned}
\tag{8.57}
$$

The calculation of C is simple if the marginal utility of income is constant. It is

$$
C = \frac{\ln\{\Sigma\exp[v_j(\boldsymbol{q}^*)]\} - \ln\{\Sigma\exp[v_j(\boldsymbol{q}^0)]\}}{\gamma}
\tag{8.58}
$$

where γ is the implicit coefficient on income. When the marginal utility of income varies by alternatives, the calculation of C becomes more difficult. It can be approximated by

$$
C \approx \frac{\Sigma\exp[v_j(q_j^0)] - \Sigma\exp[v_j(q_j^1)]}{\Sigma\gamma_j\exp[v_j(q_j^1)]}
\tag{8.59}
$$

when the marginal utility γ_j varies by alternating. For nested models, the approximation is similar to (8.59), but incorporates the σ's (Hanemann 1982).

The expression above gives the expected compensating variation for a choice occasion. If the quality change occurred at a site that has a low probability of selection even after the change, then the compensating variation will be small. And, of course, a high probability implies a larger C. These are compensating variations per choice occasion. They may be extended to the season by multiplying by the number of choice occasions, but this aggregation to the season level poses two difficulties.

Consider first the question: How many choice occasions are there, even before the quality change? In most recent studies, each trip actually observed to be taken is a choice occasion. Straightforward as it is, it prevents the estimation of changes in the demand for trips. In a few studies the researcher has attempted to define somewhat independently the number of choice occasions, allowing the decision "not to take a trip" to be one of the alternatives. Caulkins, Bishop, and Bouwes (1986) and Feenberg and Mills (1980) used the total number of days in the season; Larson (1988) used the largest number of trips in the sample. These approaches have two difficulties. First, they include "not taking a trip" in the alternative set along with taking a trip to site A, site B, etc. The independence of irrelevant alternatives assumption (IIAA) principle is clearly violated unless the participation decision is set up as a separate-stage decision. This is often made difficult by the lack of data available to explain the participation decision. In practice, this lack of information leads to a misspecified model. Factors such as time or health constraints, which might deter an individual, are not accounted for. When the participation decision is treated as a separate decision stage, the consequence is only a poorly explained participation decision. When nonparticipation is included as one alternative in many, the consequence is a violation of IIAA together with biased coefficients on the explanatory variables. These variables are generally costs and quality of site visits.

Most researchers have resorted to defining a choice occasion as a trip and multiplying welfare measures per choice occasion by the number of trips. But what happens if the change in quality (or the elimination of a site) alters the total number of trips taken? To complete the model, researchers often append to the analysis a continuous demand model for number of trips to all sites. Difficulties arise in estimating generic demand models, however; a means must be found to incorporate the "right" price and quality data. It makes little sense to regress number of trips on price and quality averaged over sites, and in many circumstances it makes equally poor sense to regress trips on the price and quality of some arbitrarily defined "best" site.

The inclusive value notion defined earlier in equation (8.53) offers a compromise. An inclusive value can be calculated from the RUM results that, in essence, weights the indirect utilities associated with different sites and functions of sites' prices and qualities by the probabilities of choosing each site. Although preferable to more arbitrary alternatives, this appended continuous model cannot be derived from the same theoretical framework as the RUM.

Though not often used for this end, the discrete choice model can also value the addition or elimination of a site. The losses from eliminating site 1, for example, could be calculated as

$$C = \frac{ln\left[\sum_{j=1}^{N} \exp(v_j) \right] - ln\left[\sum_{j=2}^{N} \exp(v_j) \right]}{\gamma} \tag{8.60}$$

where γ is the marginal utility of income. In parallel fashion, if the characteristics of a new site can be described, not only can the probability of an individual visiting that site be predicted but also a compensating variation measure can be calculated simply as

$$C = \frac{ln\left[\sum_{j=1}^{N+1} \exp(v_j) \right] - ln\left[\sum_{j=1}^{N} \exp(v_j) \right]}{\gamma} \tag{8.61}$$

where $v_{N+1}(\cdot) + \varepsilon_{N+1}$ is the indirect utility of the new site. Note that if the access cost and quality characteristics of the new site are such that an individual is not likely to use the new site (that is, the new site is dominated by existing sites) then the expected compensating variation mean will be close to zero.

8.6 Principal Problems in Recreation Valuation

The confidence and consensus that has come to be associated with welfare measurement of finite price changes for market commodities can be extended to valuing access or valuing quality changes of nonmarket commodities only with difficulty. Valuing access involves welfare areas that extend to the price access and thus exacerbate complications introduced by functional form choice, error source, truncation, and so on. Valuing quality introduces new theoretical, technical, and practical difficulties. One difficulty arises in specifying a model that incorporates variation in quality, so that behavioral responses to quality can be estimated. This usually requires a model of the demand for trips to an array of quality differentiated sites.

8.6.1 Modeling the Range of Decisions Implicit in Recreation Behavior

Both the RUM and the continuous (travel cost type) demand model offer partial solutions, but they do so in different ways. The problem, which involves participation and the demand for trips across substitutes, is too complex for either approach to handle well. The discrete choice model focuses on the choice among sites but is awkward for modeling the number of trips taken. The continuous models handle the latter well, as they are patterned after neoclassical demand functions. Because of the difficulty of incorporating substitute site prices and qualities, however, these models rarely account adequately for the possibility of substitution among sites.

When valuing a change in quality at one site, the continuous demand models will tend to overstate the gains from improvements. This will be true if other sites' qualities and prices are omitted from the estimated model but are correlated with own site qualities and prices so that the model is actually predicting behavior as if all qualities and prices are changed. It will also be true, for example, if the response to a quality improvement at site A, given a constant higher level of quality at site B, is deduced from looking at a demand function for site B (which itself is conditioned on a lower quality level at A). The gain from improving A will not be so great if B remains also at a high level of quality. Capturing the appropriate response to quality change is difficult if the researcher cannot appropriately condition the response on the quality levels of all substitutes. An earlier section showed just how difficult this is to accomplish.

In contrast, the discrete choice, or RUM, model focuses precisely on the comparison among sites and their characteristics. To value a change in quality at one site, the RUM merely predicts how that change would change the probability of choosing each site in the array. Substitution among sites is the essence of the model. If the improved site is still so inferior to other sites that no one's likelihood of visiting it is increased, then no significant welfare gain will be tallied.

Some stylized simulation experiments by Kling (1988a) give preliminary support to the notion that discrete choice models produce better benefit estimates in problems characterized by much substitution among sites, especially when a large portion of the sample is observed to choose more than one site to visit in a season. The continuous models are better suited to problems in which individuals tend always to visit the same site (although different individuals in the sample may visit different sites). In these problems, most of the action is in variations in demand for trips per season (the continuous part of the problem).

The question remains: Which type of problem most often arises in recreation? There are many types of recreation problems, especially those associated with water recreation, where individuals' time constraints are so restrictive and lumpy (due to a finite number of weekends and vacation days, weather conditions, seasonality of fisheries, and so on) that changes in environmental conditions are unlikely to produce substantial increases in the number of trips a participant in the activity can take in a year. The behavioral adjustments for existing participants may principally involve substitution among sites. For these problems, the RUM seems especially well suited.

In many applications, the participation decision is also a potentially interesting and important one. Bockstael, McConnell, and Strand, in their analysis of Chesapeake Bay recreational activities, found that a significant portion of past users had ceased recreating in the Chesapeake Bay because of the degraded water quality. Presumably, improvements in the Bay would ultimately increase the participation rate, producing much greater benefits in the long run than would be expected from estimates based solely on what current participants would do. Strictly speaking, neither the RUM nor the continuous multiple-site demand model handle this decision in a neat and consistent manner. The multiple-site continuous models can be estimated using censored (nonparticipants included) rather than truncated (nonparticipants excluded) data. But since demand functions are for sites, a nonparticipant is simply a nonparticipant at that site, not necessarily for the activity. Alternatively, a GEV discrete choice model could be estimated with two stages of decisions: whether to participate and at which site to participate. However, this model is a per choice occasion model, so the participation decision is related only to that choice occasion, not to the season.

A compromise that is not utility theoretic but approximates the nature of the problem is to append a regression of the demand for total number of trips to a RUM of choice among sites. As suggested previously, this continuous demand function might include an inclusive value variable, calculated from the RUM, reflecting the relative desirability of alternatives facing the individuals in the sample. If the sample includes nonparticipants, this continuous model could be estimated as a censored model, thus explaining seasonal participation in the recreational activity. This is not an especially appealing modeling strategy, but it has been used in the absence of any other scheme to model all the pieces of the decision. It is further complicated by the paucity of the data that are usually available to explain why some individuals are participants and others are not.

8.6.2 Multiple Price/Quality Changes: Aggregation over Sites and Activities

Throughout this chapter, valuation exercises, while set in a multiple site context, have been discussed with respect to a single recreational activity and a quality or price change at a single site. The concern has been with the individual's decision whether to participate in the activity of interest and his choice of site(s) to visit. In this context, elimination of a site or a change in quality at a site have been valued.

Consider some broader questions: A policy that affects the existence or quality of a site has ramifications for more than one recreational activity. An environmental policy changes the quality of more than one recreational site. As is usual, there is no confusion in the definition of compensating variation in either case. The confusion arises in its measurement. Whether the aggregation issue affects sites or activities or both, the welfare measure is simply

$$e(\boldsymbol{p}^0, \boldsymbol{q}^0, u) - e(\boldsymbol{p}^*, \boldsymbol{q}^*, u) \tag{8.62}$$

where $\boldsymbol{p}^0$ and $\boldsymbol{p}^*$ are the price vectors before and after the change and $\boldsymbol{q}^0$ and $\boldsymbol{q}^*$ are the quality vectors before and after the change.

Difficulties arise in capturing this measure in the continuous-demand model framework because changes in multiple markets translate into line integrals over demand functions. One example has been dealt with directly in the literature. Bockstael and Kling (1988) looked at the welfare measure where a single change in environmental quality is weakly complementary to a set of goods, which can be interpreted as a set of recreational activities. This chapter looks at one quality change — for example, water quality at a given site — but the change affects several recreational activities at that site. The welfare measure is

$$w_2 = e(\boldsymbol{p}^0, \boldsymbol{q}^0, u) - e(\boldsymbol{p}^0, \boldsymbol{q}^*, u). \tag{8.63}$$

Considering a simple example with two recreational activities, w_2 may be rewritten as

$$w_2 = \int_c [e_1(\boldsymbol{p}, \boldsymbol{q}^*, u)\mathrm{d}p_1 + e_2(\boldsymbol{p}, \boldsymbol{q}^*, u)\mathrm{d}p_2] \\ - \int_c [e_1(\boldsymbol{p}, \boldsymbol{q}^0, u)\mathrm{d}p_1 + e_2(\boldsymbol{p}, \boldsymbol{q}^0, u)\mathrm{d}p_2], \tag{8.64}$$

where c is the price path and e_i is $\partial e / \partial p_i$, the Hicksian demand for recreational activity i. Because this is an exact differential, the results depend only on the end points of the p's, and not on the path. Choose the path $(p_1^0, p_2^0) \to (\tilde{p}_1, p_2^0)$ $\to (\tilde{p}_1, \tilde{p}_2)$:

$$
\begin{aligned}
w_2 = & \int_{p_1^0}^{\tilde{p}_1} x_1^H (p_1, p_2^0, \boldsymbol{q}^*, u) \mathrm{d}p_1 + \int_{p_2^0}^{\tilde{p}_2} x_2^H (\tilde{p}_1, p_2, \boldsymbol{q}^*, u) \mathrm{d}p_2 \\
& - \int_{p_1^0}^{\tilde{p}_1} x_1^H (p_1, p_2^0, \boldsymbol{q}^0, u) \mathrm{d}p_1 - \int_{p_2^0}^{\tilde{p}_2} x_2^H (\tilde{p}_1, p_2, \boldsymbol{q}^0, u) \mathrm{d}p_2.
\end{aligned}
\tag{8.65}
$$

This expression for w_2 shows that the welfare effects of a change in quality that affects two goods is equal to the change in area under the Hicksian demand of the first good, *conditional on the current price of the second good*, plus the change in the area under the Hicksian demand for the second good, *conditional on the choke price of the first good*. Simply looking at the change in the area under the two Hicksians at current prices would yield two functions *both* conditional on the current price of the substitute good. Holding aside for a moment the problem of Hicksian versus Marshallian measures, it is clear that adding up areas under current Hicksians creates a bias. The direction and magnitude of the bias depend on how the complementarity between the two goods is influenced by changes in the environmental quality variable.

With perfectly specified demand functions, proper welfare measures can be obtained by integrating over the functions conditioned appropriately. Data problems, chiefly multicollinearity in prices, often pose difficulties that can occasionally be mitigated as discussed in Bockstael and Kling (1988). The above example can be generalized to multiple quality changes. Suppose researchers are concerned with only one activity, but a policy affects quality at two sites. The welfare measure is given by

$$
e(\boldsymbol{p}, q_1^0, q_2^0, u) - e(\boldsymbol{p}, q_1^*, q_2^*, u)
\tag{8.66}
$$

where q_1 is weakly complementary to recreation at site 1, and q_2 is weakly complementary to recreation at site 2. The following sequence of integrals over (Hicksian) demands can be shown to be equivalent:

$$
\begin{aligned}
& \int_{p_1^0}^{\tilde{p}_1} x_1^H (p_1, p_2^0, q_1^*, q_2^0, u) \mathrm{d}p_1 - \int_{p_1^0}^{\tilde{p}_1} x_1^H (p_1, p_2^0, q_1^0, q_2^0, u) \mathrm{d}p_1 \\
& + \int_{p_2^0}^{\tilde{p}_2} x_2^H (p_1^0, p_2, q_1^*, q_2^*, u) \mathrm{d}p_2 - \int_{p_2^0}^{\tilde{p}_2} x_2^H (p_1^0, p_2, q_1^*, q_2^0, u) \mathrm{d}p_2.
\end{aligned}
\tag{8.67}
$$

The sequencing is now over the two quality changes. In a sense, one quality change is evaluated as usual, but the contribution of the second quality change is only the additional contribution after the first is taken into account. Expression (8.67) suggests once again that aggregations over changes cannot be accomplished by simple addition but require proper sequential evaluation over carefully specified functions.

These aggregation conditions imply knowledge of Hicksian demand functions. Subject to certain technical difficulties in integrating back, economists can typically get Hicksians from Marshallians. But when they aggregate over Marshallian demand functions, there is even more uncertainty about the bias. Earlier it was noted that recreational demand models of the continuous sort often fail to capture price and quality substitution effects either because of modeling difficulties or multicollinearity, making the correct measurement of these multiple effects even more dubious.

Discrete choice models look appealing in this light, especially for valuing changes at more than one site. Because random utility models impose more structure on the preference function and because they do not depend on demand functions for welfare measures, they do not require evaluation of sequential integrals. The formula for welfare measurement in the discrete choice context works as well with many as with one quality change.

8.7 Conclusions

The detail of the preceding section mirrors, albeit superficially, the level of detail in the current recreational benefits literature. A perusal of this literature, including the contributions of the authors of this chapter, reveals a growing obsession with the micro details of the problem. But too often the subtleties of the models appear ludicrous juxtaposed with the simplicity of the data. Efforts toward refining individual benefit estimates are commendable, especially those that seek to identify the factors to which benefit estimates are most sensitive or to define robust measurement procedures. But in economists' untiring quest for more and more precision in individual benefit estimates, they seem to have forgotten what has always been the greatest and altogether most insurmountable difficulty in welfare evaluation — aggregating benefits or losses over time and over individuals.

How do they add up an individual's benefits over time? Calculating the present value of the loss due to the elimination of a resource requires predicting what future use (and benefits) would have been and choosing a discount rate. This is more complicated than might appear and is much affected by the dynamics of habit formation, learning by doing, and information acquisition.

Finally, how do they aggregate over individuals? Do they predict benefits by income or geographical strata and simply add up predicted benefits weighted by the size of the population in each stratum? Or are there still fundamental concerns about aggregating over individuals?

These questions remain open for debate, and those debates have considerable consequence for the prudent and thoughtful application of welfare economics to public policy questions.

8.7.1 The Dynamics of Welfare Measurement

Once a time dimension is allowed, economists can no longer turn with confidence to welfare measures based on static expenditure functions. When a public policy decision is evaluated, the resulting hypothetical change is usually a change in perpetuity barring further action. (In some extreme cases, the change may even be irreversible.) The usual procedure is to beg the question entirely and produce "annual benefits." Present values are usually avoided because the present value will, of course, be highly sensitive to the discount rate chosen.

The dynamics of behavior are at least as critical to the present value of benefits and far more interesting than debate over the discount rate. If recreational behavior can be expected to change in some systematic way over time, then the implicit assumption of constant annual benefits or losses over time cannot be sustained.

Recreation has a number of features that suggest dependence over time. Individuals are uncertain about quality of sites (be it water quality, catch rates, congestion, and so on) and this uncertainty can be resolved or substantially reduced over time by visiting the sites. This behavior produces information that reduces uncertainty. Additionally, recreational activities often involve skills that can be improved with experience. Thus, recreation may be an example of learning by doing; the more an individual participates, the more he or she enjoys participation.

Earlier economists took interest in the role of habit formation and "learning by doing" in recreational demand. Davidson, Adams, and Seneca (1966) argued that future preferences for recreational resources are influenced by current availability. Krutilla (1967) made a similar point expecting demand for unique natural resources to increase due to habit formation. Munley and Smith (1976, 1977) were among the few to model habits in empirical recreational demand analysis. McConnell, Strand, and Bockstael (1990) took up the issue again, this time in the context of nonunique resources — quality differentiated sites. They paid special attention to the effects on benefit estimates and discovered that incorporating the dynamics of habit formation in models makes a good deal of difference for benefit measures, but it also raises serious econometric problems.

8.7.2 Aggregation over Individuals

As this discussion of aggregation and the broader discussion of benefit measurement comes to a close, it is time to raise a fundamental problem of welfare economics. Evaluation of policy actions requires using the individual welfare measures that economists have sought so hard to refine to assess the effects on society as a whole. But the greatest and most controversial issue in welfare evaluation has always been adding effects over individuals.

The usual convention is to multiply the average consumer surplus of the sample by the population size. More sophisticated studies estimate benefits for categories of individuals, such as by geographical location, income strata, or boat ownership. In any event, the practice yields one value that is a simple sum of estimated consumer surpluses. If the benefits net of costs thus derived exceed zero, then in principle, gainers could compensate losers.

The long-standing criticisms, although criticisms that seem rarely to be voiced these days, are twofold. First, compensation is rarely paid, and as such, policies are rarely Pareto comparable. Second, the compensation principle presumes the initial state, that is, income distribution, to be desirable. This standard is generally not considered desirable and at best is arbitrary. Two individuals with precisely the same preference functions but different income endowments will offer different compensating variation for an increase in the quality of or access to a public resource.

These well known, but often ignored, arguments have implications for the application of welfare economics to public policies having to do with recreation. This chapter has focused on site-specific models because we believe they characterize recreational problems. Likewise, public policies have site specific implications. Individual sites, with particular characteristics located in areas with low costs to some and high costs to others, may be introduced, eliminated, upgraded or allowed to deteriorate. Clearly in this setting different policies will have different distributional results. All projects and policies have alternatives. Using a welfare economics based on simple aggregation of consumer surpluses to select among these projects has serious distributional implications over the long run.

There is no clear solution to this problem. Economists have gotten into the habit of using money bids of compensating variation even though they know compensating variation is not a measure of welfare but of money. As such it is dependent not only on preferences but on income, which itself is somewhat endogenous. A compensating variation bid in terms of any scarce resource would be equally good.[1]

Given the firmly held position that money compensating variation measures have in modern welfare theory, it is neither possible nor desirable to stem the tide. What is possible is a more careful consideration of the diversity of interests and income strata that make up the sample and the population. Credible studies should not be reduced to single-value figures but should reflect the range of bidders that are affected by resource policy. Nonmarket benefit economists have found the task of generating defensible individual consumer surplus estimates so overwhelming as to have little time or energy to devote to the bigger picture. Perhaps we should begin to reverse this trend.

---

[1] Bockstael and Strand (1987) compared implications for policy when using both time and money based compensating variation.

Measuring the Demand for Environmental Quality
John B. Braden & Charles D. Kolstad (Editors)
© Elsevier Science Publishers B.V. (North-Holland), 1991

Chapter IX

MATERIALS DAMAGES

RICHARD M. ADAMS and THOMAS D. CROCKER

Oregon State University and University of Wyoming

9.1 Introduction

Human welfare, and indeed life itself, is dependent on the biological and physical systems that provide food, shelter, and other requisite inputs. Some systems, such as climate or topography, are natural in that they are unmanaged or otherwise beyond human control; others, such as agricultural systems or physical artifacts (buildings), are managed or constructed so as to provide specific outputs of value to humans. Together, these systems comprise the "material" environment upon which much of human welfare is based. Damage to or improvement in this material environment implies welfare changes.

This chapter focuses on the application of welfare evaluation techniques to materials damages (or improvements) associated with changes in environmental quality. For the purpose of this discussion, materials damage refers to reductions in stocks of such physical assets as buildings (including the structure and interior furnishings), bridges, roads and art forms, and reductions in the flow of outputs from biological systems, such as agriculture and forestry. The emphasis of materials damages thus tends to be on use values associated with commercial, or market-based, activities. An example of an environmental change that is known to affect such materials is air pollution, which includes pollutants such as ozone and other photochemical oxidants and acid deposition. For certain pollutants, such as tropospheric ozone, the damage they cause to plants, particularly agricultural crops, is believed to be the most important of their effects on human welfare. Hence, a correct economic assessment of these effects on materials is of considerable policy importance.

A correct assessment of materials damage has three components: (1) characterizing the differential changes across space and time that environmental change causes in production and consumption opportunities; (2) determining the responses of input and output market prices to these changes; and (3) identifying the adaptations that affected agents can make so as to minimize losses or maximize gains from the changes in opportunities and prices.

Natural or physical science studies are the primary source of information for the first component. Evaluation of the latter two components represents the economics portion of the assessment task. If an environmental change causes substantial shifts in outputs, it can produce changes in price and quantity that, in turn, lead to further market-induced output changes. Moreover, even if prices remain constant, natural science information alone will still fail to provide accurate indications of output changes when producers can alter production practices and the types of outputs produced. Thus, an accurate assessment of the economic consequences of environmental change for materials effects can be attained only if the reciprocal relations between physical and biological changes and the responses of individuals and institutions are explicitly recognized.

Given the importance of physical and natural science information in materials damage assessments, its role in the economic valuation process is addressed first in this chapter. Next, section 9.3 defines three aspects of materials effects that are notable in terms of economic valuation. Section 9.4 reviews specific techniques for assessing those economic effects, and section 9.5 summarizes evidence from recent applications of those techniques to materials effects. Section 9.6 presents recommendations for future research that will improve the accuracy and robustness of materials assessments.

9.2 The Role of Natural Science Information in Economic Valuation

As noted above, an economic assessment of materials damages arising from environmental change consists of three general components. In practice, the first basic component requires two types of natural or physical science information: (1) information on the original state or condition of a natural or physical system, such as vegetation or building materials; and (2) information on the environmentally-induced changes in those features of such systems that people value. This information is typically provided by engineers or physical and natural scientists through some form of experimentation. Ideally, such experimentation is based on known natural or physical science principles and has as its objective the quantification of input-output relationships that are *technically efficient.* That is, the engineer or biologist focuses upon the design and operation of processes that are capable of producing the desired outputs with a minimum expenditure on inputs. The input prices that determine the costs of those outputs are a given. The engineer and his biological counterpart act as if they are solving a constrained optimization problem, which requires that those solving the problem have a detailed knowledge of the expected costs of material inputs. If all nonmaterial inputs, such as labor, were free, then in economic terms these engineering solutions alone would decide the optimal process design. However, nonmaterial inputs are also costly, and both material and nonmaterial costs can change. Engineering results thus identify the *technically efficient* set of materials combinations within the larger set of physical and biological opportunities for producing a given level and type of desired output.

Now, consider the economist who is also interested in defining input-output combinations for similar production processes. Although the economist also tries to minimize costs, he or she does so under the assumption that the technically efficient combinations are already defined. The economic theory of cost and production describes the effects of variable input prices upon cost-minimizing combinations of material and nonmaterial inputs. From the set of cost-minimizing combinations, the theory provides rules for identifying the combination that has the lowest material and nonmaterial costs, and it specifies how this "minimum of the minimums" will be altered as relative input prices change. It thus identifies *economically efficient* input combinations.

To understand the link between the technical and economic problems, consider a firm producing one output with multiple inputs in a static, competitive setting. In dual cost function notation, the cost minimization (short run) problem faced by the firm can be expressed as:

$$c(w,y) = \min w'x, \qquad\qquad (9.1)$$

where w is the $n \times 1$ vector of input prices (prime indicates transposition) and x is a vector of inputs that can produce y; costs are minimized subject to:

$$x \leq V(y), \qquad\qquad (9.2)$$

where $V(y)$ defines the short-run production possibility set (or input requirement set). The production possibility constraint embeds the technology selection process initially solved by the engineer. Thus, the firm's problem is to select from a set of technically efficient input combinations the one combination that will produce output y at minimum cost. If the cost function is differentiable in w, then, by applying Shephard's lemma, optimal demand for x is obtained.

Changes in environmental quality are likely to change the firm's production possibilities, including reducing the number of material input sets available to it. Outputs, too, may be affected. Environmental degradation that reduces production possibilities thus implies less choice, increased costs, and decreased profits. However, the firm's objective function and behavior toward this new set of production possibilities remain the same: it chooses the input combinations that minimize its costs. A similar situation confronts consumers: they choose utility maximizing combinations of goods and services within the context of the households production function. To the consumer, environmental degradation increases the costs of those utility maximizing combinations. Thus, even though the economic problem of selecting cost minimizing input combinations remains the same for producers and consumers, an assessment of the cost of damages to materials resulting from environmental degradation requires that the natural or physical scientist characterize accurately these changes in input and output relationships.

9.2.1 How Much Natural Science Information Is Enough?

The preceding discussion attempts to convey the importance of natural or physical science information in the economic valuation process. Given that such assessments require information on how environmental change affects material things people value, such as food, fibre, and buildings, an important issue is how that information should be collected. Ideally, the engineer or biologist will consider *all* input combinations consistent with known laws of nature (or at least all those known to occur commercially) when developing an experimental design for a particular environmental change. Unfortunately, the typical basic science, piece-by-piece experimental approach to determining technically efficient input combinations can be expensive and time consuming. In agriculture, environmental and soil factors create thousands of different input types; in manufacturing and construction, many thousands of different material inputs exist. In agriculture, each input type is embodied in one or more production processes that may appear in a variety of forms and that can be put to a number of distinctive uses. Moreover, environmental cofactors, such as temperature and moisture, act in concert with pollution to aggravate or mitigate its impact on agriculture.

One key problem, then, is to decide how much and what kind of basic science information to generate and use in any particular materials damage assessment. Most basic scientific studies of materials damage from pollution have focused upon the impacts for a *single* input combination in some production process (e.g., Nriagu 1978). Indeed, the chosen combination may have been studied independent of cost considerations. As a result, many basic science studies frequently fail to provide the range of input substitution information needed by an economist to estimate the economic consequences of the changes. This deficiency in data has led to use in some studies of highly aggregated materials damage response information.

The errors that are introduced into economic estimates by these abstractions and aggregations of natural science reality have attracted little systematic empirical attention.[1] However, the problem that has gained some attention,

---

[1] Crocker (1975), Forster (1981), and Eaton (1984) have considered some of the analytical features of the problem. Kopp and Smith (1980) look at the aggregation issue in the context of the cost of pollution control. They find that input aggregation generates substantial errors in cost estimates because it masks economically relevant input substitution possibilities.

particularly in an agricultural setting, is the influence of additional precision in natural science information upon errors in economic estimates. Adams, Crocker, and Katz (1984) showed that the contribution of additional plant science yield-response information declines rapidly. Forty or fewer yield observations per crop, similar to those found in Heck et al. (1984), adequately discriminated among the differences in economic surplus in U.S. production of corn, cotton, soybeans and wheat generated under three assumed ambient ozone levels. Feinerman and Yaron (1983) showed that the net value of information to Israeli potato growers about potato yield responses to soil salinity was maximized with as few as 15 and not more than 27 observations.

In a study of the economic benefits of ambient ozone control for corn, soybeans, and wheat, Adams and McCarl (1985) found that changes in key natural science parameters had to be substantial if they were to translate into major changes in benefit estimates. The yield response data they analyzed were part of an interdisciplinary study to assess the economic consequences of crop damages due to ozone. They found that for given pollution levels, yield responses across alternative cultivar selections and across temporal and spatial variations tended to be homogeneous, implying that additional yield-response observations do not greatly affect benefit estimates. Adams and McCarl (1985, p. 274) concluded ". . . that even a limited set of crop response data, when generated in accordance with the needs of those doing the assessments, appears adequate to measure the general benefits of pollution control."

The available evidence supports the view that economic assessments of materials damages need not await the resolution of all, or even many, interesting basic science questions. These findings, however, do not imply that correct economic assessments can be performed with faulty natural science information. As noted earlier, accurate estimates of the economic consequences of materials damage are dependent on the input and output substitutions producers and consumers make in response to pollution changes. In fact, the physical and biological phenomena that the natural scientist or engineer chooses to investigate circumscribes the economic description of the producers's decision problem. An unrealistically circumscribed representation falsely limits the producer's or the consumer's alternatives for maximizing gains or minimizing losses from an environmental change. The circumscription understates economic estimates of the gains to producers and consumers from reducing pollution and exaggerates their losses from an increase. This appears to be less of a problem for some classes of materials damages, most notably agricultural assessments, which have benefited from carefully designed monitoring and data collection, than for physical systems, such as buildings, for which there have been no systematic coordinated efforts to link response research to the needs of economists.

9.2.2 *Economics as a Guide to Natural Science Data Collection*

The preceding discussion points to the importance of generating basic science information that captures the relevant range of input and output alternatives available to the producer or consumer. However, decision problems differ greatly across economic sectors. For example, the impact of SO_2 on printing paper may be of real importance to a public curator of rare books, but be unimportant to a newsstand operator. The latter will replace his or her current inventory with next week's editions, irrespective of SO_2 levels. The curator may be able to reduce the potential damage done by SO_2; however, if the natural scientist has not collected information on how the respective alternatives affect the impact of SO_2 on the rare books, the economist cannot describe the choices the curator will make. The economist is then forced to treat the curator like the newsstand operator.

Among the economic sectors most subject to materials damage are real estate (residential, commercial, industrial, public, and regulated utility), public transport infrastructure, sanitation, commercial forestry, and agriculture. The paper example in the preceding paragraph suggests that a separate characterization of the optimizing behavior of each sector will help to identify the parameters to which that sector's adjustments are likely to be most sensitive. Thus, for example, the level of interest rates will have little or no influence on the paper used in cheap novels, but will have a major influence on the mix of materials used in building construction because high interest rates reduce builders' incentives to purchase more expensive, long-lived materials even though these materials may be strongly resistant to SO_2 or other air pollutants. To the prospective building owner, the long-lived materials are not a meaningful way of reducing the impacts of SO_2. Natural science research intended for use in economic analysis can, therefore, disregard them in order to concentrate on alternatives with shorter life spans. However, for the publishing industry, longer-lived, more SO_2-resistant paper may be a good substitute. In short, the organization of basic science research on materials by economic sector allows one to identify economically relevant gaps in that information. As a result, adaptations that are economically relevant are less likely to be overlooked. This kind of organization also happens to be the form in which most Federal, state, and private information is published, perhaps because it conforms to the natural distinctions of the marketplace.

9.3 Dimensions of Materials Effects

In addition to the frequent need for scientific information, two other dimensions of materials damages pose special analytical challenges: long-term effects that require intertemporal analysis and the need to consider risk responses of producers and consumers.

9.3.1 Materials Damages Are Intertemporal

A feature that distinguishes the economic consequences of materials damages from some other classes of environmental effects is the dominant role played by intertemporal considerations. An environmental change alters the stream of returns that assets produce, whether the assets are buildings, public infrastructure, forests, or agricultural crops. Only annual agricultural crops escape this intertemporal emphasis even though potentially long-term environmental effects, such as acidification of soil, may become an issue.

Consider the producer represented in static equations (9.1) and (9.2) who must now decide how to alter the operating, maintenance, and capital (materials) replacement program for a particular production process as the level or pattern of pollution exposures changes. This intertemporal problem can be cast in either a dual or primal specification. However, the dual version of this problem requires a set of quite restrictive assumptions, as shown in Diewert (1986). Therefore, following Crocker and Cummings (1985), the Lagrangian of the primal specification of this maximization problem for a given time interval, $t = T - \iota \geq 0$, is:

$$
L = \sum_{t=\iota}^{I} \Big([p_t(X_t)X_t - c_t(X_t,K_t) - I_t]\beta^t
$$
$$
- \mu_{t+1}\{K_{t+1} - K_t + D[X_t,M_t(A_t)] - I_t\} \Big),
$$
(9.3)

where p_t is the unit price of the output in period t, X_t is the quantity of output, c_t is operating cost, K_t is the capital (materials) stock, I_t is additions to the capital stock and is the shadow price of the capital stock constraint. D is the rate at which the capital stock depreciates or grows. It is a function of the capital stock and of the effectiveness of maintenance effort M, where $\partial D/\partial M_t \leq 0$, according to whether maintenance reduces depreciation or enhances growth. In turn, the effectiveness of any given maintenance effort is determined by pollution A_t such that $\partial M_t/\partial A_t \leq 0$. Finally, $\beta^t = (1 + r)^{-t}$, where r is the producer's subjective rate of discount. Assuming the objective function represented by the first term on the right-hand side of (9.3) is continuous and that the appropriate concavity and convexity conditions apply to all functions in (9.3), the first-order conditions for a maximum of the objective function are:

$$
\frac{\partial L}{\partial X_t} = \Big(p_t' + p_t - \frac{\partial c_t}{\partial X_t}\Big)\beta^t - \mu_{t+1}\Big(\frac{\partial D\ [X_t,M_t(A_t)]}{\partial X_t}\Big) = 0
$$
(9.4)

$$
\frac{\partial L}{\partial K_t} = -\frac{\partial c_t}{\partial K_t}\beta^t - \mu_{t+1} = 0,
$$
(9.5)

and

$$
\frac{\partial L}{\partial I_t} = -\beta^t + \mu_{t+1} = 0.
$$
(9.6)

The first term on the right-hand side of expression (9.4) is the discounted present value of the producer's marginal net revenue and the second term is the user cost. This latter term implies that the discounted value of the marginal net revenues generated by using the capital stock to produce output in period t should equal the opportunity cost, $\mu_t + 1$, of capital consumed while producing period t output, $\partial D/\partial X_t$. For commodities such as buildings, $\partial D/\partial X_t > 0$, implying that period t output depletes capital, thereby lowering the capital stocks available to future periods. However, $\partial D(A_t)/\partial X_t < 0$ for some biological resources, since thinning and harvesting can increase future stocks of trees, perennial crops, and fish.

Expression (9.5) defines the discounted present value of the marginal contribution of the capital stock to reductions in production costs in all future periods, whereas expression (9.6) says that investment during period t will be at that level that equates the marginal cost of investment, β', to the marginal value of capital, $\mu_t + 1$.

Suppose now that pollution levels rise from A_0 to A_1. Except for some biological resource cases, this implies that the effectiveness of maintenance activities declines for all future values of X. Given no change in I, capital stocks in all future periods must then be smaller. Furthermore, with concavity of the production function underlying $c(X,K)$, $\partial c/\partial K$ is positive in all future periods, implying that the shadow price of capital, $\mu t + 1$, has increased. The result of all this is that the second term on the right-hand side of (9.4) increases. It follows that the first term must also increase, given $\partial p/\partial X < 0$ and $\partial c/\partial X > 0$, which leads to a reduction in the level of output, X, and a higher price, p. In effect, higher production and user costs for the producer result in less output at a higher price. If market prices are at all sensitive to output, then consumers lose when pollution increases, but producers either lose or gain, depending on whether the percent reduction in output quantity is greater than or less than the percent increase in price.

The preceding formulation can capture a considerably wider range of producer intertemporal, spatial, input and output adjustment opportunities. For example, the producer can alter t, which is the time interval over which he produces. In fact, to the extent that pollution can be viewed as a tax on input use, the intertemporal analysis of producer behavior with respect to pollution impacts is directly analogous to the extensive economic literature on producer behavior with respect to depletable natural resources.

9.3.2 Materials Damages Introduce Risk

The strongly intertemporal features of most materials damage introduces the effects of risk on producer and consumer behavior. Treatment of risk in environmental assessments involving materials is thus potentially important. However, most materials damage assessments are *ex post* in nature. Risk is either ignored or treated imperfectly. Because an *ex post* representation establishes a number of contingent states and proceeds to treat each of them as if it were certain (see chapter 2), it is incapable of accounting for the agent's attitude toward risk; that is, it disregards the expenditures that the agent makes in preparing for states that go unrealized. The alternative, *ex ante* representation of the economic agent, addresses the consequences when the magnitude of the pollution effects is not yet known.

In a world of complete contingent claims (insurance and futures) markets, any risk attitude-induced discrepancy between *ex ante* and *ex post* measures of economic value would not exist. Complete markets enable individuals to redistribute income and associated consumption and production opportunities toward undesirable prospective states. Erlich and Becker (1972) have shown, given insurance prices that are actuarially fair and where marginal utility of income is decreasing, that insurance or futures claims would be acquired in amounts that would make individuals indifferent as to which of a set of feasible states of nature ultimately is realized. No matter what the realized state of nature, the *ex ante* premium payments and the *ex post* compensation (that the insurance supplies) maintains the *ex ante* utility level. Questions of *ex ante* versus *ex post* valuation therefore become irrelevant because the expected consequences of *ex ante* choice are always realized.

However, because fair contingent claims markets rarely exist, especially for environmental goods, due to moral hazard, adverse selection or nonindependence of risks, *ex ante* measures of value are particularly appropriate. Given incomplete markets, prospective outcomes are inherently uncertain, implying that the planned rather than the realized outcomes explain behavior (Buchanan 1969). Thus, the materials damage valuation consequences of adopting an *ex ante* perspective of the individual's decision problem can be formally deduced from the risk literature (e.g., Schmalensee 1972; Graham 1981; Chavas, Bishop, and Segerson 1986). The risk premium associated with a risk averse individual is the difference between the maximum a risk averse individual would be willing to pay to retain the option of using a future good (option price) and the expected value of *ex post* consumer surplus. The latter is a traditional Marshallian or Hicksian measure, whereas the former includes a risk premium because the individual is required to make a decision before the state of nature or its associated outcome is revealed. Most of the abundant environmental economics literature on option value has sought to establish whether it is negative, positive, or zero, which would respectively imply that the traditional measures of environmental protection are positively, negatively, or not at all biased (see chapter 10).

9.4 Alternative Assessment Techniques

The preceding sections provide an overview of some of the key natural science and economic features of materials damages. Conceptual frameworks that capture these features link the biological and physical manifestations of environmental change to the behavior of affected producers and consumers. The changes in profits and utility that accompany these behavioral responses are indications of their effects on the welfare of producers and consumers.

A conceptual framework provides an abstract representation of the economic assessment problem. Specific empirical structure is supplied by the application of economic assessment techniques. Assessment techniques are means of summarizing supply and demand behavior in particular economic sectors. Chapters 2 to 5 review the general techniques for assessing the economic consequences of environmental change. This section focuses on alternative techniques, describing their data requirements as well as their limitations, such as the kinds of error to which they are susceptible when applied to materials damages.

Before discussing alternative valuation techniques, it should be noted that there are several classification or organizational schemes that distinguish among these procedures. For example, some assessment techniques for materials damages are classified as either normative or positive. *Normative techniques,* such as mathematical programming methods, require the economist act as a representative economic agent who is to be affected by the contemplated environmental change. The economist puts himself or herself in the agent's shoes, acquires the same data as would the agent, and then solves the decision problem in an idealized way. The results of this exercise are also idealized. *Positive methods,* such as hedonics, allow the economist to infer the net effects of environmental damages from the actual choices of producers and consumers. The economic adjustments that must be explicitly modeled in normative techniques are implicit in the positive approaches.

Valuation methods can also be classified by whether they employ data on the observed or the self-reported behavior of economic agents. Contingent valuation methods form the latter category, whereas mathematical programming methods and hedonic methods comprise the former. A third classification scheme can be based on whether market price data are required to implement the method. Except for contingent valuation methods, all other methods depend to some extent on market price data. Because materials damages tend to involve marketed goods, the choice of this scheme makes available a large number of valuation techniques.

Other possible classification schemes are based on whether or not they require natural science data on the relationship between pollution dose and environmental response. Mathematical programming and other normative methods require such dose-response information. Hedonic and duality methods, which relate environmental impacts to an economic value measure, such as the effect of pollution on residential property values, do not require dose-response information. When available, the dose-response information is useful for this latter set of methods because it can be used as *a priori* information both to restrict the relationship between environmental effect and economic value and to serve as a check on the plausibility of the economic estimates obtained.

The set of techniques for valuing the materials consequences of environmental change vary in data requirements, in the degree to which they capture economic phenomena, and in the range of settings in which they are likely to provide credible economic estimates. Techniques commonly used in materials damage assessments are reviewed below.

9.4.1 Simple Multiplication

This approach uses only the physical or biological dose-response relationship to estimate the response to a change in some environmental parameter. The observed market price of the activity or entity is then multiplied by the magnitude of the physical or biological response to obtain a pecuniary measure of damage. Thus, neither behavioral adaptations nor price responses are taken into account. For example, with this approach, the impact of pollution on corn yields is estimated by multiplying biological estimates of reductions in yields per acre by acreage and the price per bushel. Or, a value for household soiling (for example, dust on exposed surfaces) can be obtained by multiplying *ex ante* cleaning frequency by the unit price of hiring cleaning services, then remultiplying by the "population-at-risk."

Simple multiplication provides an accurate estimate of economic behavior and value — in this case, changes in gross revenue — only if economic agents are limited in the ways in which they can adapt to the environmental effect and if the effect is small enough to have little or no impact on relative prices. This combination of circumstances is very unlikely. As a result, economists are critical of such assessment procedures. The technique is favored, however, by noneconomists working with the economic consequences of materials damages, perhaps because the technique maximizes the influence of noneconomic information. Two examples of the simple multiplication method are Naraganan and Lancaster (1973) on household soiling and Benedict, Miller, and Olson (1971) on agricultural damages from air pollution. The use of this approach has diminished as economic arguments about firm, consumer, and market adaptation have been widely accepted.

9.4.2 Price Imputation and Multiplication

This is a more demanding version of simple multiplication. Its operation is identical to the simple multiplication method with one exception: when an explicit market price is not readily available for the relevant entity or activity, a price that some other investigator has imputed to a unit of the activity is employed. Calish, Fight, and Teeguarden (1978) and most of Freeman (1982) are good examples of this method. Economists frequently adopt this method presumably because the imputed prices are drawn from other studies in which some account has been taken of price responses and agent adaptations. However, the transfer of imputed prices from one site, time, or allocation problem to another implicitly presumes the existence of a structure common to each setting. The refereed environmental economics literature does not contain any studies that formally test the statistical legitimacy of these transfers.[2]

9.4.3 Mathematical Programming

At the other extreme of sophistication are an assortment of methods that account more completely for the adaptations economic agents make to mitigate or exploit the opportunities of an environmental change. These methods frequently involve major commitments of technical expertise and research resources. All are normative in the previously mentioned sense that the economist tries to duplicate the decision problem of the economic agent. But because of their duplicative character, they might be called *decision duplication* methods. In principle, they can be applied to any problem for which relative price as well as dose-response information is available. The approaches derive mixes and magnitudes of production and consumption activities according to the relative prices of each activity and according to the differential impacts of the environmental change on these activities and their relative prices.

The most widely used family of decision duplication methods are mathematical programming techniques (linear, recursive, quadratic, integer, dynamic, and so on). These techniques characterize the economic problem through mathematical expressions for the decision agent's objectives, alternative ways of attaining the objectives, and resource and other restrictions upon these alternatives. Optimal (cost-minimizing or profit-maximizing) solutions are then sought for specified environmental and economic conditions that are thought to represent actual or prospective realities. The structural relations are specified rather than being estimated as a part of the exercise. Parameters within these relations may, however, be estimated. The application of mathematical programming to economic problems thus can involve a heavy dose of applied econometrics to derive an often large number of system parameters.

Mathematical programming techniques are popular in studies of the economic consequences of materials damages for three reasons: (1) their ability to assess the effects of as yet unrealized environmental or economic changes; (2) the relative ease of including output and input changes into such models; and (3) the optimization nature of the procedure and the consistency of the

---

[2] Bayesian techniques appropriate to the testing task do exist. See Lindley and Smith (1972) for the original methodology and Aigner and Leamer (1984) for an electricity pricing application.

framework with standard microeconomic theory concerning firm and market behavior.

One feature that limits the use of programming techniques is the large amount of data often required to construct an adequate representation of the economic problem. Related difficulties include the following: the treatment of more than a single objective, such as profits, as well as the handling of indivisibilities; the complex nonlinearities in the constraint system; and the stochastic phenomena.[3]

Starting with Adams, Crocker, and Thanavibulchai (1982), programming methods have been widely used to assess the economic consequences of air pollution-induced agricultural damages. The flexibility of the procedures is evident from the diverse scales found in agricultural applications, ranging from farm-level models to national-level analyses. Only rarely have the methods been applied to other classes of materials or economic sectors. An example is the McGuckin and Young (1981) study on desalination of household water supplies.

When mathematical programming models are applied to materials damage assessments, they typically start with the notion of a "representative" firm. The basic linear programming problem specification

$$\max Z = p'x \tag{9.7}$$
$$\text{s.t.} \quad Ax \le b \tag{9.8}$$
$$x \ge 0,$$

is given an economic interpretation by definition of the variables and parameters in equations (9.7) and (9.8). Typically, p is defined as an $n \times 1$ vector of (discounted) net returns resulting from a one unit increase in the jth activity, x is an $n \times 1$ vector of activity levels, A is an $m \times n$ matrix of input-output coefficients specifying the amount of the ith resource required per unit increase in the jth activity, and b is an $m \times 1$ vector of resources. In sectoral or aggregate analysis, multiple representative firms are frequently specified to account for differences in firm-level input-output relationships. To capture the behavior of each representative firm, the values assigned to the parameters are drawn from data representing the economic and physical environment of the firms in the industry.

The solutions from representative firm problems suggest how environmental changes affect levels of input use, output supply, and profits or costs. If aggregate (industry) response is needed, the second stage of a programming assessment links the response elicited from the representative firms to some measure of aggregate (industry) demand. This relationship allows the price parameter (p) to become endogenous in the system, which establishes some notion of an aggregate supply relationship (curve). Given the demand in each

---

[3] See Takayama and Judge (1971) for a detailed treatment of the analytical foundation of these methods.

market, the determination of equilibrium quantity and price is analogous to the solution of a problem of maximizing producers and consumers surplus (Samuelson 1952). The appeal of such a programming specification is obvious: the model has a firm microtheoretic foundation upon which to generate measures of social welfare that are associated with materials damage. Where data permit, programming-based evaluations of materials damage are likely to continue, particularly in those settings without complex time horizons or stochastic phenomena.

9.4.4 Simulation Models

These models provide an alternative to conventional mathematical analyses of economic systems. Naylor et al. (1968) define simulation as a numerical technique for conducting experiments on a computer. Unlike mathematical programming, there is no "theory" of simulation. Also, simulation methods are descriptive rather than normative. Like programming methods, however, simulation models can be used to assess the environmental and economic consequences of as yet unrealized events. And though they lack the optimization procedures embedded in programming techniques, they are particularly useful when theoretical issues are unresolved or when mathematical solutions are either very difficult or impossible to obtain. Simulation is particularly useful in studying dynamic systems because the systems' intertemporal relationships are key components of this technique.

In materials damages settings, simulation models have frequently been applied to dynamic and stochastic issues, where standard comparative static methods yield little insight. Examples include Thompson et al. (1979) on the spruce budworm, Ayres and Sandilya (1987) on domestic animal grazing, and Bjorndal (1988) on the herring fishery. As with the programming methods, the investigator must write down a version of the agent's objectives and constraint system. Then, rather than analytically solving for parameter values, the investigator assigns some of the values based on prior knowledge and observes how the numerical solution routine behaves when alternative parameters are used. A problem with this method is that the solution is highly specific. To obtain a more general solution, the investigator must run many simulations. Also, because simulation methods are nonanalytical, they can provide limited insight into the general structure of the system. The circumstances under which any of their results can be transferred to other settings are substantially circumscribed, if not intrinsically unknowable.

Occasionally, simulation models are built around econometric or mathematical programming models of a given system (e.g., Bjorndal 1988). In such applications, the econometric or programming models are solved to provide inputs into other (simulation) submodels of the overall systems being modeled. The submodels contain those features of the system not easily incorporated into a pure econometric or programming model; for example, sequential decisions or indivisibilities. Although the overall solution generated in these cases lacks the analytical elegance of a pure optimization technique, the embedding of econometric or programming relations improves the robustness of simulation.

9.4.5 Econometric Methods

Econometrics involves the use of numerical data to test the postulated relationships of economics (Malinvaud 1966). In one form or another, econometric approaches are frequently used in materials damage assessments. For example, one of the simplest applications employs statistical procedures to estimate the physical relationship, namely the dose-response functions between inputs and outputs, for varying levels of environmental effect. This purely biological information is then integrated with information on how input and output prices respond to relative changes in output (demand), and how producers respond to changes in input and output prices and output magnitudes (supply). These demand and supply expressions are also estimated by statistical means. Once the parameters of the market supply and demand equations have been estimated, the system is simultaneously solved by a market clearing identity which equates quantity supplied to quantity demanded. Examples of this approach include Adams, Crocker, and Katz (1984) and Brown and Smith (1984) on ozone impacts upon agriculture, Watson and Jaksch (1982) on household soiling, and Kahn and Kemp (1985) on aquatic ecosystems.

Contained within this class are some specific techniques that differ in data requirements and assumptions. For example, the econometric and other methods referred to above require dose-response information and involve separate estimation of systems of supply and demand expressions. A different set of methods infers supply and demand functions from the observed results of the actual interactions of supply and demand forces. These methods do not require dose-response information, though its availability can increase the statistical efficiency of estimators and provide a check on the plausibility of results. Nor do they involve the separate parameterizing of supply and demand functions. The econometric difficulties of hedonic and duality methods revolve around the problem of identification — that is, of distinguishing between the extent to which supply and demand respectively contribute to observed outcomes.

In principle, this latter subclass of methods consumes substantially less data and research time than do most of the assessment methods described earlier. There are two reasons for this: first, the economic agent solves the decision problem; and second, the economist is a bystander. The agent essentially does part of the work that the investigator must do with the normative methods.

It should be apparent that any method based on actual producer responses will be *ex post* relative to experienced levels of environmental change. When policies are aimed at as yet unrealized environmental changes, *ex ante* procedures, such as programming, are better alternatives. Moreover, the data on firm-level decisions are often unavailable, and the investigator has no choice but to take a normative approach.

The econometric-based approach most widely applied to materials damages is the *hedonic method* (see chapter 4). Starting with Ridker and Henning (1967) and Anderson and Crocker (1971), the method has been applied many times to residential property values. Although no analytically straightforward method exists to distinguish the economic contribution of a pollutant to materials damages from its contribution to other site-specific damages reflected in property value differences, materials damages are undoubtedly present, as household soiling studies clearly demonstrate.

Unfortunately, hedonic price theory does not say which of the hundreds of housing and neighborhood attributes are central to the determination of housing package prices. Estimates of the hedonic price locus have thus been a fertile ground for both data mining, or "specification searching," to obtain desired signs for particular attributes and selective reporting of fragile results. Because the price loci are reduced form expressions with attributes as explanatory variables, economic theory provides little guidance on selecting covariates. In addition, an absence of housing attribute "rebundling" causes many attributes to exhibit little independent variability, which implies that many hedonic price function covariates are highly collinear. When Atkinson and Crocker (1987) applied Bayesian diagnostic techniques to a Chicago property value data set, they found that the collinearity in a relatively short list of candidate covariates was so great that they could not obtain reliable estimates of the signs of most covariates. Estimates of price responses to neighborhood attributes such as air pollution were especially ambiguous. The ambiguities were uncertainties in specification and intolerances in measurement error that originated in the collinearities among candidate covariates. By employing similar Bayesian methods and a data set from Los Angeles, Graves et al. (1988) recently confirmed these ambiguities.

The other econometric-based approach to materials damages is *duality*. With the exceptions of Garcia et al. (1986) and Manuel et al. (1981), duality methods have been neglected as a means for assessing material effects of environmental change. This is unfortunate. On the producer side of the market, dual-profit function techniques are the major analytically legitimate alternative to mathematical programming methods.[4] Both methods are representations of the effort by an agent to efficiently meet objectives. For similar pollution levels, the two approaches should produce similar empirical results.

---

[4] See Diewert (1986) for an analytical treatment with immediate reference to materials effects and environmental change.

Consider agriculture, where a reasonable index of the producer's objectives is the maximization of his or her net revenues. Mathematical programming, by identifying those choices among the producer's alternatives that maximize net revenues, obtains measures of change in producer quasi rents (and consumer surpluses if cast as a regional or national-level model); the dual-profit function, by presupposing net revenue maximizing behavior, allows one to infer (recover) changes in these producer quasi rents. The former method simulates producer behavior by actually solving the producer's problem; the latter describes the producer's behavior by presuming that the producer has solved the problem. Mathematical programming provides estimates of what is best for the producer to do under environmental quality levels that may or may not have occurred historically. The dual-profit function presumes that the producer has always done what is best for him or her. Historical data on the producer's revenues and costs under different environmental regimes are then used to estimate statistically the economic consequences of his or her behavior.

In principle, the dual-profit function involves substantially less data and research time than do mathematical programming methods because the producer performs the optimization implicitly rather than explicitly, as is done by the researcher in the programming method. The researcher employing duality confronts fewer parameters and less complex interactions, thus reducing the trade-off between analytical completeness and empirical tractability found in any research effort. In fact, as Diewert (1986, 155-169) proved, the producer's dual cost function has a one-to-one correspondence with a real, unique production (dose-response) function. Dose-response functions can thus be established indirectly by applying duality methods to economic observations on producers' behavior rather than through direct natural science observations of affected materials. Mjelde et al. (1984) empirically demonstrated this for ozone impacts upon agriculture. Alternatively, duality methods can be used to assess economic consequences in the absence of immediate knowledge of dose-response functions (Garcia et al. 1986; Manuel et al. 1981).

As an illustration, recall the producer's decision problem expressed earlier in dual notation (equations 9.1 and 9.2). The profit counterpart is:

$$\pi^*(p,w) = \max \ p'y \tag{9.9}$$
$$\text{s.t.} \quad x \le V(y). \tag{9.10}$$

Assume now that in addition to prices, environmental quality z is introduced as an argument in the profit function; that is $\pi(p,w,z)$. The derivative of π with respect to z then provides some measure of the shadow price of z; the amount some producers would be willing to pay to have z increased by some marginal amount. Thus, the benefits of environmental improvement can be calculated from $\partial\pi/\partial z$.

As shown in Garcia et al. (1986), this specification can be used to measure the effect of changes in z (environmental quality) on output (supply). This is obtained by deriving the supply function for y [by differentiating $\pi^*(\cdot)$] with respect to p and then differentiating $y^*(\cdot)$ in terms of z. This process leads to a measure of the change in environmental quality that accounts for the direct affect on output of a change in z and the direct effect on output arising from adjustment of the optimal level of variable inputs. Theoretically, then, the profit function approach provides a more accurate estimate of the relationship between a change in environmental quality and output than is obtained from the simple econometric approach that uses a dose response curve in combination with aggregate supply and demand relationships.

Unfortunately, the theoretical strength of duality — of using actual producers' behavioral responses — is also one of its empirical limitations. Specifically, application of duality requires a detailed data set on producer input and output decisions over space and time. Besides confronting data limitations, dual functions typically do not address price effects (P and r are generally assumed constant). Furthermore, duality is less able to address as yet unrealized environmental states than are mathematical programming and simulation.

9.4.6 Survey Methods

Although survey methods have been used only sparingly in materials damage studies (e.g., Cummings et al. 1986), they enable researchers to develop economic value measures even in the absence of dose-response and unit price information. If the normative methods sketched above are to be properly implemented, they require large amounts of often complex information on dose-response functions, adaptation opportunities, relative prices, and standing stocks. In addition, the investigator must pretend to know the agent's objectives. Positive techniques, such as the hedonic and the duality methods, must have data on standing stocks and relative prices. Agent objectives need only conform to the basic axioms of economic choice (Deaton and Muellbauer 1980, chap. 2). Information for the positive techniques is usually more accessible than for the normative techniques; nevertheless, it will not always be available at acceptable cost. The use of survey, or contingent valuation, methods must then be considered. (These methods are discussed in greater detail in chapter 5).

Contingent valuation methods are not analytical methods; but rather, they are data-generating devices that allow the respondent's objectives to determine their choices directly rather than having them filtered through an investigator's conjectures about these objectives. More significantly, the contingent domain in which the agent responds and in which the investigator understands the agent is responding is assumed to correspond exactly (Brookshire and Crocker 1981). This common domain conforms to the strictures of an investigator-constructed analytical model rather than forcing, via a set of possibly tenuous assumptions, observed real-world data to conform to the preconceptions of the model.[5] Where dose-response information is untrustworthy or where observable market prices are lacking for goods used in activities affected by the environmental change, there may be no alternative but to use contingent valuation methods.

There is another aspect of materials damage assessments where contingent valuation methods are required. As noted in a previous section, the intertemporal features of materials damages justify more serious attention to the effects of risk on producer and consumer behavior. The presence of risk, combined with incomplete contingent claims expressed among *ex ante* lotteries of outcomes, suggests the potential application of contingent valuation methods.

These *ex ante* issues have received little attention in empirical efforts to assess the economic consequences of materials damages. They are especially important if a major objective is to predict agents' behavioral responses to environmental change. Contingent valuation techniques provide a ready forward-looking means to assess these responses as well as their *ex ante* value consequences.

9.5 Empirical Evidence

The empirical literature on the economic consequences of materials damages from environmental change is dominated by two classes of applications: air pollution damages to annual agricultural crops and household soiling from air pollution. As table 9.1 shows, these studies of household soiling and agriculture encompass three valuation techniques. All are atemporal estimates that disregard uncertainty. They therefore miss what are some of the more important and challenging features of materials damages. In addition, the transferability of these estimates over time and space, as frequently required for policy analysis of environmental changes, is unclear. Nevertheless, in spite of their restricted scope, the agricultural studies of annual crops exhibit several regularities in behavioral responses and sensitivities to imposed conditions and to data accuracies and precisions (Adams and Crocker 1989). These regularities may allow subsequent researchers to restrict the dimensionality of their analytical and empirical problems.

Comparisons of the results of studies can provide little meaningful insight unless the conditions under which the studies were done are compared. Where regularities appear, they generally are derivable directly from economic theory; others are more subtle and perhaps arise from unique characteristics of the agricultural setting. Together, the regularities illustrate the current understanding of the behavioral responses of economic agents to materials effects induced

[5] Except for contingent valuation, all of the previously described methods employ observed rather than stated behavior: they implicitly presume there is no discrepancy between planned and realized outcomes.

TABLE 9.1
Recent Studies of the economic effects of ozone and other pollutants on materials.

| Effect category | Region | Pollutant and concentration | Model features | | | | Material | Results (annual 1980 U.S. dollars) | | |
			Price changes	output substitutions	Input substitutions	Quality changes		Consumer benefits	Producer benefits	Total benefits
Agriculture										
Howitt, Gossard, and Adams (1984)	Corn Belt	Ozone, universal reduction to 40 ppb[a]	Yes	Yes	Yes	No	18 crops	$17×10^6$	$28×10^6$	$45×10^6$
Adams and McCarl (1985)	Corn Belt	Ozone, reduction NAAQS from 120 ppb to 80 ppb[b]	Yes	Yes	Yes	No	Corn, soybeans, wheat	$2079×10^6$	$2079×10^6$	$668×10^6$
Mjelde et al. (1984) Garcia et al. (1986)	Illinois	Ozone, 10% increase from 46.5 ppb[a]	No	Yes	Yes	No	Corn, soybeans	None	$226×10^6$	$226×10^6$
Adams and Crocker (1985)	U.S.	Ozone, universal reduction from 53 ppb to 40 ppb[a]	Yes	Yes	Yes	No	Corn, soybeans, cotton	Not reported	Not reported	$220×10^6$
Adams, Crocker, and Katz (1984)	U.S.	Ozone, universal reduction from 48 ppb to 40 ppb[a]	Yes	No	No	No	Corn, soybeans, cotton, wheat	Not reported	Not reported	$2400×10^6$
Adams, Hamilton, and McCarl (1986)	U.S.	Ozone, 25% reduction from 1980 level for each state[a]	Yes	Yes	Yes	No	Corn, soybeans, cotton, wheat, sorghum, barley	$1160×10^6$	$550×10^6$	$1700×10^6$
Kopp, et al. (1985)	U.S.	Ozone, universal reduction from 53 ppb to 40 ppb[a]	Yes	Yes	Yes	No	Corn, soybean, wheat, cotton, peanuts	Not reported	Not reported	$1300×10^6$
Shortle, Dunn, and Phillips (1986)	U.S.	Ozone, universal reduction from 53 ppb to 49 ppb[a]	Yes	No	No	Yes	Soybeans	$880×10^6$	$-90×10^6$	$790×10^6$
Adams, Callaway, and McCarl (1986)	U.S.	Acid deposition, 50% reduction in wet acidic deposition	Yes	Yes	Yes	No	Soybeans	$172×10^6$	$-30×10^6$	$142×10^6$
Adams et al. (1989)	U.S.	Ozone, seasonal standard of 50 ppb with 95% compliance[c]	Yes	Yes	Yes	No	Corn, soybeans, cotton, wheat, sorghum, rice, hay, barley	$905×10^6$	$769×10^6$	$1674×10^6$

TABLE 9.1 *Continued*

Effect category	Region	Pollutant and concentration	Price changes	Model features output substitutions	Input substitutions	Quality changes	Material	Consumer benefits	Results (annual 1980 U.S. dollars) Producer benefits	Total benefits
Forests										
Callaway, Darwin, and Neese (1985)	Eastern U.S.	Forest growth reductions of 10%, 15% and 20%	No	No	Limited spatial only	No	Hardwoods and softwoods	Not represented	$-270×10^6$ to $-563×10^6$	$-270×10^6$ to $-563×10^6$
Building materials										
Lareau, Horst, Manuel, and Sipfert (1986)	Northeastern and North Central U.S. urban areas	Reduction of SO_2 to background levels	No	No	Yes	Yes	Buildings	Not represented	$1600×10^6$	$1600×10^6$
Household soiling										
Manuel, Horst et al. (1981)	Northeastern and North Central U.S. urban areas	Universal reduction of SO_2 and particulates to secondary standards	No	Yes	Yes	Yes	Soiling	$434×10^6$	Not represented	$434×10^6$

[a] Seven-hour growing season geometric mean. Given a log-normal distribution of air pollution events, a 7-hour seasonal ozone level of 40 ppb is approximately equal to an hourly standard of 80 ppb, not to be exceeded more than once a year [Heck et al. (1984)].

[b] Averaging time of one hour; not to be exceeded more than once a year.

[c] Seven and twelve-hour growing season geometric means. Analysis includes both fixed roll-backs (e.g. 25 percent) and seasonal standards (with variable compliance rates).

by environmental change. They also indicate which important features of materials damages should be included in such evaluations and which assessment techniques are appropriate. The regularities are as follows:

1. *Producers can gain from increases in air pollution.* Adams and McCarl (1985) found that substantial increases in ozone increased the quasi rents of Corn Belt growers. Shortle, Dunn, and Phillips (1986) also estimated that Corn Belt growers benefited from increases in ozone. A condition for these findings is that the pollution-induced percent reduction in output for a given crop is less than the associated increase in market price; that is, inelastic demand and an upward shift in the supply curve. When secondary effects on livestock producers are included in national assessments, the net effect of higher levels of ozone on all producers is a loss in welfare (Adams, Hamilton, and McCarl 1986). Even then, however, producers in some regions benefit while others lose. Early studies generally concluded that growers always lost from increased air pollution. However, because those studies used simple multiplication and fixed prices, they guaranteed producer losses with increased pollution.

2. *Losses to consumers are a significant portion of the total agricultural losses due to air pollution.* Of the studies that accounted for air pollution-induced changes in the prices of agricultural outputs, the percentage of total losses due to losses of consumer surplus ranged from 50 (Adams, Hamilton, and McCarl 1986) to 100 percent (Adams and McCarl 1985). Given that producers can sometimes benefit from air pollution increases, methods which disregard consumer impacts can understate total losses in surplus and grossly misstate the distribution of these welfare effects.

3. *As the air becomes more polluted, the percent changes in total economic surplus are less than the percent changes in biological yields that triggered these losses.* Any study that accounts for price effects of crop and input substitutions will obtain this result. The result is a direct outgrowth of microeconomic theory. Substitutions allow producers and consumers to attenuate the losses they would otherwise suffer from declines in air quality.

4. *Changes in air pollution affect both the productivity and the aggregate demand for factors of production.* Several studies have demonstrated that changes in air pollution will change the productivity of and the demand for specific inputs. Mjelde et al. (1984) estimated that a 10 percent increase in ozone in Illinois will result in a 4 percent decline in the demand for variable inputs such as labor, water, and fertilizer. Three other studies have shown that increased ozone increases the demand for land inputs: Howitt, Gossard, and Adams (1984); Kopp et al. (1985); and Brown and Smith (1984). Crocker and Horst (1981) found that ambient ozone conditions prevailing in southern California during the mid-1970s reduced the economic productivity of selected agricultural workers by an arithmetic mean of 2.2 percent. Some workers were unaffected, while others suffered productivity declines of more than 7 percent. These "input demand" effects were influenced by two interdependent

processes. The first process affects the marginal productivity of a given input. For example, if changes in air pollution are viewed as a neutral technological change, then reductions in air pollution will increase the productivity of all inputs proportionally. And as the productivity of agricultural inputs increase, growers become willing to pay more for the inputs. The second process that results in increased demand for some inputs refers to the "crowding out" phenomenon noted above. That is, reductions in air pollution will increase the yields of some crops and regions more than others. Consequently, the amount of acreage planted in the more productive crops or regions may expand. This will result in an increase in aggregate input use in these more favored crops and regions. Conversely, total crop acreage of crops or of regions that do not realize net relative productivity gains from pollution control may decrease, thus causing a reduction in aggregate input use. The exact nature of these input demand changes will be a function of the input mixes used for each crop or region and the relative effects of pollution changes within and across regions.

5. *Changes in air pollution have differential effects on the comparative advantage of agricultural production regions.* Several studies have focused on the consequences of air pollution across large geographical areas, (Adams and Crocker 1985; Kopp et al. 1985; and Adams, Hamilton, and McCarl 1986). To correctly represent aggregate supply response across large geographical regions, economists typically define the regions as consisting of distinct subregions representing different factors, such as crop production alternatives and resource and environmental conditions. A useful feature of such spatial equilibrium models is that economists can use them to measure the aggregate economic consequences of different policies as well as their effects on each subregion. The literature demonstrates that economic effects can vary sharply. For example, studies by Adams, Hamilton, and McCarl (1986) and Kopp et al. (1985) indicated that local producers benefited economically when air pollution was reduced in areas characterized by high ambient levels of pollution and pollution sensitive crops. In these areas, the percent gain in yields was greater than the overall reduction in agricultural prices nationwide, thus increasing regional net income. This lead to an expansion of crop acreage. Conversely, the studies showed that producers in areas with the opposite characteristics experienced reductions in revenues and acreages of some crops, and hence lost a "market share" of these crops. Thus, although these analyses showed a net gain to society from reductions in air pollution, they also found that some subregions gained at the expense of others.

6. *Air pollution alters international trade flows with attendant gains and losses to exporters and importers.* The economic gains from free trade are well recognized. International trade in agricultural commodities has profound effects on the welfare of consumers and producers in exporting and in importing countries. A few recent studies of U.S. agriculture incorporate a trade component into their models (Adams, Hamilton, and McCarl 1986; Shortle,

Dunn, and Phillips 1986). These studies found that reductions in air pollution altered the supply of export commodities because much of the increase in agricultural output was exported. The net effect was that foreign consumers of these commodities captured some of the consumer gains while foreign producers lost. To illustrate, Adams, Hamilton, and McCarl estimated that of the total consumer gains from ambient U.S. ozone reductions, 60 percent accrued to non-U.S. consumers. Conversely, increases in air pollution implied a reduction in the welfare of importing countries. As a result, national environmental policies can readily have economic transboundary implications even in the absence of transboundary pollution.

7. *Outside the annual agricultural crop sector, there is evidence that losses in surplus increase at a decreasing rather than an increasing rate over some intervals.* Manuel et al. (1981) and Watson and Jaksch (1982) noted that damages from household soiling were concave with respect to pollution. This finding has a psychophysical basis in Fechner's law, which states that the strength of a sensation is proportional to the negative of the logarithm of the stimulus. Perception thresholds for soiling are highly plausible, but they have not been incorporated in any published soiling study. Self-protection, in which potential sufferers act to reduce their pollution exposures to near-zero, also implies that there is an approximate upper bound to pollution damages. Moreover, Crocker and Forster (1981) and Repetto (1987) reviewed several examples of natural system dose-response functions in which the standard premise of nondecreasing marginal physical damages was clearly violated. The examples encompass a broad spectrum of effects, including effects on soils, aquatic ecosystems, forests, and atmospheric visibility. In fact, any natural system dose-response function with a damage threshold is likely to exhibit the largest marginal reduction in service flows in the immediate neighborhood of the threshold. Such thresholds are especially likely to exist wherever the natural system is able to protect itself. Many biological systems, such as fisheries, possess this property.

9.6 Areas for Future Research

As the preceding section indicates, existing empirical results capture only a portion of the plausible economic consequences of materials damages. Capturing the larger set of consequences will require more than an extension of existing analytical formulations to an intertemporal, sequential setting. Specifically, what are needed are formulations with substantially greater generality. This section identifies three neglected, but potentially significant, economic dimensions of materials damages. Development of appropriate analytical frameworks for each is an important research agenda.

9.6.1 The Environment as Wealth

In economic analysis, people's ability to achieve their objectives is restricted by their wealth. Environmental economists generally identify endowed wealth with perfectly fungible, that is, freely exchangeable, commodities to which individuals have open access and fully secure property claims. In a conventional benefit-cost analysis, this definition is likely to result in undervaluations of environmental assets when the appropriate measure of value for a harmful impact is the amount of compensation individuals must receive to keep their welfare intact (Knetsch 1984). Undervaluation occurs when individuals treat environmental commodities that are not freely exchangeable as an integral part of their wealth. Examples include access to a national forest or an urban public park. The economic manifestations of legal terms of access and ability to exclude others are important, but subtle, dimensions of this issue. For example, Merrill (1985) pointed out that the right to exclude others' intrusions can be defined in terms of nuisance or of trespass. Nuisance implies zero liability for the intruder ". . . absent a showing of substantial harm." With trespass, strict liability applies and the owner can obtain an injunction against the intruder. A nonactionable nuisance allows B to intrude offensively upon A; trespass allows A to exclude B's offending intrusions. The distributional and incentive features of the choice of the applicable law clearly differ. Nuisance law involves a "balancing of the equities"; trespass law does not.

Though people may not be able to trade these environmental commodities to prevent an undesired change, the loss of the commodities can affect the consumers' welfare as much as can any reduction in income or inventory of exchangeable commodities, as implied by the consumer analogue to equations (9.1) and (9.2).[6]

The following simple formulation demonstrates the place such commodities have in individual behavior. Let $U = U(Z,X)$ be a quasi-concave, twice differentiable utility function that is monotonically increasing and weakly separable in Z and X. Both Z and X are choice variables. The quantity Z represents the consumer's expenditures on goods or activities that are not sensitive to variations in the environment. Its price is normalized at unity. The quantity X refers to an activity that is environmentally sensitive; that is, the unit cost, c, of engaging in it increases with declines in environmental quality Q, $c' < 0$ (prime now indicating a derivative). Engagement costs include costs of access and of exclusion with respect to exchange possibilities

---

[6] Consider the following example from Crocker and Cummings (1985). The structures and the statuary of a public square provide an aesthetically pleasing backdrop for private musical performances and artwork. The backdrop reduces the artists' costs of attracting an audience of potential customers. Away from the square, the unit costs of attracting an audience exceed any artists' reservation price for producing art. In short, the art would not be supplied in the absence of the square. The structures and the statuary reduce the individual's costs of transacting and, by making markets less thin, expand trading opportunities. Thin markets reduce demands for goods, thereby discouraging production and raising holding costs (Clower and Howitt 1978).

and production possibilities. The utility function is maximized subject to a budget constraint

$$Z + c(Q) \cdot X + g(Q) \le y, \tag{9.11}$$

where y is income, and g, $g' > 0$, are any rents due to the capitalization of the material environment into any immobile and durable assets the individual uses. These rents may have nonpecuniary as well as pecuniary components. Both $c(Q)$ and $g(Q)$ are exogenous.

Utility maximization yields the constraint (9.11) and the following additional first-order conditions:

$$U_z - \lambda = 0, \tag{9.12}$$

$$U_x - \lambda c(Q) = 0 \tag{9.13}$$

$$g' = -Xc' \tag{9.14}$$

where a subscript represents a partial derivative. Expressions (9.12) and (9.13) together imply

$$U_x/U_z = c(Q), \tag{9.15}$$

which states that the consumer undertakes those amounts of environmentally insensitive expenditures Z and environmentally sensitive activities X such that the marginal utility per dollar spent is the same for all activities. Note from (9.13) that the shadow price λ of the sensitive activities is not a constant but is determined by the effect of the environment upon the cost of undertaking these activities. If there are multiple sensitive activities, it follows that each such activity will have its own shadow price reflecting differences in the ease of converting the activities into money, and hence, into the expected marginal utility of income. Moreover, since $c < 0$, a decline in environmental quality will increase the cost of engaging in sensitive activities, implying utility with a given amount of income will fall. These increases in c cause people to substitute toward commodities, the Z, that are not environmentally sensitive.

The point here is that although individuals cannot trade an entitlement to a nonfungible environmental commodity, its loss can still have a major impact upon the decisions they make based upon their income.[7] If the Z goods are perfectly fungible while the X goods or activities are imperfectly so, household budget constraints will vary according to the forms in which their initial entitlements were accumulated. People may be able to readily trade fungible goods for access to less fungible commodities, but the act of engaging in the opposite trade may be much more costly. The substitution effects of price changes are, therefore, asymmetrical, and price effects of entitlement changes can exist. For example, individuals can readily trade money in the form of

---

[7] Sen (1983, p. 755) states: "The power of the market force depends on relative prices and, as the price of some good rises, the hold of income on the corresponding entitlement weakens. With nonmarketability, it slips altogether."

travel cost and time to a public park, but they are unable to reverse the order so as to trade access for money. A loss in access would reduce their rents, necessitating, by expression (9.15), an increase in their expenditures on substitute environmentally sensitive goods. Because individuals' costs of engaging in trade differ according to the goods combination in their original entitlement, each combination will result in a different, generally nonlinear, budget constraint (Crocker 1986). From the individual's perspective, a dollar is thus not an invariant pecuniary measure: instead, its value depends on the form of an income change; that is, on the good and the entitlement in which the increment is embodied.

Given the nature of the material environment, the implication of this simple formulation for materials damage assessments is straightforward: declines in environmental quality are likely to impose direct costs on users beyond those recorded in current assessments. This notion of the environment as wealth would also apply to assessments of other categories of effects, such as recreation. Although the magnitude of this undervaluation is unclear, the likelihood of such undervaluation will increase in the presence of pervasive, chronic forms of environmental degradation, including most forms of air pollution and acid deposition.

9.6.2 Derived Demand

A common observation in popular, nontechnical literature on environmental issues is that the behaviors of ecosystems and economies are fused. Human economic activities modify the services provided by the ecosystem, and in turn, the responses of ecosystems affect individual and societal behaviors. Such reciprocal relations are an important dimension of some forms of materials damage, such as to forests or other perennials. Some of the technical literature in ecology and economics recognizes these reciprocal adaptations.[8] However, as Forster (1981) demonstrated, these formulations lump together many disparate organisms and materials and therefore do not generate propositions that economists can use to predict and evaluate which ecosystem services are to be usurped by humans, which waste flows are to be allowed in ecosystems, and which organisms are to be maintained. Arguably, the contributions of ecosystems to human well-being depend as much on the *details* of environmental and economic states as they do on the *gross relationships* between natural and economic sectors.

The need for a more detailed understanding of the ecosystem-economic linkages becomes apparent as economists recognize that many environmental changes have no immediate influence upon these components of an ecosystem that have direct, intrinsic commercial or utilitarian value. Rather, environ-

---

[8] d'Arge and Kogiku (1973) and Cropper (1976) are worthy early examples in economics. Hannon (1973) is the best early example in ecology.

mental changes influence the components that in turn nourish those components with intrinsic value. The "demand" for environmental quality conducive to these components (inputs) is thus derived from the value of the components with intrinsic value. For example, acid deposition is known to harm benthic organisms, which are food for freshwater recreational fish. In principle, the value of the benthic organisms is measured by their marginal product in producing the recreational fish stock. Though the point of examples like this is obvious, its analytical development and empirical implementation is likely to be complex, especially when the marginal physical product of the ecosystem is endogenous to the economic system. Building upon original work of Hannon (1976), Tschirhart and Crocker (1987) suggested a structure whereby concepts common to bioenergetics and economic models of choice may be used to develop a valuation framework that explicitly accounts for the reciprocal adaptations of ecological and economic systems.

The framework assumes that ecosystem components behave "as if" they maximize stored energy. A physiology function F_i, having the same properties as a production function in economics, describes the maximum stored energy R_i that the ith component can obtain from X_j kinds and biomasses of inputs, $i, j = 1, \ldots, n$. Solving for the first-order necessary conditions and applying the implicit function theorem yields species i's demand for the biomass of species j:

$$X_{ji}(\overline{E}, \overline{X}), \qquad i, j = 1, \ldots, n, \qquad\qquad (9.16)$$
$$i \neq j$$

where $\overline{E}$ is a vector of net energy content per unit biomass, and $\overline{X}$ is a vector of mass flows from the ith species to the jth species, while X_{ji} is a mass flow from the jth species to the ith species. In the absence of human intervention, the system will attain a "natural equilibrium" in which all species are maximizing their assimilated energy, all biomass markets are cleared, and an energy budget equation is satisfied.

When human beings use human labor L or capital to manage ecosystems, Tschirhart and Crocker (1987) showed that expression (9.17) becomes:

$$X_{ji}(\overline{E}, Eji, \overline{X}, \overline{L}) \qquad\qquad (9.17)$$

where $\overline{E}$ is the net energy content of the biomass flow from humans to the ith ecosystem component and $\overline{L}$ is human effort expenditure. In short, human decisions about effort influence the "as if" choice behavior of ecosystem components.

Humans allocate their labor inputs among manufactured goods and ecosystem management activities until the values of the marginal products of labor in each set of activities is equated. However, Tschirhart and Crocker (1987) showed that this value is in part determined by the "as if" choice behavior of ecosystem components. In short, the ecologist, in order to comprehend the behavior of ecosystems, must grasp the decision processes of humans. Similarly, the economist, in addition to recognizing that humans respond to environmental change, must also be aware that these responses influence the "as if" choices of ecosystem components.

Failure to recognize the link between environmental quality and the derived demand for ecosystem components will lead to incorrect estimates of materials damages from environmental degradations under at least two conditions. The first condition occurs when long lags exist between the reduced performance of an ecosystem component(s) and the output of the ecosystem component of value, such as commercial timber or long-lived fish species. The second condition occurs when an ecosystem's responses to environmental degradations are characterized by either threshold effects or discontinuities over some combination of inputs. Other conditions may also exist. The magnitude of the misestimates that these conditions create depends on the number of direct and cross-product relationships in the ecosystem production function.

9.6.3 Endogenous Risk[9]

With many types of prospective materials damages, consumers and producers must decide how much they are willing to spend to reduce the chances that an undesirable event will occur, and then if it does, how much they are willing to spend to mitigate its impact. For example, producers (and ultimately consumers) must weigh the costs of earthquake-proofing a building versus repairing it after an earthquake, or rust proofing an automobile versus *ex post* body work and painting. In general, the optimal allocation of scarce inputs between prevention and cure is unclear when it is doubtful if the desirable state can ever be realized. As noted earlier, the relevance of this risk problem to materials damages is obvious, but is not considered in the contemporary literature; that is, it is assumed that economic agents are helpless when confronted by risk. The literature, as in Lareau et al. (1986), assumes that at best, consumers can do little more than expend scarce resources to return damaged goods to a semblance of their former state. The consumer is assumed to lack the ability to anticipate these damages.

Marshall (1976) showed that exogenous risk requires a complete set of Arrow-Debreu contingent claims contracts. Because writing a contract is costly, complete contracts rarely, if ever, exist; therefore, individuals must choose either to contractually define states of nature or to make an effort to alter them. Although it is always possible to redefine a problem so that the state of nature is independent of human actions, the redefinition frequently will be economically irrelevant. Erlich and Becker (1972) asked the reader to consider the probability that a bolt of lighting will burn down a house. The probability of this event will be altered if the owner places a lightning rod

---

[9] Jason Shogren, Iowa State University, is a co-author of this subsection.

upon the roof. In this example, the state of the world can be redefined to be independent of the owner's actions if the event is thought of in terms of the probability of lightning striking the house. The owner has no control over the probability of a strike. However, this probability is not economically relevant. The probability that is important to the owner is whether or not the house will burn when struck by lightning. The owner is able to exercise some control over this event.

Another example of the implications of risk concerns the decisions individuals make to protect their welfare when they learn that one of their valuable personal assets, such as their house, may soon be exposed to pollution. Assume that they cannot move for the period in question and that they face a particular liability regime. In this instance, a dilemma arises because the individuals' expenditures for self-protection reduce the likelihood and the severity of damage. Hence, the individuals' costs for *ex post* damages will also reduce their *ex ante* personal consumption. And because of adverse selection, moral hazard, and nonindependence of risks, the individuals either will not or cannot acquire enough market insurance to completely avoid the dilemma. Given their insurance purchases and given that their utility is intemporally separable, a minimal formulation appropriate to most atemporal materials damages problems is:

$$\max_{x,s} \text{E} U - \int_a^b U(x,Q)f(Q;s,A)\mathrm{d}Q\,|\,x + s + C(Q;s,A) \le y, \qquad (9.18)$$

where E is the expectations operator and U is a von Neumann-Morgernstern utility index.

Expression (9.18) says that given a full income, y, individuals' decision problems are to choose that combination of expenditures on personal consumption x and on self-protection s that maximize their expected utility. Their probability-weighted utility is a function of their personal consumption and environmental quality Q where $U_x > 0$, $U_Q > 0$, $U_{xx} < 0$, and $U_{QQ} < 0$. Subscripts refer to partial derivatives.

The probability weights in (9.18) are represented by a symmetrical subjective probability density function f defined over the minimum a and the maximum b environmental quality that the natural and the developmental history of the asset allows. Presume that the interval $[a,b]$ is independent of self-protection. Let $F_s > 0$ and $F_A < 0$. Although individuals acting alone may be unable to influence the extent of pollution, they may use self-protection to reduce their exposure, thus influencing the cumulative probability distribution $F(\cdot)$ of asset-specific environmental quality states. The probability density function of asset states or exposures is dependent upon self-supplied protection s from prospective pollution A. No restrictions need be placed on the signs of F_{ss}, F_{AA}, and F_{sA} in the immediate neighborhood of the expected utility maximizing level of self-protection, s^*.

For each environmental quality outcome that individuals realize, they select a minimum cost combination of *ex post* curative expenditures. Their *ex ante* efforts to protect themselves from the pollution will influence these costs C such that $C_s < 0$, $C_A > 0$, and $C_{ss} > 0$. The signs of C_{AA} and C_{sA} have no restrictions. The absence of signs for F_{sA} and C_{sA} reflects the possibility that these responses depend on the environmental concentration (quality) of contamination as well as the extent to which individuals choose to reduce their exposures.

The implications of simultaneously accounting for risk, the severity of *ex post* effects (where severity is measured as a continuum rather than binary event) and the relationship between *ex ante* protection and the cost of *ex post* cures have been explored elsewhere (Shogren and Crocker forthcoming). This formulation extends the intertemporal dimensions of materials damages as represented in equation (9.3), into a formal risk setting. The purpose here is simply to suggest that in the presence of incomplete contingent claims markets, traditional evaluations of the economic consequences of materials damages have neglected large chunks of economic reality. Consideration of this matter may have significant consequences for economic measures of materials damages.

9.7 Conclusions

This chapter has focused on issues in estimating the materials consequences of environmental change. Specific empirical studies of the effects of air pollution on agricultural crops, forests, and building materials substantiate the importance of these issues. The studies also conveyed a general sense of the magnitudes of such damages. However, much of the empirical and analytical work necessary to obtain a complete picture of the economic consequences of materials damages remains to be done.

Although this chapter has identified shortcomings in current economic evaluations, the studies are precise enough for some effects categories to distinguish among the consequences of most policy options. Precision here refers to the statistical variability associated with those estimates; that is, the less variable the estimates are, the greater the precision. Precision in these economic estimates derives from the underlying precision of biological or physical science parameters, which in turn are driven by the quality of the biological and physical science data used in most economic evaluations. Precision is enhanced, and conversely, variability is reduced, by the addition of observations or data on these estimated relationships. As Adams, Crocker, and Katz (1984) have shown, the precision in existing estimates of agricultural crop damage is sufficient to discriminate between some alternative regulatory policies; that is, the underlying crop response-ozone pollution data are sufficient to answer economic policy questions. It is unclear, however, whether this conclusion applies to other categories of effects, such as forestry, where biologic data are ambiguous.

The precision claimed for some existing studies does not guarantee that these estimates are accurate or complete. Current assessments typically ignore many dimensions on specific materials effects. Inadequate model formulations may lead to inaccurate estimates of materials damage. Furthermore, the current economic assessments tend not to include the full set of materials consequences associated with environmental change. This implies an under-valuation of such economic effects. Future research directed at refining and extending some of the model formulations discussed here thus seems worth-while.

Measuring the Demand for Environmental Quality
John B. Braden & Charles D. Kolstad (Editors)
© Elsevier Science Publishers B.V. (North-Holland), 1991

Chapter X

TOTAL AND NONUSE VALUES

ALAN RANDALL

Ohio State University

10.1 Introduction

The idea of nonuse benefits is motivated by a concern that even after all of the various benefits associated with using an environmental amenity have been estimated and entered into the benefit calculation, something important might be missed. At first glance, the idea of benefits without explicit, observable use seems radical, just as did the earlier notion that use benefits could be unaccompanied by direct and observable commercial transactions. However, both of these ideas are well within the purview of the standard economic model of benefits. What counts as a benefit is delimited by neither the act of use nor the commercial transaction. Rather, an action has a *prima facie* economic benefits if an action increases the availability of something that is scarce at the margin and if that "something" is desired by someone; that is, if it is at least potentially a source of human utility.

As for what might be overlooked in a benefit account confined to use values, there has been no shortage of candidates. Those that are still frequently proposed include existence value, first suggested by Krutilla (1967), vicarious use value (Krutilla 1967), option value (Weisbrod 1964), and quasi-option value (Arrow and Fisher 1974; Henry 1974). Each of these nonuse value concepts was introduced to deal with a legitimate concern. *Existence value* addresses the idea that, in Krutilla's much-quoted example, there are people who value the existence of wilderness even though they would be appalled by the prospect of exposure to it. *Vicarious use value* addresses the possibility that people who cannot visit unusual environments nevertheless gain pleasure from pictures, broadcasts, written accounts, and so on, of these places. Although Randall and Stoll (1983) argued that vicarious use value is a form of use value, they conceded that it is difficult to distinguish by observation from existence values. *Option value* is applied to circumstances in which individuals may be willing to pay a premium to ensure the future availability of an amenity. This scenario, in which significant option value emerges for environmental amenities, is analogous to purchased options in real estate and financial instruments. *Quasi-option value* concerns unique natural terrain, species, and ecosystems, and their susceptibility to irreversible change. Some economists argue that a complete account of preservation benefits must include a benefit associated with the possibility that future discoveries may make fragile natural resources more or less valuable than they seem to be now.

The economic literature on nonuse benefits is almost a quarter-century old and has become quite voluminous. Unfortunately, from the practitioner's perspective, this literature was contentious almost from the beginning. New controversies flare-up faster than old issues are resolved. Economists debate the validity of important categories of nonuse value, as well as the relevance of others. They differ on how the various categories of use and nonuse value may add up to the total benefits of a resource service or amenity. Furthermore, their answer to a major question for practitioners — whether the analyst is better advised to estimate total value directly or by aggregating estimates of all the various components of use and nonuse value — depends on which side they take in the controversy about the merits of alternative estimation methods.

The intent of this chapter is first, to provide a theoretical basis for conceptualizing total economic value and its nonuse components, and second, to suggest some pragmatic methods for estimating benefits. Both objectives are essential because pragmatic strategies for benefit estimation must be built on valid conceptual foundations.

10.2 A Total Value Framework

The first step in laying this conceptual foundation is developing a simple model of total economic value in a deterministic framework. This model has important implications: total economic value is uniquely defined; it decomposes readily into existence value and several categories of use value; and in general, it misstates total value if the independently-estimated values for the benefit categories are simply summed. When uncertainty is introduced into the model, all benefit values become *ex ante*; that is, they represent the individual's expected benefits based on what is known at the time valuation takes place rather than in retrospect after she has experienced the consequences of her choice. A decomposition of *ex ante* total economic value reveals no new categories of value. Instead, *ex ante* existence value and various kinds of *ex ante* use value emerge. Furthermore, the concepts underlying *ex ante* use value, as a concept, adequately addresses the legitimate concerns that led to proposals that option value and quasi-option value be included in benefit accounts.

An illustration of the problem of valuation will be helpful before formally modeling total economic value. The Tongass National Forest covers approximately twenty-seven thousand square miles and accounts for more than 97 percent of the land area in southeastern Alaska. In and near Tongass the diversity of landscape, biota, and natural-resource-based human activities, both commercial and nonmarket, is unusually rich. Some sixty-five thousand residents and several hundred thousand annual visitors enjoy the services and amenities produced on Tongass. Due to its topography, southeastern Alaska has few roads connecting it with the rest of North America, and as a result, most visitors arrive by sea or air. Visits, therefore, typically are expensive, have several destinations, and include several activities.

To develop a total value framework for the environmental resources of such a place, it makes sense to consider three mutually exclusive classes of individuals who may value these resources: residents of southeastern Alaska, visitors to southeastern Alaska, and nonresident nonvisitors. All three classes of individuals would have nonuse values. For example, they all might value the continued existence of a nondegraded Tongass environment independently of any concern about actually using it. Residents and visitors would have, in addition, values deriving from prospective use.

It is customary for resource management agencies to conceptualize use values in terms of values derived from particular activities: for example, motorized water-based recreation and hunting are among the activities recognized in the Tongass National Forest. In addition to these specific activities in which both residents and visitors may participate, there is also a more amorphous kind of use. To give just two examples, every southeastern Alaska resident lives in an environment dominated by the Tongass National Forest in terms of scenery, wildlife, air and water quality, and so on. Therefore, they may well experience a positive residential value independent of participation in any specific activities. Similarly, visitors who arrive on the many cruise ships that travel the inner navigation passage can enjoy being immersed in an unusual environment, often without participating in any explicit activities within the forest. For a situation such as the Tongass National Forest, which is among the more complex valuation problems likely to be encountered, a total valuation framework should include existence values and two kinds of use values: site experience values and the values attributable to explicit activities.

10.2.1 Total Value in a Deterministic Context

Consider an individual with the utility function

$$u = u(x_e, x_s, x_1, x_2, Q, z) \tag{10.1}$$

where x is environmental services, Q is the state or condition of the environment, and z is a vector of ordinary goods and services. The subscripts denote, respectively: e, existence; s, site experience; 1, activity 1; and 2, activity 2. At this stage, x_e, x_s, x_1, and x_2 are each single elements; however, it would be simple to extend the analysis to consider a vector of existence services, a vector of site experience services, and n activities rather than just two.

The solution to the problem

$$\min \ (p_e x_e + p_s x_s + p_1 x_1 + p_2 x_2 + pz) \tag{10.2}$$
$$\text{s.t.} \ \ u(\cdot) \geq u^o$$

is the expenditure function

$$e = e(p_e, p_s, p_1, p_2, Q, p, u^o). \tag{10.3}$$

If one assumes, purely for notational convenience, that prices of ordinary goods and services are exogenous, the expenditure function can be written with p implicit: $e(p_e, p_s, p_1, p_2, Q, u^o)$.

The total value (TV) of environmental services is then defined, in terms of Hicksian compensating welfare change measures, as

$$\text{TV} = e(p_e^*, p_s^*, p_1^*, p_2^*, Q^o, u^o) - e(p_e^o, p_s^o, p_1^o, p_2^o, Q^o, u^o) \tag{10.4}$$

where p^* is a choke price (that is, a price so high that quantity demanded is zero), p^o is a baseline price, and Q^o is the baseline level of resource quality. Existence services generally are unpriced: thus, p_e is the virtual price of x_e. Virtual price, which is synonymous with shadow price, is defined as the price that would be efficient where efficient price is absent; that is, virtual price would clear efficient markets if they existed. The choke price p_e^* indicates a situation where existence services are not provided. When other kinds of environmental services are unpriced, for institutional or other reasons, the virtual price interpretation of price also applies.

Expression (10.4) suggest a one-shot, or holistic, measure of total value. However, it is possible to break down total value into its components, and to provide a common sense interpretation of each component.

$$\text{TV} = \quad e(p_e^*, p_s^*, p_1^*, p_2^*, Q^o, u^o) - e(p_e^o, p_s^*, p_1^*, p_2^*, Q^o, u^o) \tag{10.5a}$$
$$+ \ e(p_e^o, p_s^*, p_1^*, p_2^*, Q^o, u^o) - e(p_e^o, p_s^o, p_1^*, p_2^*, Q^o, u^o) \tag{10.5b}$$
$$+ \ e(p_e^o, p_s^o, p_1^*, p_2^*, Q^o, u^o) - e(p_e^o, p_s^o, p_1^o, p_2^*, Q^o, u^o) \tag{10.5c}$$
$$+ \ e(p_e^o, p_s^o, p_1^o, p_2^*, Q^o, u^o) - e(p_e^o, p_s^o, p_1^o, p_2^o, Q^o, u^o) \tag{10.5d}$$

Equation (10.5) suggests a sequented valuation that captures total value as the sum of various value components. Of the many possible valuation sequences, one particular sequence is chosen and elaborated here. Line (10.5a) defines existence value, (10.5b) defines site experience value, and (10.5c) and (10.5d) define the activity values for activities 1 and 2, respectively. Alternatively, line (10.5a) identifies existence value, while the subsequent lines define particular kinds of use value. Total value is, in a deterministic framework, the sum of existence and use values evaluated sequentially. The conventional wisdom, that nonuse value is the residual between total value and use value, is seen to be true but only with the proviso that all value components are evaluated in some valid sequence.

In contrast to the holistic or sequential TV frameworks, there is a common procedure that calculates total value by aggregating component values estimated independently. This procedure is denoted *independent valuation and summation* (IVS). The IVS value measure analogous to (10.5) is

$$\text{IVS} = \quad e(p_e^*,p_s^*,p_1^*,p_2^*,Q^o,u^o) - e(p_e^o,p_s^*,p_1^*,p_2^*,Q^o,u^o) \tag{10.6a}$$

$$+ \; e(p_e^*,p_s^*,p_1^*,p_2^*,Q^o,u^o) - e(p_e^*,p_s^o,p_1^*,p_2^*,Q^o,u^o) \tag{10.6b}$$

$$+ \; e(p_e^*,p_s^*,p_1^*,p_2^*,Q^o,u^o) - e(p_e^*,p_s^*,p_1^o,p_2^*,Q^o,u^o) \tag{10.6c}$$

$$+ \; e(p_e^*,p_s^*,p_1^*,p_2^*,Q^o,u^o) - e(p_e^*,p_s^*,p_1^*,p_2^o,Q^o,u^o) \tag{10.6d}$$

Existence value and each of the use values is evaluated independently of the others, which is not consistent with (10.5).

Hoehn and Randall (1989) demonstrated that, in general, IVS $\neq$ TV; this result is not unexpected, given that the consumer surplus value of a bundle of goods is generally not the same as the sum of consumer surpluses for individual goods. Furthermore, they showed that TV is unique, whether defined in holistic (10.4) or sequential (10.5) terms; and each component value is not unique but depends on its position in the valuation sequence. As the number of components becomes large, the error in IVS becomes systematic: IVS overestimates the benefits of policy change, and non-net-beneficial policy changes may be misidentified as net beneficial.

Now, consider some proposed policy that would affect the condition of the environment, changing Q^o to Q^1. The total value for the policy change is

$$\text{TV}(Q^1;Q^o) = e(p_e^*,p_s^*,p_1^*,p_2^*,Q^1,u^o) - e(p_e^o,p_s^o,p_1^o,p_2^o,Q^1,u^o)$$
$$- \; [e(p_e^*,p_s^*,p_1^*,p_2^*,Q^o,u^o) - e(p_e^o,p_s^o,p_1^o,p_2^o,Q^o,u^o)] \tag{10.7}$$
$$= e(p_e^o,p_s^o,p_1^o,p_2^o,Q^o,u^o) - e(p_e^o,p_s^o,p_1^o,p_2^o,Q^1,u^o).$$

This expression may be decomposed:

$$\text{TV}(Q^1;Q^o) = \quad e(p_e^o,p_s^*,p_1^*,p_2^*,Q^o,u^o) - e(p_e^o,p_s^*,p_1^*,p_2^*,Q^1,u^o) \tag{10.8a}$$

$$+ \; e(p_e^o,p_s^*,p_1^*,p_2^*,Q^1,u^o) - e(p_e^o,p_s^o,p_1^*,p_2^*,Q^1,u^o) \tag{10.8b}$$
$$- \; [e(p_e^o,p_s^*,p_1^*,p_2^*,Q^o,u^o) - e(p_e^o,p_s^o,p_1^*,p_2^*,Q^o,u^o)]$$

$$+ \; e(p_e^o,p_s^o,p_1^*,p_2^*,Q^1,u^o) - e(p_e^o,p_s^o,p_1^o,p_2^*,Q^1,u^o) \tag{10.8c}$$
$$- \; [e(p_e^o,p_s^o,p_1^*,p_2^*,Q^o,u^o) - e(p_e^o,p_s^o,p_1^o,p_2^*,Q^o,u^o)]$$

$$+ \; e(p_e^o,p_s^o,p_1^o,p_2^*,Q^1,u^o) - e(p_e^o,p_s^o,p_1^o,p_2^o,Q^1,u^o) \tag{10.8d}$$
$$- \; [e(p_e^o,p_s^o,p_1^o,p_2^*,Q^o,u^o) - e(p_e^o,p_s^o,p_1^o,p_2^o,Q^o,u^o)]$$

Expression (10.7) defines a holistic total valuation of the proposed policy change, whereas (10.8) defines a sequential total valuation. Again, total value is unique. Expression (10.8a) identifies the change in existence value attributable to the policy change, whereas (10.8b) through (10.8d) define, sequentially, the changes in various components of use value. It is, of course, possible also to define an IVS measure of the total value for the policy change, but such a measure would be misleading in general, and systematically upward biased for policies with many components (Hoehn and Randall 1989).

Most applied benefit-cost analyses evaluate proposed policy changes, and therefore expressions in (10.7) and (10.8) define the appropriate value measures. Nevertheless, for notational brevity and consistency, all further definitions and analyses will address the baseline valuation case as defined in (10.4) and (10.5).

10.2.2 Total Valuation Under Uncertainty

Uncertainty is introduced in two steps. First, define J possible states of the world, S_j, each independent of policy. Thus, S is a J-element vector, and S_j has probability π_j, such that $\Sigma_{j=1}^{J} \pi_j = 1$. Each category of environmental services is now a J-element vector of contingent commodities, each with its contingent price or virtual price. Thus, these are now J-element vectors: x_e, x_s, x_1, x_2, p_e, p_s, p_1, and p_2. This formulation permits policy-independent uncertainty that may affect the benefits of environmental services.

Second, uncertainty may pertain to baseline Q^o or to the effects of policy on Q^1. So, let q_k be the probability that the environment is in some condition Q_k. Thus, Q is a K-element vector of states of the environment and q is a K-element vector of probabilities, $\Sigma_{k=1}^{K} q_k = 1$. Uncertainty with respect to Q addresses what the standard literature calls *supply uncertainty*. In contrast to the tone of most of the option value and quasi-option value literature, neither the passage of time nor the introduction of policy necessarily decreases or eliminates uncertainty. New information emerging with time may increase uncertainty about states of the world, S, and a new policy may make Q^1 more uncertain than Q^o.

To define total value under uncertainty, economists use the *ex ante* or *planned expenditure function* (Helms 1985; Simmons 1984; Smith 1987a) *circa* equation (2.37): $\tilde{e} = \tilde{e}(p_e, p_s, p_1, p_2, \pi, q, EU)$, where π and q are independent and EU is expected utility.

Ex ante total value for the baseline policy (or environmental condition) is

$$\text{TV} = \tilde{e}(p_e^*, p_s^*, p_1^*, p_2^*, \pi^o, q^o, EU^o) - \tilde{e}(p_e^o, p_s^o, p_1^o, p_2^o, \pi^o, q^o, EU^o), \tag{10.9}$$

and

$$\begin{aligned}
\text{TV} = &\ \tilde{e}(p_e^*, p_s^*, p_1^*, {}_2, \pi^o, q^o, EU^o) &- \tilde{e}(p_e^o, p_s^*, p_1^*, p_2^*, \pi^o, q^o, EU^o) &\tag{10.10a} \\
&+ \tilde{e}(p_e^o, p_s^*, p_1^*, p_2^*, \pi^o, q^o, EU^o) &- \tilde{e}(p_e^o, p_s^o, p_1^*, p_2^*, \pi^o, q^o, EU^o) &\tag{10.10b} \\
&+ \tilde{e}(p_e^o, p_s^o, p_1^*, p_2^*, \pi^o, q^o, EU^o) &- \tilde{e}(p_e^o, p_s^o, p_1^o, p_2^*, \pi^o, q^o, EU^o) &\tag{10.10c} \\
&+ e(p_e^o, p_s^o, p_1^o, p_2^*, \pi^o, q^o, EU^o) &- \tilde{e}(p_e^o, p_s^o, p_1^o, p_2^o, \pi^o, q^o, EU^o) &\tag{10.10d}
\end{aligned}$$

Observe how closely the total valuation framework under uncertainty parallels the framework for the deterministic situation. Either a holistic valuation (10.9) or a sequenced valuation (10.10) may be specified. Furthermore, the categories that emerge from the sequenced valuation under uncertainty (10.10) are the same as in the deterministic case: that is, existence value, site experience value, activity-1 value and activity-2 value. In this general total value formulation for the uncertainty case, no new categories of value emerge (see also Smith 1987a): there are no categories for demand-side option value, supply-side option value, or quasi-option value.

Uncertainty causes the emergence of no new components of value. Rather, it now permeates all of the value categories that were present in the deterministic case. Holistic total value and all of the value components are now *ex ante*, based on the planned expenditure function and dependent on the probabilities associated with the policy-independent states of the world (π) and the impacts of policy on the environment (q). Hicksian compensation tests are performed subject to maintaining the initial level of expected utility.

It is important to explain how the total value framework — which includes no terms for the traditional categories of supply-side option value, demand-side option value, and quasi-option value — nevertheless addresses the concerns that initiated the introduction of these terms into the standard literature.

Supply Uncertainty. Supply uncertainty is modeled by assuming that Q is a K-element vector of the states of the environment and q is a K-element vector of probabilities associated with these states. Uncertainty would be eliminated if one particular q_k were to be set equal to 1 and all q_h, if $h \neq k$, equal to 0. The earliest literature on supply uncertainty assumed that a policy action could be taken to change Q^o to Q^1_k; that is, to eliminate baseline uncertainty by taking action to ensure a unique post-policy environmental condition, Q^1_k. The model permits valuation of such a proposed policy.

More generally, both Q^o and Q^1 can be uncertain, and either can be "more uncertain" (that is, the probabilities associated with various elements of either the Q^o or Q^1 vector can be more dispersed) than the other. Policy may amplify rather than reduce uncertainty. The model permits empirical analysis of all the various possibilities.

Demand Uncertainty. One valid interpretation of S and π is that, independent of policy, demand for environmental services might be uncertain. Again, the uncertainty would be eliminated if one π_j took the value of 1 and all π_i, $i \neq j$, were set at 0. This appears to represent the traditional formulation of demand uncertainty, which states that if the potential user is required to enter into a use contract in advance, or if she can choose to enter into such a contract to reduce supply uncertainty, then uncertainty about what her demand will be when the time of use arrives may reduce her maximum offer for the advance contract. This uncertainty about future demand could arise from incomplete knowledge of one's own future consumption technology or income and from uncertainty about the value of demand shifters, such as the weather, when future use becomes available. The *ex ante* total value framework allows for these various forms of uncertainty.

Quasi-Option Value. Scenarios from which positive quasi-option value is said to emerge have something of a special-case quality. Within these scenarios there are two alternatives: development, $d = 1$, and preservation, $d = 0$; and two time periods: $t = 1$, and $t = 2$. Development is irreversible, such that $d_1 = 0$ permits d_2 to take either value, 0 or 1; whereas $d_1 = 1$ permits only $d_2 = 1$. If there is a chance that new information will emerge after $t = 1$ but before $t = 2$, information that will shift rightward the demand for services produced by $d_2 = 0$ but not by $d_2 = 1$, then this possibility generates a quasi-option value that should be counted as a benefit of choosing $d_1 = 0$. The typical example concerns period-1 species preservation, which might permit future benefits from as yet undiscovered uses of a species. The theoretical developments and numerical examples in the literature typically assume discrete time periods, discrete choices ($d = 1$ or $d = 0$), and strict irreversibility. Without these assumptions, the analysis becomes untidy and the theoretical results are more ambiguous (Fisher and Hanemann 1986).

A more general analysis of quasi-option value within this framework would allow: (1) reversion ($d_1 = 1$ and $d_2 = 0$) at some cost, which might range from quite small to very large, depending on the particular case; (2) d to be continuous in the range $0 \le d \le 1$; and (3) development (as well as preservation) to serve as a precondition for the usefulness of emerging information that might shift later-period demands rightward (or leftward).

What modifications would it take, for the total value framework to address quasi-option value? The only significant change necessary would be to extend the time frame to allow a time sequence of decision points, and to allow the earlier-period decisions to affect the costs attached to the later-period alternatives. Information emerging over time, independently of policy, would change π_t as t increases. However, prior policy decisions may render some kinds of new information less useful, by reducing the scope for later-period decisions that benefit from the new information. In concept, the necessary modifications to the *ex ante* total value framework are straightforward, but in execution they may be quite messy. That is the problem Fisher and Hanemann found when attempting to generalize their theory of quasi-option value.

10.3 Strategies for Estimating Total Value and Nonuse Values

For nonmarket valuation, the standard techniques are contingent valuation (chapter 5), the household production/travel cost method (chapter 3), and hedonic price analysis (chapter 4). In designing the research plan for any particular empirical study, the researchers must answer a number of questions.

First, do they want to start from scratch to design and execute a study that will estimate either total value holistically or total and component values in a valid piecewise sequential valuation framework? If so, they will be pushed toward contingent valuation. And, is it important to include in their research design some opportunities for using travel cost or hedonic methods to estimate use values? And last, as studies designed from scratch tend to be expensive, do they want to use at least some "typical" unit values that studies such as that of Sorg and Loomis (1984) have compiled? If their answer to either of the last two questions is positive, then they must resolve three additional problems.

The first concerns the forms of the values. Component values that are estimated by travel cost or hedonic methods or are gleaned from compilations of "typical" unit values, are usually in IVS form. That is, for the example of activity 1, such values usually conform to (10.6c) rather than (10.5c). To use them within a valid total value framework, the researchers would need to find a method of approximating a valid piecewise valuation using independent piecewise value estimates.

Second, an eclectic valuation strategy of using different methods (and perhaps even different teams of researchers) to estimate the various value components exposes the overall research effort to the problems of multiple-counting and its obverse, omissions. The discipline of the total value framework is absent and in its place the researcher must substitute painstaking examinations for consistency: for example, the researcher must determine if the existence value estimate includes some value that should be attributed to use value or option value, or if any important sources of value been omitted.

Third, when uncertainty is a concern, the total value framework calls for *ex ante* values (see equations 10.9 and 10.10), but travel cost, hedonic, and "typical" unit values are usually considered *ex post*; that is, values estimated from decisions made after the uncertainty has been resolved. An eclectic strategy for valuation under uncertainty seems to require procedures for translating between *ex ante* and *ex post* values.

The remainder of this section will elaborate on these issues, and so far as possible, resolve them.

10.3.1 Contingent Valuation Approaches

The total value framework developed in the previous section permits economists to measure total value either directly and holistically or piecewise and sequentially. Total value should be unique, whether measured holistically or sequentially. The components of total value will include existence value and various kinds of use value, and if estimates of particular component values are desired, a sequential piecewise estimation strategy may be used. The magnitudes of particular component values will depend on their places in the valuation sequence, which may be a problem when there is no reason to chose any particular sequence. If, instead of a piecewise sequential strategy, a piecewise independent estimation strategy was chosen, the estimates obtained would be misleading with respect to both total value and component values. The introduction of uncertainty changes none of these conclusions. Total value and its components are all defined in *ex ante* terms and measured with respect to a baseline level of expected utility. These modifications to the deterministic framework provide a valid response to the problem of uncertainty; it is unnecessary to introduce new components of value, such as option value and quasi-option value.

Total value and its components can be estimated *de novo* (for example, in an exercise that starts from scratch) via the contingent valuation method (CVM). Scenarios can be constructed to elicit total value holistically, or total value and component values in a valid piecewise sequence. The match between the total value framework and CVM is very close: the total value framework has a sound intuitive appeal; CVM is flexible in that it permits the valuation of a wide variety of plausible constructed policy scenarios, and CVM scenarios constructed to implement a total value approach, are as plausible and intuitive as the total value framework.

To implement CVM in a total value framework, all of the standard considerations and caveats pertinent to CVM (chapter 5) apply. The focus here is on constructing scenarios. For holistic total value, the scenario should follow the logic of equation (10.4) to estimate baseline value and equation (10.7) to evaluate a proposed policy change under certainty, and equation (10.9) to estimate baseline value under uncertainty. A CVM scenario based on equation (10.4) requires the respondent to compare two situations: $(p_e^o, p_s^o, p_1^o, p_2^o, Q^o)$, in which the particular environment exists, and site experience and the various activities are available at the current quality levels and prices; and $(p_e^*, p_s^*, p_1^*, p_2^*, Q^o)$, in which the environment does not exist and hence is unavailable for use. A Hicksian compensating approach to valuation would suggest questions designed to elicit — directly or by binary choice, perhaps in referendum format — willingness to accept compensation (WTA) to permit $(p_e^*, p_s^*, p_1^*, p_2^*, Q^o)$ given a reference situation of $(p_e^o, p_s^o, p_1^o, p_2^o, Q^o)$. A Hicksian equivalent approach would suggest eliciting WTP to avoid the less-desired situation.

Scenario construction for eliciting the holistic total value of a proposed policy (10.7), is even simpler because the environmental quality level changes, whereas the prices do not. The scenario must emphasize that existence and the opportunity for site experience and the various activities are to be valued simultaneously. If a proposed policy would change prices and quality levels, this would need to be reflected in the scenarios. To elicit *ex ante* holistic value under uncertainty (10.9), the basic scenarios for (10.4) must be modified to communicate the uncertainties that exist with respect to demand and to environmental quality. Recent developments in communicating risk and uncertainty in CVM are reviewed in chapter 5.

For valid piecewise total and component values, the researcher must develop scenarios that conform to, for example, equations (10.5), (10.8), and (10.10). As an example, consider (10.8). First, choose a valuation sequence. Intuition suggests that existence comes first since it is a prerequisite for any kind of use. (Previous studies have often treated existence value as an afterthought, perhaps worth considering after the use values have been estimated.) Similarly, intuition suggests that site experience has a claim to second position. The specialized use activities (hiking, boating, sightseeing, and so on) must also be sequenced, but there may be no intuitive reason to prefer any particular sequence.

Once a sequence has been chosen, as in (10.8), scenario construction proceeds according to the logic of sequential valuation. For $Q^1 > Q^o$, the first step is to elicit the value (again directly or via a binary choice) of the improvement from Q^o to Q^1 given that site experience and use activities are unavailable (Bennett 1984). The second step is to elicit the value of the Q^o to Q^1 upgrade in site experience given that the existence upgrade has already been secured and paid for. For example, a researcher might ask a respondent: Now, how much more would (the upgrade to Q^1) be worth to you if you were free to visit the site at a cost of p_e^o but not to engage in any special activities like hiking or boating? The third through nth steps (for an n-item valuation sequence) introduce an activity upgraded to Q^1 assuming all previous upgraded components have been secured and paid for. In general, a valid valuation requires that the respondent determine his or her valuation while considering fully the array of substitutes, complements, and alternatives, as well as budget constraints. For example, when he or she is considering existence values for particular species or for small areas with unusual ecosystems, a multitude of alternative potential projects may compete for (contingent) payments. CVM scenarios should communicate to the respondent that the proposed project or policy is just one of many potential efforts to preserve specics or natural areas.

Clearly, researchers can obtain valid estimates of total value, existence value, and various kinds of use value (in a deterministic or uncertain context) by conducting, *de novo,* a CVM exercise. However, research sponsers and some analysts could object to a valuation research strategy that was confined to *de novo* CVM exercises. They could argue that a *de novo* effort is not always essential, and that there are circumstances in which they could adapt value data that had been assembled for some other valuation purpose. An insistence on *de novo* valuation efforts would unnecessarily increase the costs of routine benefit estimation. They could also argue that methods for obtaining value data other than CVM — for example, revealed preference methods, such as hedonics or travel cost — could be used in ways that approximate a valid piecewise sequential valuation framework. It is difficult, however, to see how these methods could capture existence value or total value holistically. To those practitioners who prefer revealed preference methods to CVM, and to those who believe that a sound benefit estimation strategy should include tests for cross-technique consistency in the estimates for at least some components of total value, it is important to find a legitimate role for the methods in a valid piecewise valuation strategy.

Are De Novo Value Estimates Essential? *De novo* valuation efforts can be designed to implement the valid holistic valuation framework (10.4) and the piecewise sequential framework (10.5). However, many agencies that routinely undertake benefit-cost analysis have invested substantial effort in assembling an inventory of "typical" unit values for various beneficial uses.[1] The idea is that the up-front costs of carefully assembling these "typical" unit values can be recouped by using them repeatedly in routine benefit-cost analysis.

The objection to this procedure is that it usually implies an IVS, or piecewise independent, valuation procedure (10.6) that Hoehn and Randall (1989) showed produces misleading estimates of total value and component values. The problem is that an IVS valuation procedure does not take proper account of resource scarcity and the interactions, such as substitution or complementarity among various kinds of environmental services. Still, given the considerable savings in research costs that might result from using "typical" unit value estimates taken "off the shelf," it is reasonable for researchers to try to devise methods that approximate the valid valuations using "shelf" data.

Randall, Hoehn, and Sorg Swanson (1990) and Hoehn (1989) have suggested methods for approximating the valid total and component values while economizing on the need for *de novo* value estimates. Hoehn started with a set of CVM values for a three-component policy in which each component could be implemented at one of three levels (the baseline and two alternative levels). He used this data to estimate a multidimensional value function, using quadratic and translog functional forms. He was able to test empirically for substitution relationships among policy components and to estimate the total value of any level or combination of policy impacts within the range of the sample data. Based on this result, one avenue of economizing on *de novo* value estimation may be to elicit values for a sample of policy combinations rather than for every relevant combination of policy components, and to econometrically estimate the values of other policy combinations within the range of the data.

If a large body of empirical results such as Hoehn's (1989) is assembled, it is conceivable that certain empirical patterns might emerge. Then, economists may start to get a feel for the magnitudes of the substitution relationships and the errors introduced by IVS procedures in various environmental valuation contexts. If this happened, the accumulated empirical experience

---

[1] Consider, for example, the work of Sorg and Loomis (1984) on recreation values, that was sponsored by U.S. Forest Service.

would provide correction factors that would enable economists to use "typical" unit values, such as those assembled by Sorg and Loomis (1984), to construct approximately valid total and component values for complex policies. This procedure could be justified only in relatively routine policy evaluations where economizing on research expense is an overriding consideration.

Opportunities for Using Revealed Preference Methods. Revealed preference methods, such as the travel cost method for estimating recreation benefits and hedonic price analysis for valuing various kinds of amenities that have spatial dimensions, are widely used in environmental benefit estimation (see chapters 3 and 4). Some practitioners prefer them to CVM because of their empirical basis in historical transactions. Others argue that as long as CVM and revealed preference methods remain roughly equal competitors, it is wise to make opportunities to use both kinds of methods in the same situation to check for cross-technique comparability of results.

However, the total value framework imposes some important impediments to the use of revealed preference methods. First, as mentioned above, revealed preference methods are generally inappropriate for measuring total value and its existence value component. So, incorporation of these methods in a benefit estimation strategy will force the practitioner into implementing some kind of procedure to approximate a valid piecewise sequential benefit estimate. (The present status of such procedures was discussed above.) Second, it can be argued that the long and tortuous controversy about option value, when benefits are uncertain, is an artifact of the use of revealed preference methods.

Revealed preference methods are typically used to estimate the amount of consumers' surplus implicit in a historical pattern of transactions. Using projections of future transactions, the analyst projects this surplus forward in *ex ante* benefit analysis and labels it the expected surplus (ES) from future use. Under uncertainty, it has long been argued that, in addition to expected surplus based on projections from historical experience, there is an option value (OV) representing the premium that prospective users would pay to reduce the risks associated with future supply. Therefore, option value emerges only as a result of a decision to estimate total value in piecewise fashion by applying revealed preference methods and historical transactions data to generate estimates of expected surplus for some kinds of use value.

There has been a long controversy about the sign and size of option value, and by the early 1980s a consensus emerged that, in general, sign is ambiguous while there is little evidence that size is typically substantial (see, e.g., Freeman 1984, 1985b). However, *option price — ex ante* willingness to pay to secure an option for future use — may be defined: option price (OP) = expected surplus (ES) + option value (OV), where OV $\gtreqless$ O. Recently, several empirical studies have documented substantial option prices for various environmental resources (e.g., Brookshire, Eubanks, and Randall 1983; Desvousges, Smith, and Fisher 1987; Greenley, Walsh, and Young 1981; Walsh, Loomis, and Gillman 1984).

However, two recent developments have shattered this consensus. First, OP, ES, and OV are sure payments; that is, the amount paid is independent of which state of the world eventually emerges. Graham (1981) introduced the concept of an *ex ante* contract for an array of state-conditional payments. With perfect markets in contingent commodities, a researcher could define a fair bet point (FB), which is the set of state-conditional payments whose expected value (EF) is the maximum feasible, so that EF $\geq$ OP, and EF $\geq$ ES. This led some researchers, such as Cory and Saliba (1987), to question the relevance of OP and OV.

Second, Smith (1987a) introduced a formulation that combines the concepts of state-conditional payments and planned expenditures in order to define *ex ante* use values. As in the equation set (10.10), no separate category for option value emerges. Smith was adamant that *ex ante* use value, $\tilde{U}V$ is fundamentally noncomparable with ES because $\tilde{U}V$ is defined with respect to *ex ante* planned expenditures whereas ES is defined with respect to actual expenditures *ex post* (after the uncertainty has been resolved). It can be readily shown that OP is simply a restricted case of $\tilde{U}V$ (compare with Smith 1987a). Thus, it follows that OP is noncomparable with ES and that the traditional claim of OP = ES + OV is meaningless.

Given that *ex ante* benefit-cost analysis of proposed projects and policies requires *ex ante* measures of use value, such as $\tilde{U}V$ or OP, the noncomparability provides a powerful argument against using revealed preference methods to estimate ES and then adding in OV to estimate total use value. This argument clearly tilts the choice of method toward CVM, which is readily adaptable to measuring $\tilde{U}V$ and OP.

However, in response to an earlier version of Smith (1987a), Bishop (1986) argued that Smith overstated the distinction between *ex ante* and *ex post*. His point was that Smith's distinction was based on the relative timing of the value-revealing decision and the resolution of uncertainty. Smith stated that if the decision occurs while things are still uncertain, the valuation it reveals is *ex ante*; however, if the decision occurs after the uncertainty has been resolved, the valuation is *ex post*. Bishop argued that many revealed preference data sets are based on decisions that were made before all the uncertainties pertaining to use and enjoyment could possibly have been resolved. How, for example, can people be sure what the weather will be like during their vacation at the time they reserve their plane tickets and accommodations? Bishop believed that rather than overplaying the *ex ante/ex post* distinction, analysts should be alert to any differences among the information bases that underly the various value measures.

Obviously, several of the issues raised in this subsection await final resolution. In particular, the optimism engendered by Cory and Saliba that ES can serve validly as a lower-bound estimate of *ex ante* use value — and therefore that revealed preference methods have a valid role in a piecewise benefit evaluation strategy — has been questioned by Smith's analyses of the noncomparability of EF and OP to ES. For now, there seem to be serious impediments to incorporating benefit measures based on revealed preference methods into a valid measure of *ex ante* total value. However, there is little guidance as to the likely magnitude of the empirical errors that may be involved.

10.4 Empirical Estimates of Nonuse Values

During the 1980s, a number of empirical studies for estimating nonuse values were published. Yet, given that the conceptual understanding of nonuse values has been in a state of flux and (as has often been the case in environmental economics) that the earliest empirical applications preceded the systematic development of the relevant theory, several conceptual controversies remain unresolved. Thus, it is not surprising that the conceptual foundations of these various empirical studies are often inconsistent with one another. In some cases, there are — when examined from hindsight — clear conceptual errors (Greenley, Walsh, and Young 1981). Nevertheless, a body of literature has accumulated that shows, first, that estimation of nonuse values is often feasible, and second, that nonuse values are frequently substantial and sometimes exceed current use values by a considerable margin (Majid, Sinden, and Randall 1983; Schulze, Brookshire, et al. 1983).

Some general observations about the empirical literature may be helpful. First, all of the studies cited in this section used CVM. This is to be expected since existence value, option price, and option value are not readily accessible via revealed preference methods. Second, given that little more than a dozen empirical works are cited, the range of resources addressed is quite broad: air quality (Schulze, Brookshire, et al. 1983); water quality (Desvousges, Smith, and Fisher 1987; Edwards 1988; Greenley, Walsh, and Young 1981; Sutherland and Walsh 1985); wilderness and wildlands (Bennett 1984; Majid, Sinden, and Randall 1983; Walsh, Loomis, and Gillman 1984); hunting (Brookshire, Eubanks, and Randall 1983); fisheries (Bishop, Boyle, and Welsh 1987); and nonconsumptive wildlife use and existence (Bowker and Stoll 1988; Boyle and Bishop 1987; Samples, Dixon, and Gowen 1986; Stoll and Johnson 1984). Third, the studies vary with respect to the categories of value estimated. Some address total value or preservation value; and some of these attempt to break-out various components of value. Others are focused on OP. Still others are primarily concerned with existence value. Those cited in this paragraph and summarized in the table 10.1 include several of each type.

In an empirical experiment concerning existence values, Samples, Dixon, and Gowen (1986) specifically addressed the issue of how information provided before or in the course of the CVM exercise affects the value estimates obtained. Values were sensitive to the amount and kind of information provided. In one experiment, positive values were obtained for preserving

TABLE 10.1

A summary of published studies emphasizing nonuse values.

Publication	Range of value estimates ($/household/year)				
	Total[a]	Preservation[a]	Use[a]	Option[a]	Existence[a]
Bennett (1984) Preservation of Nadgee Nature Reserve, Australia	—	—	—	—	0-2.00
Bowker and Stoll (1988) Whooping cranes, Texas	5-149	—	—	—	—
Boyle and Bishop (1987) and Bishop, Boyle, and Welsh (1987) Bald eagles and	6.50-75.31	—	—	—	4.92-28.38
stripped shiners, Wisconsin	—	—	—	—	1.00-5.66
Brookshire, Eubanks, and Randall (1983) Grizzley Bear and	—	—	—	9.70-21.50	—
Bighorn Sheep, Wyoming	—	—	—	11.18-22.90	—
Desvousges, Smith, and Fisher (1987) Water quality for boating, fishing, swimming, Monongahela River	—	—	—	54.1-117.60	—
Edwards (1988) Potable groundwater protection, Cape Cod, Massachusetts	—	—	—	4,930-24,850[d]	—

TABLE 10.1 *Continued*

Publication	Range of value estimates ($/household/year)				
	Total[a]	Preservation[a]	Use[a]	Option[a]	Existence[a]
Majid, Sinden, and Randall (1983) Incremental value of additional parks, Australia	1.5-5.3[b]	—	—	—	—
Schulze and Brookshire, et al. (1983) Visibility preservation in S.W. parklands	—	2.89-4.50	3.16-4.93	—	—
Stoll and Johnson (1984) Whooping cranes, Texas	—	—	2.42-3.15	28.90-41.96[e]	-0.29-39.48
Sutherland and Walsh (1985) Distance and water preservation values, Flathead Lake, Montana	—	8,183.70[c]	—	2,770.50[c]	5,413.20[c]
Walsh, Loomis, and Gillman (1984) Colorado wilderness levels, 10.0 million acres protected	93,200[c]	35,000[c]	58,000[c]	10,200[c]	24,800[c]

[a] Consult original for the definition employed.
[b] Increments to total value from addition of new parks to system.
[c] Aggregated annual values of all households x 1000.
[d] Present value of a 30-year stream of benefits.
[e] Option price *plus* existence value for households expecting to visit in future.

plausible but imaginary species. This study barely hints at the broad influences information might have on CVM-estimated existence values concerning resources about which the public, and even the specialists, have very incomplete knowledge. The issue of exactly how much and what kind of information to provide respondents is unresolved.

10.5 Conclusions

Despite several key developments in the 1980s — including the Hoehn-Randall work on valid holistic and piecewise sequential benefit estimation structures, Graham's theory of benefits under uncertainty in an environment of complete contingent markets, and Smith's elaboration of the distinction between *ex ante* and *ex post* value measures — the researcher is still faced with ambiguity about correct methods of evaluating nonuse benefits. What is known and not known about these methods is summarized below.

1. In either a deterministic framework or under uncertainty, total value is uniquely defined and can be decomposed into existence value and various kinds of use values. Under uncertainty, all value concepts are *ex ante*: ex ante total value, *ex ante* existence values, and *ex ante* use value.

2. Total value may be defined holistically or in a piecewise sequential framework. If a piecewise sequential framework is used, the value of a particular component is dependent on its place in the valuation sequence.

3. Piecewise independent valuation is a common procedure, and it has the virtue of permitting some economies in benefit estimation through the use of "typical" values for various kinds of use. Unfortunately, this procedure is misleading with respect to both total value and component values. Although some researchers are developing procedures that will use piecewise independent data for approximately valid benefit evaluation, as yet none exist. Furthermore, little is known about the magnitude of the errors due to piecewise independent valuation.

4. Contingent valuation methods are readily adaptable for measuring, *de novo,* all of the *ex ante* concepts of total, existence, and use value. Researchers are working on approximately valid methods that will economize on de novo benefit estimation by extending the scope of a sample of *de novo* benefit estimates.

5. Some researchers prefer revealed preference methods to CVM; others use both CVM and revealed preference methods to facilitate cross-technique comparisons. In a deterministic framework, revealed preference methods may be used to evaluate some kinds of use benefits, although care must be taken because of the piecewise independent framework in which these methods typically are used.

6. When uncertainty about future supply, demand, or knowledge is explicitly

recognized, CVM methods are readily adaptable to the benefit estimation task. However, theoretical developments in the 1980s have not resolved the option value controversy. Rather, additional complexities have emerged. These developments have tended to make it more, rather than less, difficult to define a valid role for revealed preference methods in benefit estimation under uncertainty.

10.5.1 Must the Practitioner Always Be Concerned with Existence Values and Uncertainty?

It seems clear that a concern with total value, existence value, and uncertainty tends to swing the balance, with respect to valuation techniques, toward CVM. On the other hand, an analyst who is satisfied to evaluate use benefits from a historical perspective will have much greater scope for reliance on revealed preference methods. Analysts who would prefer to reserve a substantial role for the latter methods have good reason to ask whether it is always necessary to be concerned with existence value and uncertainty.

It seems to be a truism that everything has an existence value; however, existence value at the margin approaches zero when existence is not scarce. Projects that modify small amounts of resources that are in large supply or have many substitutes are unlikely to cause substantial losses in existence value. This suggests that, in many routine benefit estimation contexts, existence values will be unimportant. Nevertheless, this conclusion should be approached cautiously; the burden of proof should always lie upon the analyst who claims existence value does not matter. Randall and Stoll (1983) have argued that, for environmental goods with a spatial dimension, important local and regional existence values may be at stake, even if existence at the global level is not threatened.

With respect to uncertainty and the concomitant obligation to estimate *ex ante* total (or existence and use) value, it is perhaps appropriate to observe that uncertainty is often recognized but treated as academic: often the analysis proceeds with scant attention to uncertainty. Where there is nothing especially unusual about the kinds and degrees of uncertainty involved, this may well be a justifiable strategy. However, where the uncertainties are great and it would be very costly to reverse a wrong decision or mitigate its damage, these uncertainties should be explicitly considered in the benefit analysis.

Measuring the Demand for Environmental Quality
John B. Braden & Charles D. Kolstad (Editors)
© Elsevier Science Publishers B.V. (North-Holland), 1991

Chapter XI

SUMMARY AND CONCLUSIONS

JOHN B. BRADEN and CHARLES D. KOLSTAD

University of Illinois at Urbana-Champaign

11.1 Introduction

As Bockstael, McConnell, and Strand stated in chapter 8, the basic motivation for placing economic values on preferences for environmental goods and services is "the failure of the market to provide efficient quantities and qualities of natural resources" for recreation, aesthetics, health, and other aspects of the environment that people value. Prior to the 1970s, it was the demand for quantities of environmental commodities such as recreational sites, scenic rivers, and wildlife that received most attention from economists. The early 1970s witnessed a shift toward interest in environmental quality. Quality presented new theoretical challenges, but more importantly, it required new methods of measurement and estimation. The challenge was met with adaptations of hedonic methods, which were developed to evaluate quality differentiation of market goods, and of constructed market methods, which were in common use for new product development.

If the 1970s were a decade of discovering new strategies for measuring the demand for environmental goods and services, then scholarship in the 1980s has emphasized tactics. By the beginning of the 1980s, basic economic theories of demand and welfare measurement had been thoroughly integrated into the field. In some instances, notably weak complementarity and the design of constructed markets, this field has been a source of new insights into general demand and welfare theories. The subsequent literature is largely concerned with issues of specification and design.

These nearly two decades of sustained professional effort have produced a diverse portfolio of analytical methods and a rich inventory of applications. By and large, the methods now in widespread use produce reasonable estimates of the demand for and value of quantities and qualities of environmental attributes, and they do so with some regularity and consistency (Smith and Kaoru 1989). The methods have been compared in a number of settings and there is an emerging professional sense of how they relate to one another. The methods are sufficiently diverse that one or another can usually be adapted to a particular environmental context. The tactical literature provides guidance on the needed adaptations.

Although the field has matured to the point of tactical refinement, it still resists simplification or generalization. Economists are as far now from placing a single dollar value on fishable water, unimpaired scenery, or other environmental phenomena as they were two decades ago, perhaps farther.

The extensive understanding of technical issues has revealed just how dependent are demand and benefit estimates on specific circumstances and analysts' professional judgments. The specification of a regression equation, the nature of the sample data, or the way research questions are posed can greatly affect the results. Judgments about these issues must be made case by case; there are very few simple rules of thumb that analysts can confidently apply. Thus, artful application continues to play a large role in the science of environmental demand estimation.

11.2 Theory and Methods

Chapter 2 covered demand theory as it is used for environmental goods and services. Two points deserve to be repeated. One is the demise of "ad hocery" in specifying demand relationships. When deriving willingness-to-pay measures from observable data, special theoretical conditions must be satisfied. Failing to specify ordinary demand in a theoretically appropriate way can undermine the rest of the analysis.

The second point to draw from chapter 2 is that the frontier of environmental demand theory is dominated by considerations of risk. As yet, no consensus has been reached on the best way to treat risk in a demand framework. Most approaches are based on expected utility theory or its offspring, decision theory, and these approaches have much in common with demand estimation for quantities of goods and services. However, because of a number of anomalies in expected utility theory, other approaches also are being explored. There is much to gain from more scholarly attention to evaluating risk.

Chapter 3 covered the household production function (HPF) framework. This framework supports two of the major empirical methods for nonmarket commodity valuation: the travel cost model and averting expenditure analysis. The HPF approach emphasizes the interrelationships between consumer goods and services, both those purchased and not purchased in markets. Establishing the basic relationships is essential for using observable market transactions data to infer the worth of nonmarket commodities. Such inference is theoretically valid and meaningful only under certain conditions, such as weak complementarity and perfect substitutability. The HPF reveals those conditions, and analysis based on this model shows how to derive valid nonmarket value from transactions data.

In chapter 3, Smith argues that considerable scope remains for extending the household production function approach and related valuation methods. In particular, he contends that the approach can be translated directly into a behavioral model that includes environmental quality variables as arguments. The model would involve a complete, simultaneous system of demand equations. This extension would allow the direct estimation of preferences, rather than indirect estimation through the travel cost model or averting expenditures. Detailed household-level data would be required to apply the direct approach. The general lack of such data may present difficulties, but might be circumvented through the use of partial demand systems.

The other major "revealed-preference" approach to demand measurement is the hedonic method. The hedonic method infers environmental demand from the prices of closely related goods or services. As is evident from chapter 4, this approach has become another work horse of demand analysis techniques. It is specifically designed to evaluate quality, which is the leading issue for contemporary environmental demand analysis.

The main promise of hedonic methods is that it becomes theoretically possible to infer demand for nonmarketed commodities from markets for related commodities. The way air quality affects the value of a house is determined by underlying preferences for air quality.

Unfortunately, it seems as if the promise of hedonics has been accompanied by an even greater number of problems, including some very fundamental general equilibrium questions on the extent to which stocks of environmental quality are reflected in wages and land prices. Progress over the last few decades has been steady: chipping away at conceptual and implementation problems.

The third approach — direct elicitation of preferences — was covered in chapter 5. If there has been a breakthrough in demand evaluation during the 1980s, it is in the increased professional acceptance of these techniques. Contingent valuation techniques are not appreciably younger than hedonics, but their very different ancestry has slowed their acceptance. They stem from marketing surveys rather than analysis of actual consumption patterns. Stated preferences frequently are not borne out in observed consumption, and many environmental economists have been skeptical of demand estimates based on intentions or ideals rather than actual commitments.

Some direct elicitation studies have succeeded in simulating real markets and eliciting actual consumption decisions. But even survey approaches have advanced greatly, to the point where they often produce highly credible demand estimates that withstand comparison to measures based on revealed preference approaches. The so-called referendum approach, in which respondents choose between two outcomes that differ in environmental attributes and in cost, seems to perform particularly well. In general, the great flexibility of these approaches, their relative transparency, and the fact that they avoid some demanding theoretical gymnastics by getting directly at willingness to pay, have kindled great interest among public agencies faced with performing environmental demand studies.

Other advantages of constructed markets are: (1) they measure demand and welfare *ex ante,* as theory indicates is appropriate for proposed policies; (2) they easily handle risk and nonuse demand, rather than measuring option values separately; and (3) they circumvent problems of aggregating across commodity classes, as emphasized by Randall in chapter 10.

With growing acceptance of the basic direct elicitation approaches, scholarly effort in this area also has turned increasingly to tactical issues. There is considerable agreement that the design of a simulated market or a survey — the conditions under which it operates and the messages it conveys about the commodity being evaluated — can substantially affect the outcome. Ways of minimizing possible distortions have received extensive attention. Various design options have been proposed, each leading to a number of sampling and econometric challenges.

One of the key technical issues identified by Carson in chapter 5 is whether to value specific changes in commodity quality or quantity, or to derive a function that relates value to quality or quantity in a general way. The former approach has been the norm in contingent valuation studies, and the latter in hedonics or travel cost method (TCM), although this is not dictated by the structures of the approaches. Clearly, value functions are better suited to generalization. Unfortunately, the theoretical baggage is much heavier, even for constructed markets. Given the professional reluctance to generalize estimates from revealed preference studies, there may not be a lot gained from turning contingent valuation methods (CVM) studies toward valuation functions.

11.3 Classes of Environmental Commodities

11.3.1 Health

One of the most basic and important effects of environmental degradation is in terms of health, both mortality and morbidity. Placing monetary value on health frequently raises hackles but is implicit in virtually every economic decision that affects health. Several fundamental philosophical questions are raised here in valuing health effects, particularly the question of the value of life and the distinction between voluntary and involuntary risks. Two major methods have been used for inferring values for health effects — averting behavior and wage studies. Unfortunately the disagreement among studies purporting to measure the same effect are uncomparably large. Cropper and Freeman, in chapter 6, suggest that more refined models of the value of mortality risks can reduce the disagreement. They also suggest that there are opportunities for progress in using averting behavior by identifying behavior that is uniquely identified with the risk in question, such as symptom-specific medication. Another avenue for future research suggested by these authors is to better connect human perceptions of risk with actual risks. Finally, a direction for future work that is a common thread throughout this book, is to apply diverse techniques to the same environmental insult in order to cross-validate techniques for valuing environmental effects.

11.3.2 Aesthetics

The major problem confronting any valuation of aesthetics is the difficulty in unraveling aesthetic aspects of a good or experience from other aspects. It is for this reason that, in chapter 7, Graves identifies contingent valuation as a necessary, though not totally satisfactory, method for eliciting preferences. He indicates that hedonic methods circumvent the arbitrary nature of CVM but introduce a host of different and equally vexing problems. Graves indicates that household production methods are rarely used to value aesthetics because of the insurmountable problem of separating aesthetic characteristics from other characteristics of recreation areas.

In reviewing the CVM, Graves notes a plethora of potential problems with the method, ranging from difficulty in clearly defining an aesthetic good to bias in the survey. Graves is more concerned with fundamental problems associated with hedonic methods. He is particularly concerned about the extent to which amenity values are capitalized in land as opposed to reflected in wage rates.

All in all, both CVM and hedonic methods have a role to play in valuing aesthetics although each is far from perfect. Future research will focus on refining each method and in providing cross-method validity checks.

11.3.3 Recreation

The travel cost method covered in chapter 8 by Bockstael, McConnell, and Strand has dominated recreation demand estimation, so the limitations of TCM are in many respects the limitations of measuring recreation demand. Bockstael and her colleagues note that recreation studies are most successful when they are site-specific. This illustrates the difficulty of generalizing, even when using approaches like TCM that produce value functions.

Bockstael, McConnell, and Strand also cast considerable doubt on the attempts to date to merge hedonic and travel cost methods in order to measure the demand for recreation site attributes. The problem is that the supply of recreation sites that draw on unique physical amenities (unlike artificial game parks or water slides) does not operate as supposed in the hedonic model. As a result, the second stage of the hedonic estimation has no obvious interpretation. Moreover, part of the true prices for recreation consists of travel and time costs. These costs vary among individuals, so there are really many hedonic functions.

The effect of quality on recreation demand is an important policy issue. Recreation is often a major impact category. The current difficulty of dealing with site quality attributes is a real limitation of the recreation demand models based on the TCM. The hedonic travel cost model has not solved the problem, but it has helped define the issues.

Studies of recreation demand have provided much insight into the theoretical challenges facing revealed preference approaches. For example, the effects on welfare estimates of different functional forms and estimation techniques have been studied largely in the context of recreation demand. Models and empirical techniques with equal claim to theoretical validity sometimes produce widely varying welfare estimates. Analysts are left without clear guidelines about the most appropriate estimation strategies. Bockstael, McConnell, and Strand argue that the rigor in these approaches ensures that we at least know what the results mean, even if we have difficulty choosing between them. However, the insights have also encouraged interest in alternatives. Contingent valuation and other constructed market approaches are used increasingly because they avoid some of the theoretical restrictions and empirical dilemmas. They also can address quality dimensions, although these can be troublesome to measure and to describe, they capture both use and nonuse values, and they develop data where frequently no behavioral data are available.

Experience with direct elicitation studies of recreation has been encouraging. Several comparative studies have produced quite reasonable relationships between value estimates derived from surveys and estimates based on actual behaviors. Advances have been made in drawing survey studies closer to actual choices, for example, with the referendum approach. Moreover, the strategic responses that economists expected in survey responses appear not to be terribly important. Nevertheless, Bockstael and colleagues remain skeptical about constructed market approaches. In their view, because of a lack of behavioral theory underlying constructed markets, we really do not know what the participants are revealing in their responses. From behavioral methods, we have at least the comfort of knowing that we are analyzing real behaviors, even if we cannot unambiguously decipher them.

11.3.4 Materials Damages

The tremendous complications that arise in assessing human preferences are least troublesome, although not absent, in dealing with materials damages.

This is because many materials are intermediate goods rather than direct sources of utility. Individuals probably do not care what kind of mortar is used in a lovely brick building as long as the mortar is durable. In evaluating the effects of increased air pollution that destabilizes mortar, the additional cost of a more durable mortar (for new buildings) or more frequent maintenance (for old buildings) is an appropriate measure of the damage.

Rather that wrestling with preferences, most studies of materials damages are concerned with producer and consumer decision sets. That is, they seek to learn how the final good can be provided most cheaply given an anticipated change in environmental quality. In most cases, there are substitutes and complements for the directly affected material. Producers or consumers may be able to moderate the effect of the change by selecting a different mix of inputs. Clearly, this is fertile ground for production function analysis (household production function analysis in the case of consumers).

A variety of techniques, including mathematical programming, dose-response studies, econometric estimation, and simulation can be used to capture the decision environment surrounding materials damage problems. Adams and Crocker point to the application of dual econometric models as an especially promising approach. Dual models have the same advantages of interpretation as do mathematical programming models, but are less onerous in their data requirements and may be less restrictive in imposing structure on production relationships.

In addition to production function analysis, which focuses on the selection of quantities of inputs, hedonic methods also have been used to evaluate damages to real property. Property values have been the most common target of analysis. Carson and Crocker are skeptical of these studies on several counts. One is that hedonics provide no *a priori* guidance about which qualitative influences should be important. Second, in the case of real estate, they expect high covariation among a number of important quality characteristics, and high covariation significantly reduces the statistical resolution of relative effects.

As for contingent valuation techniques, Carson and Crocker view them as holding much potential for future work. Their application in materials damage would greatly reduce the need to characterize technologies, just as they get around the need to characterize preferences for environmental commodities that are directly consumed. The advantage of *ex ante* rather than *ex post* analysis also carries over.

Relatively few types of materials damage have been addressed with modern techniques; the old studies of the impacts of fetid air on buildings and sanitation were largely simple cases of multiplying prices times quantities without considering possible adjustments in the mix of goods and services or resulting income effects on welfare. Agricultural damages of air pollution have dominated the recent studies. What is especially interesting about the agricultural studies is that some have been extended to general equilibrium analysis, as has happened with few other classes of environmental commodities. When crop yields are reduced over a wide area, as might occur for example from regional ozone build-up or acid precipitation, overall output may be affected enough to change relative prices. Thus, welfare effects may need to be traced through several markets. This raises a number of interesting issues, but chiefly the issue of aggregation, which cuts through all commodity classes.

11.3.5 Nonuse Values

We see in all impact categories the dilemma of evaluating things that do not pass through markets. The dilemma is obviously most difficult where nothing is even consumed or experienced directly.

Revealed preference approaches have a very difficult time getting at these nonuse benefits because they simply are not systematically revealed. To a great degree, the literature is dominated by controversy over definitional issues. There have been few efforts to measure nonuse demand through behavioral models. It would seem that charitable contributions are one of the few ways of tackling this, and the limitations of the charitable sector make this an unlikely avenue for capturing the full scope of demand.

As Randall argues in chapter 10, nonuse demand appears to be an area in which there are few good alternatives to constructed markets — more specifically, contingent valuation. Moreover, with contingent valuation, use and nonuse demands can be evaluated together rather than being treated as distinct. This has substantial virtues when it comes to aggregation, a point that Randall emphasizes and that is common to all areas of environmental demand analysis.

11.4 The Direction of Future Research

One impression the reader will undoubtedly take away from this book is that estimating demand for environmental goods and services is a very active area of research. All of the right characteristics are there — real policy relevance with many unanswered questions, a solid methodological infrastructure on which to build, and nearly as many deficiencies with current empirical techniques as strengths. In other words, we know a lot, but there is a lot more to learn and the learning could be very valuable for policy purposes.

Household production methods are probably the oldest of the three major techniques discussed as they are based in the work of Hotelling and Becker. Recent research has focused on developing more sophisticated theoretical structures, including random utility models, that can extract more meaningful information from data.

Hedonic methods have been applied successfully in many environmental contexts. However, as we have seen, problems remain both in terms of econometrics and in terms of interpreting partial equilibrium hedonic prices.

Contingent valuation methods are the youngest methods and thus, not surprisingly, have matured the fastest in recent years. CVM has moved from the status of often-maligned to being of considerable professional interest. The potential problems with CVM have been well articulated and researchers are gradually addressing these although the hypothetical nature of CVM will be a lasting drawback. The overriding virtue that CVM has, and that assures its longevity, is that it can be applied in virtually any situation, albeit at some cost.

Our expectation is that the coming decade will be as productive as the last two in developing and refining techniques for measuring demand for environmental quality. It is clearly one of today's most important areas of applied economics. Nonmarket demand estimation has achieved a good deal of credibility. From a scholarly perspective, the issues raised by nonmarket demand have enriched economics with new theoretical and econometric insights. From a practical perspective, government agencies routinely call for demand studies when considering significant changes in environmental or resource management policies, and valuation studies are presented with increasing frequency in litigation over alleged environmental damages, such as those caused by oil spills.

But credibility should not breed complacency. Many challenges remain to be faced in applying nonmarket demand estimation methods. Furthermore, not all of those challenges are technical in nature. As Bockstael and her colleagues have eloquently stated in this volume: "... in our untiring quest for more and more precision in individual benefit estimates, we seem to have forgotten what has always been the greatest and altogether most insurmountable difficulty in welfare evaluation — aggregating benefits or losses over time and over individuals." Scholarship in this field is in some danger of losing perspective on the big questions of relevance and strategy in the pursuit of small questions of technique and tactic. It is important that researchers remember and that users of our work understand the limitations of what we have to offer. Our estimates of the value of changes that play out over many years will probably be more affected by our assumptions about future demand pattern than about which functional form is used. Yet, we foresee the future without noticeably greater skill than others. Furthermore, our estimates are based on an accepted economic status quo. The worth that individuals attach to environmental commodities directly reflect their relative income positions. The consequence is that some individuals' preferences count a great deal and others' count almost not at all. And, despite our careful efforts to derive income compensated measures of welfare change, compensation almost never occurs in practice. Gainers rarely pay and losers rarely are compensated, so the basic premises of our analyses are rarely, if ever, realized.

Far from suggesting a ship that is sinking or deserves to be sunk, these limitations surround an area of scholarship that is not only stimulating and fast moving, but also surrounded by some of the most compelling and fundamental philosophical questions in economics.

BIBLIOGRAPHY

Acton, J.P. 1973. Evaluating public programs to save lives: the case of heart attacks. Research report R-73-02, Rand Corporation, Santa Monica, CA.

Adams, R.M., L.M. Callaway, and B.A. McCarl. 1986. Pollution agriculture and soil welfare: the case of acid deposition. *Canadian Journal of Agricultural Economics* 34:3-19.

Adams, R.M., and T.D. Crocker. 1985. Economically relevant response estimation and the value of information: acid deposition. In *Economic perspectives on acid deposition control*, ed. T.D. Crocker, pp. 35-64. Boston: Butterworth Publishers.

————. 1989. The agricultural economics of environmental change: some lessons from air pollution. *Journal of Environmental Management* 28:295-307.

Adams, R.M., T.D. Crocker, and R.W. Katz. 1984. Assessing the adequacy of natural science information: a Bayesian approach. *Review of Economics and Statistics* 66:568-575.

Adams, R.M., T.D. Crocker, and N. Thanavibulchai. 1982. An economic assessment of air pollution to selected annual crops in southern California. *Journal of Environmental Economics and Management* 9:42-58.

Adams, R.M., J.D. Glyer, S.L. Johnson, and B.A. McCarl. 1989. A reassessment of the economic effects of ozone on U. S. agriculture. *Journal of the Air Pollution Control Association* 39:960-968.

Adams, R.M., S.A. Hamilton, and B.A. McCarl. 1986. The benefits of pollution control: the case of ozone and U.S. agriculture. *American Journal of Agricultural Economics* 68:886-893.

Adams, R.M., and B.A. McCarl. 1985. Assessing the benefits of alternative oxidant standards on agriculture: the role of response information. *Journal of Environmental Economics and Management* 12:264-276.

Aigner, D.J., and E.E. Leamer. 1984. Estimation of time-of-use pricing response in the absence of experimental data. *Journal of Econometrics* 26:205-227.

Äkerman, J., L. Bergman, and F.R. Johnson. 1989. Paying for safety: voluntary reduction of residential radon risks. Unpublished paper, Office of Policy Analysis, U.S. Environmental Protection Agency, Washington, DC.

Alberini, A., and R.T. Carson. 1990. The relative efficiency of simple discrete choice estimation. Draft manuscript, Department of Economics, University of California, San Diego, CA.

Amemiya, T., and J.L. Powell. 1981. A comparison of the Box-Cox maximum likelihood estimator and the nonlinear two-stage least squares estimator. *Journal of Econometrics* 17:351-382.

Anderson, A.B., A. Basilevsky, and D.P.J. Hum. 1983. Missing data. In *Handbook of survey research,* eds. P.H. Rossi, J.D. Wright, and A.B. Anderson. New York: Academic Press.

Anderson, R.J., and T.D. Crocker. 1971. Air pollution and residential property values. *Urban Studies* 8:171-180.

Arnould, R.J., and L.M. Nichols. 1983. Wage-risk premiums and worker's compensation: a refinement of estimates of compensating wage differential. *Journal of Political Economy* 91:332-340.

Arrow, K.J. 1986. Comments. In *Valuing environmental goods,* eds. R.G. Cummings, D.S. Brookshire, and W.D. Schulze. Totowa, NJ: Rowman and Allanheld Publishers.

Arrow, K.J., and A.C. Fisher. 1974. Environmental preservation, uncertainty, and irreversibility. *Quarterly Journal of Economics* 88:313-319.

Arthur, W.B. 1981. The economics of risks to life. *American Economic Review* 71:54-64.

Atkinson, S.E., and T.D. Crocker. 1987. A Bayesian approach to assessing the robustness of hedonic property value studies. *Journal of Applied Econometrics* 2:27-45.

Ayres, R.U., and M.S. Sandilya. 1987. Utility maximization and catastrophe aversion: a simulation test. *Journal of Environmental Economics and Management* 14:337-370.

Bailey, M.J. 1980. *Reducing risks to life: measurement of the benefits.* Washington: American Enterprise Institute.

Bajac, V. 1985. Housing market segmentation and demand for housing attributes: some empirical findings. *Journal of the American Real Estate and Urban Economics Association* 13:58-75.

Ball, M.J., and R.M. Kirwan. 1977. Accessibility and supply constraints in the urban housing market. *Urban Studies* 14:11-32.

Barnett, W.A. 1977. Pollak and Wachter on the household production function approach. *Journal of Political Economy* 85:1073-1082.

Baron, D.P. 1970. Price uncertainty, utility and industry equilibrium in pure competition. *International Economic Review* 11:463-480.

Bartik, T.J. 1986. Neighborhood revitalization's effects on tenants and the benefit-cost analysis of government neighborhood programs. *Journal of Urban Economics* 19:234-248.

————. 1987a. Estimating hedonic demand parameters with single market data: the problems caused by unobserved tastes. *Review of Economics and Statistics* 69:178-180.

————. 1987b. The estimation of demand parameters in hedonic price models. *Journal of Political Economy* 95:81-88.

————. 1988a. Evaluating the benefits of non-marginal reduction in pollution information on defensive expenditures. *Journal of Environmental Economics and Management* 15:111-127.

————. 1988b. Measuring the benefits of amenity improvements in hedonic price models. *Land Economics* 64:172-183.

Bartik, T.J., and V.K. Smith. 1987. Urban amenities and public policy. In *Handbook of regional and urban economics,* ed. E.S. Mills, vol. 2, pp. 1207-1254. Amsterdam: North-Holland Publishing Company.

Batra, R.N., and A. Ullah. 1974. Competitive firm and the theory of input demand under price uncertainty. *Journal of Political Economy* 82:537-548.

Becker, G. 1965. A theory of the allocation of time. *Economic Journal* 75:493-517.

Belsley, D.A., E. Kuh, and R.E. Welsch. 1980. *Regression diagnostics.* New York: John Wiley & Sons, Inc.

Bender, B., T.J. Gronberg, and H.-S. Hwang. 1980. Choice of functional form and the demand for air quality. *Review of Economics and Statistics* 62:638-643.

Benedict, H.M., C.J. Miller, and R.E. Olson. 1971. *Economic impact of air pollution on plants in the United States.* Menlo Park, CA: Stanford Research Institute.

Bennett, J.W. 1984. Using direct questioning to value existence benefits of preserved natural areas. *Australian Journal of Agricultural Economics* 28:136-152.

Berger, M.C., G.C. Blomquist, D. Kenkel, and G.S. Tolley. 1987. Valuing changes in health risks: a comparison of alternative measures. *Southern Economic Journal* 53:967-984.

Bergstrom, T.C. 1982. When is a man's life worth more than his human capital? In *The value of life and safety,* ed. M.W. Jones-Lee, pp. 3-26. Amsterdam: North-Holland Publishing Company.

Bergstrom, T.C., D.L. Rubinfeld, and P. Shapiro. 1982. Micro-based estimates of demand functions for local school expenditures. *Econometrica* 50:1183-1205.

Bergstrom, J.C., B.L. Dillman, and J.R. Stoll. 1985. Public environmental amenity benefits of private land: the case of prime agricultural land. *Southern Journal of Agricultural Economics* 17(1):139-149.

Bishop, G. 1981. Survey research. In *Handbook of political communication,* eds. D.D. Nimmo and K.R. Sanders. Beverly Hills: Sage.

Bishop, R.C. 1986. Resource valuation under uncertainty: theoretical principles for empirical research. In *Advances in applied micro-economics,* ed. V.K. Smith, vol. 4, pp. 133-152. Greenwich, CT: JAI Press.

Bishop, R.C., and K.J. Boyle. 1985. The economic value of Illinois Beach State Nature Preserve. Report to the Illinois Department of Conservation. Madison, WI: Heberlein and Baumgartner Research Series.

Bishop, R.C., K.J. Boyle, and M.P. Welsh. 1987. Toward total economic valuation of Great Lakes fishery resources. *Transactions of the American Fisheries Society* 116:339-345.

Bishop, R.C., and T.A. Heberlein. 1979. Measuring values of extramarket goods: are indirect measures biased? *American Journal of Agricultural Economics* 61:926-930.

———. 1980. Simulated markets, hypothetical markets, and travel cost analysis: alternative methods of estimating outdoor recreation demand. Staff paper series no. 187, Department of Agricultural Economics, University of Wisconsin, Madison, WI.

———. 1986. Does contingent valuation work. In *Valuing environmental goods,* eds. R. Cummings, D. Brookshire, and W. Schulze, pp. 123-147. Totowa, NJ: Rowman and Allanheld Publishers.

Bishop, R.C., T.A. Heberlein, M.P. Welsh, and R.A. Baumgartner. 1984. Does contingent valuation work? A report on the Sandhill Study. Paper presented to the American Economics Association annual meeting, August, Ithaca, NY.

Bjorndal, T. 1988. The optimal management of North Sea herring. *Journal of Environmental Economics and Management* 15:9-29.

Blair, R.D. 1974. Random input prices and the theory of the firm. *Economic Inquiry* 12:214-225.

Blomquist, G. 1979. Value of life savings: implications of consumption activity. *Journal of Political Economy* 87:540-558.

Blomquist, G., M. Berger, and J. Hoehn. 1988. New estimates of quality of life in urban areas. *American Economic Review* 78:89-107.

Blomquist, G., and L. Worley. 1981. Hedonic prices, demands for urban housing amenities, and benefit estimates. *Journal of Urban Economics* 9:212-221.

Bockstael, N.E., W.M. Hanemann, and C.L. Kling. 1987. Modeling recreational demand in a multiple site framework. *Water Resources Research* 23:951-960.

Bockstael, N.E., and C.L. Kling. 1988. Valuing environmental quality: weak complementarity with sets of goods. *American Journal of Agricultural Economics* 70:654-662.

Bockstael, N.E., and K.E. McConnell. 1980. Calculating equivalent and compensating variation for natural resource facilities. *Land Economics* 56:56-62.

———. 1981. Theory and estimation of the household production function for wildlife recreation. *Journal of Environmental Economics and Management* 8:199-214.

———. 1983. Welfare measurement in the household production framework. *American Economic Review* 73:806-814.

———. 1987. Welfare effects of changes in quality: a synthesis. Working paper, Department of Agricultural and Resource Economics, University of Maryland, College Park, MD.

Bockstael, N.E., K.E. McConnell, and I.E. Strand. 1988. *Benefits from improvements in Chesapeake Bay water quality.* Report for USEPA, Cooperative agreement CR-811043-01-0, U.S. Environmental Protection Agency, Washington, DC.

Bockstael, N.E., and I.E. Strand. 1987. The effect of common sources of regression error on benefit estimates. *Land Economics* 63:11-20.

Bockstael, N.E., I.E. Strand, and A. Graefe. 1986. *Economic analysis of artificial reefs: a pilot study of selected valuation methodologies.* Technical Report no. 6. Washington: Artificial Reef Development Center, Sportfishing Institute.

Bockstael, N.E., I.E. Strand, and W.M. Hanemann. 1987. Time and the recreational demand model. *American Journal of Agricultural Economics* 69:293-302.

Bohm, P. 1972. Estimating demand for public goods: an experiment. *European Economic Review* 3:111-130.

———. 1984. Revealing demand for an actual public good. *Journal of Public Economics* 24:135-151.

Bowen, H. 1943. The interpretation of voting in the allocation of economic resources. *Quarterly Journal of Economics* 58:27-48.

Bowker, J.M., and J.R. Stoll. 1988. Use of dichotomous choice nonmarket methods to value the whooping crane resource. *American Journal of Agricultural Economics* 70:372-381.

Box, G.E.P., and N.R. Draper. 1987. *Empirical model-building and response surfaces.* New York: John Wiley & Sons, Inc.

Boyle, K.J., and R.C. Bishop. 1987. Valuing wildlife in benefit cost analysis: a case study involving endangered species. *Water Resources Research* 23:943-950.

Bradburn, N.M. 1982. Question wording effects in surveys. In *Question framing and response consistency,* ed. R.M. Hogarth. San Francisco: Jossey-Bass.

Bradford, D., and G. Hildenbrandt. 1977. Observable preferences for public goods. *Journal of Public Economics* 8:111-131.

Brookshire, D.S., and T.D. Crocker. 1981. The advantages of contingent valuation methods for benefit cost analysis. *Public Choice* 36:235-252.

Brookshire, D.S., L.S. Eubanks, and A. Randall. 1983. Estimating option prices and existence values for wildlife resources. *Land Economics* 59:1-15.

Brookshire, D.S., B.C. Ives, and W.D. Schulze. 1976. The valuation of aesthetic preferences. *Journal of Environmental Economics and Management* 3:325-346.

Brookshire, D.S., A. Randall, and J.R. Stoll. 1980. Valuing increments and decrements in natural resource service flows. *American Journal of Agricultural Economics* 62:478-488.

Brookshire, D.S., R.C. d'Arge, W.D. Schulze, and M.A. Thayer. 1981. Experiments in valuing public goods. In *Advances in Applied Microeconomics,* ed. V.K. Smith, vol. 1, pp. 123-172. Greenwich, CT: JAI Press.

Brookshire, D.S., M.A. Thayer, W.D. Schulze, and R.C. d'Arge. 1982. Valuing public goods: a comparison of survey and hedonic approaches. *American Economic Review* 72:165-178.

Brookshire, D.S., M.A. Thayer, J. Tschirhart, and W.D. Schulze. 1984. A test of the expected utility model: evidence from earthquake risks. Unpublished manuscript, Department of Economics, University of Wyoming, Laramie, WY.

Broome, J. 1978. Trying to value a life. *Journal of Public Economics* 9:91-100.

————. 1982. Uncertainty in welfare economics, and the value of life. In *The value of life and safety,* ed. M.W. Jones-Lee. Amsterdam: North-Holland Publishing Company.

————. 1985. The economic value of life. *Economica* 52:281-294.

Brown, B.W., and M.B. Walker. 1989. The random utility hypothesis and inference in demand systems. *Econometrica* 57:815-29.

Brown, C. 1980. Equalizing differences in the labor market. *Quarterly Journal of Economics* 94:113-134.

Brown, D., and M. Smith. 1984. Crop substitution in the estimation of economic benefits due to ozone reduction. *Journal of Environmental Economics and Management* 11:327-346.

Brown, G., Jr., and R. Mendelsohn. 1984. The hedonic travel cost method. *Review of Economics and Statistics* 66:427-433.

Brown, J.N., and H.S. Rosen. 1982. On the estimation of structural hedonic price models. *Econometrica* 50:765-768.

Buchanan, J. 1969. *Cost and choice.* Chicago: Markham Publishing Co.

Burt, O.R., and D. Brewer. 1971. Estimation of net social benefits from outdoor recreation. *Econometrica* 39:813-828.

Butler, R.V. 1980. Cross-sectional variation in the hedonic relationship for urban housing markets. *Journal of Regional Science* 20:439-453.

————. 1982. The specification of housing indexes for urban housing. *Land Economics* 58:96-108.

Calish, S., R.D. Fight, and D.E. Teeguarden. 1978. How do nontimber values affect Douglas fir rotations. *Journal of Forestry* 76:217-223.

Callaway, J.M., R.F. Darwin, and R.J. Nesse. 1985. Economics valuation of acidic deposition: preliminary results from the 1985 NAPAP assessment. Draft report for National Acid Precipitation Assessment Program, U.S. Environmental Protection Agency, Washington, DC.

Cameron, T.A. 1988. Auto-validation: empircal discrete continuous choice modeling with contingent valuation referendum survey data. Unpublished paper, University of California, Los Angeles, CA.

Cameron, T.A., and D. Huppert. 1987. Non-market resource valuation: assessment of value elicitation by "payment card" versus "referendum" methods. Paper presented at the Western Economic Association Meetings, July 1987, Vancouver, British Columbia.

Cameron, T.A., and M.D. James. 1987. Efficient estimation methods for 'closed-ended' contingent valuation surveys. *Review of Economics and Statistics* 69:269-276.

Carmines, E.G., and R.A. Zeller. 1979. *Reliability and validity assessment.* Beverly Hills: Sage.

Carson, R. 1988. Economic value of a reliable water supply. Draft manuscript, University of California, San Diego, CA.

Carson, R., M. Hanemann, and R. Mitchell. 1986. Determining the demand for public goods by simulating referendums at different tax prices. Unpublished manuscript, University of California, San Diego, CA.

Carson, R., M. Hanemann, and D. Steinberg. 1988. *A discrete choice contingent valuation estimate of the value of Kenai King Salmon.* Unpublished manuscript, University of California, San Diego, CA.

Carson, R.T., and R.C. Mitchell. 1988. Value of clean water: the public's willingness to pay for boatable, fishable, swimmable quality water. Staff paper 88-13, Department of Economics, University of California, San Diego, CA.

Carson R., and P. Navarro. 1988. Fundamental issues in natural resource damage assessment. *Natural Resources Journal* 28:815-836.

Carson, R., and D. Steinberg. 1989. Estimation of demand curves via survival analysis. Paper presented at the Annual American Statistical Association Winter Conference, January, San Diego, CA.

Cassel, E., and R. Mendelsohn. 1985. The choice of functional forms for hedonic price equations: comment. *Journal of Urban Economics* 18:135-142.

————. The demand for housing characteristics: a spatial hedonic approach. Undated working paper.

Caulkins, P.P., R.C. Bishop, and N.W. Bouwes. 1986. The travel cost model for lake recreation: a comparison of two methods for incorporating site quality and substitution effects. *American Journal of Agricultural Economics* 68:291-297.

Cesario, F.J. 1976. Value of time in recreation benefit studies. *Land Economics* 52:32-41.

Chavas, J.P., R.C. Bishop, and K. Segerson. 1986. *Ex ante* consumer welfare evaluations in cost benefit analysis. *Journal of Environmental Economics and Management* 13:255-268.

Chestnut, L.G., S.D. Colome, L.R. Keller, W.E. Lambert, B. Ostrow, R.D. Rowe, and S.L. Wojciechowski. 1988a. *Heart disease patients' averting behavior, costs of illness, and willingness-to-pay to avoid angina episodes.* Report to the Office of Policy Analysis. U.S. Environmental Protection Agency, Washington, DC.

————. 1988b. Risk to heart disease patients from exposure to carbon monoxide: assessment and evaluation. Draft report for U.S. Environmental Protection Agency, Washington, DC.

Chestnut, L.G., R.D. Rowe, and J.C. Murdoch. 1986. *Review of establishing and valuing the effects of improved visibility in eastern United States.* Report for USEPA contract no. 68-01-7033. U.S. Environmental Protection Agency, Washington, DC.

Cicchetti, C.J., and V.K. Smith. 1973. Congestion, quality deterioration, and optimal use: wilderness recreation in the Spanish peaks primitive area. *Social Science Research* 2:15-30.

Ciracy-Wantrup, S.V. 1947. Capital returns from soil-conservation practices. *Journal of Farm Economics* 29:1181-1196.

————. 1952. *Resource conservation: economics and policies.* Berkeley: University of California Press.

Clawson, M. 1959. Methods of measuring the demand for and value of outdoor recreation. Reprint no. 10. Resources for the Future, Inc., Washington, DC.

Clawson, M., and V.L. Knetsch. 1966. *Economics of outdoor recreation.* Baltimore: Johns Hopkins University Press.

Clower, R.W., and P.W. Howitt. 1978. The transactions theory of the demand for money: a reconsideration. *Journal of Political Economy* 86:449-456.

Cobb, S.A. 1977. Site rent, air quality, and the demand for amenities. *Journal of Environmental Economics and Management* 4:214-218.

————. 1984. The impact of site characteristics on housing cost estimates. *Journal of Urban Economics* 15:26-45.

Cobb, S., A. Barkume, and P. Shapiro. 1978. Amenties and property values in a model of an urban area. *Journal of Public Economics* 9:107-110.

Cocheba, D.J., and W.A. Langford. 1978. Wildlife valuation: the collective good aspect of hunting. *Land Economics* 54:490-504.

Cochran, W. 1977. *Sampling techniques,* 3rd. ed. New York: John Wiley & Sons, Inc.

Conley, B.C. 1976. The value of human life in the demand for safety. *American Economic Review* 66:45-55.

Converse, J.M., and S. Presser. 1986. *Survey questions: handcrafting the standard questionnaire.* Beverly Hills: Sage.

Cook, P.J. 1978. The value of human life in the demand for safety: comment. *American Economic Review* 68:710-711.

Cook, P.J., and D.A. Graham. 1977. The demand for insurance and protection: the case of irreplaceable commodities. *Quarterly Journal of Economics* 91:143-156.

Cooper, B.S., and D.P. Rice. 1976. The economic cost of illness revisited. *Social Security Bulletin* 39:21-36.

Cory, D.C., and B.C. Saliba. 1987. Requiem for option value. *Land Economics* 63:1-10.

Courant, P., and R. Porter. 1981. Averting expenditures and the cost of pollution. *Journal of Environmental Economics and Management* 8:321-329.

Coursey, D.L., J.J. Hovis, and W.D. Schulze. 1987. The disparity between willingness to accept and willingness to pay measures of value. *Quarterly Journal of Economics* 102:679-690.

Coursey, D.L., and W.D. Schulze. 1986. The application of laboratory experimental economics to the contingent valuation of public goods. *Public Choice* 49:47-68.

Cragg, J.G. 1971. Some statistical models for limited dependent variables with application to the demand for durable goods. *Econometrica* 39:829-844.

Craig, C.S., and J.M. McCann. 1978. Item nonresponse in mail surveys: extent and correlates. *Journal of Marketing Research* 15:285-289.

Crocker, T.D. 1975. Cost-benefit analyses of cost-benefit analysis. In *Cost-benefit analysis and water pollution policy,* eds. H.M. Peskin and E.P. Seskin, pp. 341-360. Washington: The Urban Institute.

————. 1984. On the value of the condition of a forest stock. Unpublished manuscript, Department of Economics, University of Wyoming, Laramie, WY.

————. 1986. Economic effects of materials degradation. In *Materials degradation caused by acid rain,* ed. R. Baboian, pp. 369-383. Washington: American Chemical Society.

Crocker, T.D., and R.G. Cummings. 1985. On valuing acid deposition-induced materials damages: a methodological inquiry. In *Acid deposition,* eds. D.D. Adams and W.P. Page, pp. 359-384. New York: Plenum Press.

Crocker, T.D., and B.A. Forster. 1981. Decision of problems in the control of acid precipitation: nonconvexities and irreversibilities. *Journal of the Air Pollution Control Association* 31:31-37.

Crocker, T.D., and R.L. Horst, Jr. 1981. Hours of work, labor productivity, and environmental conditions: a case study. *Review of Economics and Statistics* 63:361-368.

Cronin, F.J., and K. Herzeg. 1982. *Valuing nonmarket goods through contingent markets.* Report no. PNL-4255. Richland, WA: Pacific Northwest Laboratory.

Cropper, M.L. 1976. Regulating activities with catastrophic environmental effects. *Journal of Environmental Economics and Management* 3:1-15.

————. 1981a. Measuring the benefits from reduced morbidity. *American Economic Review* 71:235-240.

————. 1981b. The value of urban amenities. *Journal of Regional Science* 21:359-374.

Cropper, M.L., and A.S. Arriaga-Salinas. 1980. Inter-city wage differentials and the value of air quality. *Journal of Urban Economics* 8:236-254.

Cropper, M.L., L.B. Deck, and K.E. McConnell. 1988. On the choice of functional form for hedonic price functions. *Review of Economics and Statistics* 70:668-675.

Cropper, M.L., and F.G. Sussman. 1988a. Families and the economics of risks to life. *American Economic Review* 78:255-260.

————. 1988b. Valuing future risks to life. Working paper no. 88-30. University of Maryland, Baltimore, MD.

Cummings, R.G., D.S. Brookshire, and W.D. Schulze. 1986. *Valuing environmental goods: a state of the arts assessment of the contingent valuation method.* Totawa, NJ: Rowland and Allanheld Publishers.

Cummings, R.G., H.S. Burness, and R.D. Norton. 1981. Measuring household soiling damages from suspended particulates: a methodological inquiry. Research report for USEPA. Department of Economics, University of New Mexico, Albuquerque, NM.

Cummings, R.G., W.D. Schulze, S.D. Gerking, and D.S. Brookshire. 1982. Measuring the elasticity of substitution of wages for municipal infrastructure: a comparison of the survey and wage hedonic approach. Working paper, Department of Economics, University of New Mexico, Albuquerque, NM.

————. 1986. Measuring the elasticity of substitution of wages for municipal infrastructure: a comparison of the survey and wage hedonic approahces. *Journal of Environmental Economics and Management* 13:269-276.

d'Arge, R.C., and K.C. Kogiku. 1973. Economic growth and the environment. *Review of Economic Studies* 40:61-78.

Dardis, R. 1980. The value of a life: new evidence from the marketplace. *American Economic Review* 70:1077-1082.

Daubert, J.T., and R.A. Young. 1981. Recreational demands for maintaining instream flows: a contingent valuation approach. *American Journal of Agricultural Economics* 63:666-676.

Davidson, P., F.G. Adams, and J. Seneca. 1966. The social value of water recreational facilities resulting from an improvement in water quality: the Delaware estuary. In *Water research,* eds. A.V. Kneese and S.C. Smith, pp. 175-211. Baltimore: Johns Hopkins University Press for Resources for the Future.

Davis, R.K. 1963. The value of outdoor recreation: an economic study of the Maine woods. Unpublished Ph.D. dissertation, Harvard University, Cambridge, MA.

————. 1964. The value of big game hunting in a private forest. In *Transactions of the twenty-ninth North American wildlife conference.* Washington: Wildlife Management Institute.

————. 1980. Analysis of the survey to determine the effects of water quality on participation in recreation. Davis to John Parsons, National Park Service, July 28, 1980.

Deaton, A., and J. Muellbauer. 1980. *Economics and consumer behavior.* New York: Cambridge University Press.

Desvousges, W.H., V.K. Smith, D.H. Brown, and D.K. Pate. 1984. The role of focus group in designing a contingent valuation survey to measure the benefits of hazardous waste management regulations. Draft report for the US Environmental Protection Agency. Research Triangle Park, NC: Research Triangle Institute.

Desvousges, W.H., V.K. Smith, and A. Fisher. 1987. Option price estimates for water quality improvements: a contingent valuation study for the Monongahela River. *Journal of Environmental Economics and Management* 14:248-267.

Desvousges, W.H., V.K. Smith, and M.P. McGivney. 1983. *A comparison of alternative approaches for estimation of recreation and related benefits of water quality improvements.* USEPA Report no. EPA-230-05-83-001. Washington: U.S. Environmental Protection Agency.

Desvousges, W.H., V.K. Smith, and H.H. Rink III. 1989. *Communicating radon risk effectively: radon testing in Maryland.* Washington: USEPA, Office of Policy, Planning, and Evaluation.

Deyak, T.A., and V.K. Smith. 1974. Residential property values and air pollution: some new evidence. *Quarterly Review of Economics and Business* 14:93-100.

Diamond, D.B., Jr., and B.A. Smith 1985. Simultaneity in the market for housing characteristics. *Journal of Urban Economics* 17:280-292.

Dickens, W.T. 1984. Differences between risk premiums in union and nonunion wages and the case for occupational safety regulation. *American Economic Review* 74:320-323.

Dickie, M., and S. Gerking. 1989. Valuing nonmarket goods: a household production approach. Paper presented at the Association of Environmental and Resource Economists conference on Estimating and Valuing Morbidity in a Public Context, June 8-9, 1989, Research Triangle Park, NC.

Dickie, M., S. Gerking, D. Brookshire, D. Coursey, W. Schulze, A. Coulson, and D. Tashkin. 1987. Reconciling averting behavior and contingent valuation benefit estimates of reducing symptoms of ozone exposure draft. In *Improving accuracy and reducing costs of environmental benefit assessments.* Washington: U.S. Environmental Protection Agency.

Dickie, M., S. Gerking, W. Schulze, A. Coulson, and D. Tashkin. 1986. Values of symptoms of ozone exposure: an application of the averting behavior method. In *Improving accuracy and reducing costs of environmental benefit assessments.* Washington: U.S. Environmental Protection Agency.

Diewert, W.E. 1986. *The measurement of the economic benefits of infrastructure services.* New York: Springer-Verlag.

Dillingham, A.E. 1985. The influence of risk variable definition on value-of-life estimates. *Economic Inquiry* 24:227-294.

Dillman, D.A. 1978. *Mail and telephone surveys — the total design method.* New York: John Wiley & Sons, Inc.

Dillman, D.A. 1983. Mail and other self-administered questionnaires. In *Handbook of Survey Research,* eds. P.H. Rossi, J.D. Wright, and A.B. Anderson. New York: Academic Press.

Domencich, T., and D. McFadden. 1975. *Urban travel demand: a behavioral analysis.* Amsterdam: North-Holland Publishing Company.

Duan, N. 1983. Smearing estimate: a nonparametric retransformation method. *Journal of the American Statistical Association* 78:605-610.

Eaton, J.S. 1984. Theoretically optimal environmental metrics and their surrogates. *Journal of Environmental Economics and Management* 11:18-27.

Edmonds, R.G., Jr. 1985. Some evidence on the intertemporal stability of hedonic price functions. *Land Economics* 61:445-451.

Edwards, S.F. 1988. Option prices for groundwater protection. *Journal of Environmental Economics and Management* 15:475-487.

Edwards, S.F., and G.D. Anderson. 1987. Overlooked biases in contingent valuation surveys: some considerations. *Land Economics* 63:168-178.

Ellickson, B. 1981. An alternative test of the hedonic theory of housing markets. *Journal of Urban Economics* 9:56-79.

Ellickson, B., B. Fishman, and P.A. Morrison. 1977. Economic analysis of urban housing markets: a new approach. Report no. R-2024-NSF. Santa Monica, CA: Rand.

Epple, D. 1984. Closed form solutions to a class of hedonic equilibrium models. Working paper, Carnegie-Mellon University, Pittsburgh, PA.

————. 1987. Hedonic prices and implicit markets: estimating demand and supply functions for differentiated products. *Journal of Political Economy* 95:59-80 (unpublished Appendix, 1984).

Epstein, L.G. 1975. A disaggregated analysis of consumer choice under uncertainty. *Econometrica* 43:877-891.

————. 1981. Generalized duality and integrability. *Econometrica* 49:655-678.

Erlich, I., and G. Becker. 1972. Market insurance, self-insurance and self-protection. *Journal of Political Economy* 80:623-648.

Ervin, D.E., and J.W. Mill. 1985. Agricultural land markets and soil erosion: policy relevance and conceptual issues. *American Journal of Agricultural Economics* 67:938-942.

Feenberg, D., and E.S. Mills. 1980. *Measuring the benefits of water pollution abatement.* New York: Academic Press.

Feinerman, E., and D. Yaron. 1983. The value of information on the response function of crops to soil salinity. *Journal of Environmental Economics and Management* 10:72-85.

Ferejohn, J.A., and R. Noll. 1976. An experimental market for public goods: the PBS Station Program Cooperative. *American Economic Review* 66.

Fienberg, S.E., and J.M. Tanur. 1985. A long and honorable tradition: intertwining concepts and constructs in experimental design and sample surveys. Paper presented at the International Statistical Institute Meeting, Amsterdam.

Finney, D.J. 1971. *Probit Analysis,* 3rd ed. Cambridge: Cambridge University Press.

Fischhoff, B., and L. Furby. 1987. *A review and critique of Tolley, Randall, et al.: establishing and valuing the effects of improved visibility in the eastern United States.* ERI Technical Report 87-6. Eugene, OR: Eugene Research Institute.

————. 1988. Measuring values: a conceptual framework for interpreting transactions with special reference to contingent valuation of visibility. *Journal of Risk and Uncertainty* 1:147-184.

Fischhoff, B., S. Lichtenstein, P. Slovic, S.L. Derby, and R.L. Keeney. 1981. *Acceptable risk.* New York: Cambridge University Press.

Fisher, A.C., and W.M. Hanemann. 1986. Option value and the extinction of species. In *Advances in applied micro-economics,* ed. V.K. Smith, vol. 4, pp. 169-190. Greenwich: JAI Press.

Fisher, A., D. Violette, and L. Chestnut. 1989. The value of reducing risks of death: a note on new evidence. *Journal of Policy Analysis and Management* 8:88-100.

Fisher, W.L. 1984. Measuring the economic value of fishing and hunting: a conceptual overview. Paper presented at the Annual Meeting of Omicron Delta Epsilon, Lycoming College, April 16, Williamsport, PA.

Follain, J.R., and E. Jimenez. 1985. Estimating the demand for housing characteristics: a survey and critique. *Regional Science and Urban Economics* 15:77-107.

Forster, B.A. 1981. Separability, functional structure and aggregation for a class of models in environmental economics. *Journal of Environmental Economics and Management* 8:118-134.

Frankel, M. 1979. Opportunity and the valuation of life. Preliminary report for the Department of Economics, University of Illinois, Urbana, IL.

Freeman, A.M., III. 1971. Air pollution and property values: a methodological comment. *Review of Economics and Statistics* 53:415-416.

————. 1974a. Air pollution and property values: a further comment. *Review of Economics and Statistics* 56:554-556.

————. 1974b. On estimating air pollution control benefits from land value studies. *Journal of Environmental Economics and Management* 1:277-288.

————. 1975. Spatial equilibrium, the theory of rents, and the measurement of benefits from public programs: a comment. *Quarterly Journal of Economics* 89:470-473.

————. 1979a. *The benefits of environmental improvement: theory and practice.* Baltimore: Johns Hopkins University Press.

————. 1979b. Hedonic prices, property values and measuring environmental benefits: a survey of the issues. *Scandinavian Journal of Economics* 81:155-173.

————. 1982. *Air and water pollution control: a benefit-cost assessment.* New York: John Wiley & Sons, Inc.

————. 1984. The sign and size of option value. *Land Economics* 60:1-13.

————. 1985a. Methods for assessing the benefits of environmental programs. In *Handbook of natural resource and energy economics,* eds. A.V. Kneese and J.L. Sweeney, vol. 1, pp. 223-270. Amsterdam: North-Holland Publishing Company.

————. 1985b. Supply uncertainty, option price, and option value in project evaluation. *Land Economics* 61:176-181.

————. 1988. Valuing changes in risks. Unpublished manuscript, Bowdoin College, Brunswick, ME.

Frey, J.H. 1983. *Survey research by telephone.* Beverly Hills: Sage.

Gallagher, D.R., and V.K. Smith. 1985. Measuring values for environmental resources under uncertainty. *Journal of Environmental Economics and Management* 12:132-143.

Gallant, A.R. 1981. On bias in flexible functional forms and an essentially unbiased form: the fourier functional form. *Journal of Econometrics* 15:211-245.

Garbacz, C., and M. Thayer. 1983. An experiment in valuing senior companion program services. *Journal of Human Resources* 18:147-153.

Garcia, P., B.L. Dixon, J.W. Mjelde, and R.M. Adams. 1986. Measuring the benefits of environmental change using a duality approach: the case of ozone and Illinois cash grain farms. *Journal of Environmental Economics and Management* 13:69-80.

Gardner, K., and R. Barrows. 1985. The impact of soil conservation investments on land prices. *American Journal of Agricultural Economics* 67:943-947.

Gegax, D., S. Gerking, and W. Schulze. 1985. Perceived risk and the marginal value of safety. Working paper. U.S. Environmental Protection Agency, Washington, DC.

Gerking, S., M. DeHaan, and W. Schulze. 1988. The marginal value of job safety: a contingent valuation study. *Journal of Risk and Uncertainty* 1:185-199.

Gerking, S., and L.R. Stanley. 1986. An economic analysis of air pollution and health: the case of St. Louis. *Review of Economics and Statistics* 68:115-121.

Giannias, D.A. 1988. A structural approach to hedonic equilibrium models. Working paper. University of Guelph, Guelph, Ontario.

Gilbert, C.C.S. 1985. *Household adjustment and the measurement of benefits from environmental quality improvements.* Unpublished Ph.D. dissertation, University of North Carolina, Chapel Hill, NC.

Goldberger, A.S. 1968. The interpretation and estimation of Cobb-Douglas functions. *Econometrica* 35:464-482.

Goodman, A.C. 1978. Hedonic prices, price indices and housing markets. *Journal of Urban Economics* 5:471-484.

Graham, D.A. 1981. Cost benefit analysis under uncertainty. *American Economic Review* 71:715-725.

Gramlich, F.W. 1977. The demand for clean water: the case of the Charles River. *National Tax Journal* 30:183-194.

Graves, P.E., J.C. Murdoch, M.A. Thayer, and D. Waldman. 1988. The robustness of hedonic price estimation: urban air quality. *Land Economics* 64:220-233.

Graves, P.E., and D. Waldman. 1989. A test of the multimarket amenity capitalization hypothesis. Manuscript, Department of Economics, University of Colorado, Boulder, CO.

Green, H.A.J. 1980. *Consumer Theory.* Revised ed. New York: Academic Press.

Greenley, D.A., R.G. Walsh, and R.A. Young. 1981. Option value: empirical evidence from a case study of recreation and water quality. *Quarterly Journal of Economics* 96:657-673.
————. 1982. *Economic benefits of improved water quality.* Boulder, CO: Westview Press.
Grieson, R.E., and J.R. White. 1989. The existence and capitalization of neighborhood externalities: a reassessment. *Journal of Urban Economics* 25:68-76.
Griliches, Z., ed. 1971. *Price indexes and quality change.* Cambridge, MA: Harvard University Press.
Gross, D.J. 1988. Estimating willingness to pay for housing characteristics: an application of the Ellickson bid-rent model. *Journal of Urban Economics* 24:95-112.
Grossman, M.. 1972. On the concept of health capital and the demand for health. *Journal of Political Economy* 80:223-255.
Groves, R.M., and R. Kahn. 1979. *Comparing telephone and personal interview surveys.* New York: Academic Press.
Hageman, R. 1985. Valuing marine mammal populations: benefits valuations in multi-species ecosystem. Administrative report no. LJ-85-22, Southwest Fisheries Center, National Marine Fisheries Service, La Jolla, CA.
Halvorsen, R., and R. Palmquist. 1980. The interpretation of dummy variables in semilogarithmic equations. *American Economic Review* 70: 474-475.
Halvorsen, R., and H.O. Pollakowski. 1981. Choice of functional form for hedonic price equations. *Journal of Urban Economics* 10:37-49.
Hamermesh, D.S. 1985. Expectations, life expectancy, and economic behavior. *Quarterly Journal of Economics* 100:389-408.
Hammack, J., and G.M. Brown, Jr. 1974. *Waterfowl and wetlands: toward bioeconomic analysis.* Baltimore: Johns Hopkins University Press for Resources for the Future.
Hammerton, M., M.W. Jones-Lee, and V. Abbott. 1982. The consistency and coherence of attitudes to physical risk: some empirical evidence. *Journal of Transport Economics and Policy* May:181-199.
Hammit, J.K. 1986. Estimating consumer willingness to pay to reduce food-borne risks. R-3447- EPA, report to the U.S. Environmental Protection Agency and Rand Corporation, Santa Monica, CA.
Hanemann, W.M. 1978. A methodological and empirical study of the recreation benefits from water quality improvement. Ph.D. dissertation, Department of Economics, Harvard University, Cambridge, MA.
————. 1980a. Measuring the worth of natural resource facilities: a comment. *Land Economics* 56:482-486.
————. 1980b. Quality changes, consumer's surplus, and hedonic price indices. California Agricultural Experiment Station working paper no. 116.
————. 1982. Applied welfare analysis with qualitative response models. California Agricultural Experiment Station working paper no. 241. Berkeley, CA: University of California.
————. 1984a. Discrete/continuous models of consumer demand. *Econometrica* 52:541-561.
————. 1984b. Welfare evaluations in contingent valuation experiments with discrete responses. *American Journal of Agricultural Economics* 66:332-341.
————. 1989. Willingness to pay and willingness to accept: how much can they differ? Unpublished paper, Department of Agricultural and Resource Economics, University of California, Berkeley, CA.
Hannon, B. 1973. The structure of ecosystems. *Journal of Theoretical Biology* 41:535-546.
————. 1976. Marginal product pricing in the ecosystem. *Journal of Theoretical Biology* 56:253-267.
Harford, J.D. 1984. Averting behavior and the benefits of reduced soiling. *Journal of Environmental Economics and Management* 11:296-302.
Harrington, W., A.J. Krupnick, and W.O. Spofford, Jr. 1989. The economic losses of a waterborne disease outbreak. *Journal of Urban Economics* 25:116-137.
Harrington, W., and P.R. Portney. 1987. Valuing the benefits of health and safety regulation. *Journal of Urban Economics* 22:101-112.
Harrison, D., Jr., and D.L. Rubinfeld. 1978. Hedonic housing prices and the demand for clean air. *Journal of Environmental Economics and Management* 5:81-102.
Haurin, D. 1980. The regional distribution of population, migration, and climate. *Quarterly Journal of Economics* 95:293-308.

Hausman, J.A. 1981. Exact consumer's surplus and deadweight loss. *American Economic Review* 71:662-676.

Hausman, J.A., and D.A. Wise. 1978. A conditional probit model for qualitative choice: discrete decision recognizing interdependence and heterogeneous preferences. *Econometrica* 46:403-26.

Heberlein, T.A., and R. Baumgartner. 1978. Factors affecting response rates to mailed questionnaires: a quantitative analysis of the published literature. *American Sociological Review* 43(4):447-462.

Heck, W.W., W.W. Cure, J.O. Rawlings, L.J. Zaragoza, A.S. Heagle, H.E. Heggestad, R.J. Kohut, L.W. Kress, and P.J. Temple. 1984. Assessing impacts of ozone on agricultural crops. *Journal of the Air Pollution Control Association* 34:729-735, 810-817.

Heckman, J. 1976. The common structure of statistical models of truncation, sample selection and limited dependent variables and a simple estimation of such models. *Annals of Economic and Social Measurement* 5:475-492.

Helms, L.J. 1985. Expected consumer's surplus and the welfare effects of price stabilization. *International Economic Review* 26:603-617.

Henry, C. 1974. Option values in the economics of irreplaceable assets. In *Symposium on the economics of exhaustible resources* 41(S):89-104.

Hicks, J.R. 1939. *Value and capital.* Oxford: Clarendon Press.

Hirshleifer, J., and J.G. Riley. 1979. The analytics of uncertainty and informationan expository survey. *Journal of Economic Literature* 17:1375-1421.

Hoehn, J.P. 1989. *Valuing environmental policy alternatives in the presence of substitutes.* Staff paper, Department of Agricultural Economics, Michigan State University, East Lansing, MI.

Hoehn, J.P., M.C. Berger, and G.C. Blomquist. 1987. A hedonic model of interregional wages, rents, and amenity values. *Journal of Regional Science* 27:605-620.

Hoehn, J.P., and A. Randall. 1989. Too many proposals pass the benefit cost test. *American Economic Review* 79:544-551.

Hoffman, E., and M.L. Spitzer. 1982. The coase theorem: some experimental tests. *Journal of Law and Economics* 25:73-98.

Hori, H. 1975. Revealed preferences for public goods. *American Economics Review* 65:978-991.

Horowitz, J.L. 1985. Inferring willingness to pay for housing amenities from residential property values. Report for USEPA Cooperative agreement no. CR-810753-01-0. U.S. Environmental Protection Agency, Washington, DC.

―――. 1986. Bidding models of housing markets. *Journal of Urban Economics* 20:168-190.

―――. 1987. Identification and stochastic specification in Rosen's hedonic price model. *Journal of Urban Economics* 22:165-173.

Horowitz, J., and R. Carson. 1988. Discounting statistical lives. Staff paper 88-x, University of California, San Diego, CA.

Hotelling, H. 1949. Letter to the National Park Service (dated 1947). In *An economic study of the monetary evaluation of recreation in the national parks.* Washington: U.S. Department of the Interior, National Park Service and Recreational Planning Division.

Houthakker, H.S. 1952. Compensated changes in quantities and qualities consumed. *Review of Economic Studies* 19:155-164.

Howitt, R.E., T.W. Gossard, and R.M. Adams. 1984. Effects of alternative ozone levels and response data on economic assessments: the case of California crops. *Journal of the Air Pollution Control Association* 34:1122-1127.

Hurwicz, L., and H. Uzawa. 1971. On the integrability of demand functions. In *Preferences, utility and demand,* ed. John S. Chipman, pp. 114-148. New York: Harcourt Brace Jovanovich, Inc.

Ippolito, P.M., and R.A. Ippolito. 1984. Measuring the value of life saving from consumer reactions to new information. *Journal of Public Economics* 25:53-81.

Johansson, P.-O. 1987. *The economic theory and measurement of environmental benefits.* Cambridge: Cambridge University Press.

Jones, D. 1980. Location and the demand for nontraded goods: a generalization of the theory of site rent. *Journal of Regional Science* 20:331-342.

Jones-Lee, M.W. 1974. The value of changes in the probability of death or injury. *Journal of Political Economy* 99:835-849.

————. 1976. *The value of life: an economic analysis.* Chicago: University of Chicago Press.

————. 1979. Trying to value a life: why Broome does not sweep clean. *Journal of Public Economics* 12:249-256.

————. 1986. Personal communication to M.L. Cropper, November 1986.

Jones-Lee, M.W., M. Hammerton, and P.R. Philips. 1985. The value of safety: results of a national sample survey. *Economic Journal* 95:49-72.

Jorgenson, D.W., L.J. Lau, and T.M. Stoker. 1982. The transcendental logarithmic model of aggregate consumer behavior. In *Advances in Econometrics,* eds. R.L. Basmann and G.F. Rhodes, vol. I, pp. 97-238. Greenwich, CT: JAI Press.

Just, R., D. Hueth, and A. Schmitz. 1982. *Applied welfare economics and public policy.* Englewood Cliffs, NJ: Prentice-Hall, Inc.

Kahn, F.R., and W.M. Kemp. 1985. Economic losses associated with the degradation of an ecosystem: the case of submerged aquatic vegetation in Chesapeake Bay. *Journal of Environmental Economics and Management* 12:246-263.

Kahn, S., and K. Lang. 1988. Efficient estimation of structural hedonic systems. *International Economic Review* 29:157-166.

Kahneman, D. 1986. Comments. In *Valuing environmental goods,* eds. R.G. Cummings, D.S. Brookshire, and W.D. Schulze, pp. 185-194. Totowa, NJ: Rowman and Allanheld Publishers.

Kahneman, D., and A. Tversky. 1979. Prospect theory: an analysis of decisions under risk. *Econometrica* 47:263-291.

Kalton, G. 1983. *Compensating for missing survey data.* Ann Arbor, MI: University of Michigan Press.

Kanemoto, Y. 1988. Hedonic prices and the benefits of public projects. *Econometrica* 56:981-989.

Kanemoto, Y., and R. Nakamura. 1986. A new approach to the estimation of structural equations in hedonic models. *Journal of Urban Economics* 19:218-233.

Kennedy, P.E. 1981. Estimation with correctly interpreted dummy variables in semilogarithmic equations. *American Economic Review* 71:801.

Kirsch, I.S., and A. Jungeblut. 1986. *Literacy: profiles of America's young adults.* National Assessment of Education Progress Report no. 16-PI-02, Educational Testing Service, Princeton, NJ.

Kling, C.L. 1988a. Comparing welfare estimates of environmental quality changes from recreation demand models. *Journal of Environmental Economics and Management* 15:331-340.

————. 1988b. The reliability of estimates of environmental benefits from recreation demand models. *American Journal of Agricultural Economics* 70:892-901.

Knetsch, J.L.. 1974. *Outdoor recreation and water resources planning.* American Geographical Union, Water Resources Monograph no. 3.

————. 1984. Legal rules and the basis for evaluating economic losses. *International Review of Law and Economics* 4:5-13.

Knetsh, J.L., and R.K. Davis. 1965. Comparisons of methods for recreation evaluation. In *Water Research,* eds. A.V. Kneese and S.C. Smith, pp. 125-142. Baltimore: Johns Hopkins University Press.

Knetsch, J.L., and J.A. Sinden. 1984. Willing to pay and compensation demanded: experimental evidence of an unexpected disparity in measures of value. *Quarterly Journal of Economics* 94:507-521.

Knez, M., and V.L. Smith. 1989. Hypothetical valuations and preferences reversals in the context of asset trading. Draft manuscript, University of Arizona, Tucson, AZ.

Kohlhase, J.E. The impact of toxic waste sites on housing values. *Journal of Urban Economics,* forthcoming.

Kopp, R.J., and V.K. Smith. 1980. Measuring factor substitution with neoclassical models: an experimental evaluation. *Bell Journal of Economics* 11:631-655.

————. 1989. Benefit estimation goes to court: the case of natural resource damage assessments. *Journal of Policy Analysis and Management* 8:593-612.

Kopp, R.J., W.J. Vaughn, M. Hazilla, and R. Carson. 1985. Implications of environmental policy for U.S. agriculture: the case of ambient ozone standards. *Journal of Environmental Management* 20:321-331.

Krueger, R.A. 1988. *Focus groups: a practical guide for applied research.* Beverly Hills, CA: Sage.

Krutilla, J.V. 1967. Conservation reconsidered. *American Economic Review* 57:777-786.

LaFrance, J., and W.M. Hanemann. 1989. The dual structure of incomplete demand systems. *American Journal of Agricultural Economics* 71:262-274.

Lancaster, K.J. 1966. A new approach to consumer theory. *Journal of Political Economy* 74:132-157.

Landefeld, J.S., and E.P. Seskin. 1982. The economic value of life: linking theory to practice. *American Journal of Public Health* 72:555-66.

Lang, J.R., and W.H. Jones. 1979. Hedonic property valuation models: are subjective measures of neighborhood amenities needed? *Journal of the American Real Estate and Urban Economics Association* 7:451-465.

Lareau, T.J., R.L. Horst, Jr., E.H. Manuel, Jr., and F.W. Sipfert. 1986. Model for economic assessment of acid damage to building materials. In *Materials degradation caused by acid rain,* ed. R. Baboian, pp. 397-410. Washington: The American Chemical Society.

Larson, D.M. 1988. *Choice and welfare measurement under risk: theory and applications to natural resources.* Ph.D. dissertation, Department of Agricultural and Resource Economics, University of Maryland, College Park, MD.

Lave, L.B., and E.P. Seskin. 1971. Air pollution and human health. *Science* 169:723-731.

―――――. 1977. *Air pollution and human health.* Baltimore: Johns Hopkins University Press for Resources for the Future.

Leland, H.E. 1972. Theory of the firm facing uncertain demand. *American Economic Review* 62:278-291.

Lerman, S.L., and C.R. Kern. 1983. Hedonic theory, bid rents, and willingness-to-pay: some extension of Ellickson's results. *Journal of Urban Economics* 13:358-363.

Lind, R.C. 1973. Spatial equilibrium, the theory of rents, and the measurement of benefits from public programs. *Quarterly Journal of Economics* 87:188-207.

―――――. 1975. Spatial equilibrium, the theory of rents, and the measurement of benefits from public programs: reply. *Quarterly Journal of Economics* 89:474-476.

―――――. 1982. A primer on the major issues relating to the discount rate for evaluating national energy options. In *Discounting for time and risk in energy policy,* ed. R.C. Lind. Baltimore: Johns Hopkins University Press for Resources for the Future.

Lindley, D.V., and A.F.M. Smith. 1972. Bayes estimates for the linear model. *Journal of the Royal Statistical Society* Series B, 34:1-41.

Linneman, P. 1980. Some empirical results on the nature of the hedonic price function for the urban housing market. *Journal of Urban Economics* 8:47-68.

―――――. 1981. The demand for residence site characteristics. *Journal of Urban Economics* 9:129-148.

Linnerooth, J. 1982. Murdering statistical lives . . . ? In *The value of life and safety,* ed. M.W. Jones-Lee. Amsterdam: North-Holland Publishing Company.

Lippman, S.A., and J.J. McCall. 1981. Competitive production and increases in risk. *American Economic Review* 71:207-211.

Loehman, E.T. 1984. Willingness to pay for air quality: a comparison of two methods. Staff paper 84-18, Department of Agricultural Economics, Purdue University, West Lafayette, IN.

Loehman, E.T., and V.H. De. 1982. Application of stochastic choice modeling to policy analysis of public goods: a case study of air quality improvements. *Review of Economics and Statistics* 64:474-480.

Loehman, E.T., S.V. Berg, A.A. Arroyo, R.A. Hedinger, J.M. Schwartz, M.E. Shaw, R.W. Fahien, V.H. De, R.P. Fishe, D.E. Rio, W.F. Rossley, and A.E.S. Green. 1979. Distributional analysis of regional benefits and cost of air quality control. *Journal of Environmental Economics and Management* 6:222-243.

Loomis, J.B. 1983. A review of the suitability of the U.S. Fish and Wildlife Service 1980 national survey of fishing, hunting, and wildlife associated recreation in national assessments and forest planning. Unpublished manuscript, U.S. Forest Service, Fort Collins, CO.

―――――. 1987. An economic evaluation of the public trust resources of Mono Lake. Institute of Ecology Report no. 30, College of Agriculture and Environmental Sciences, University of California, Davis, CA.

MacLean, D. 1986. *Values at risk.* Totowa, NJ: Rowman and Allanheld Publishers.

Maddala, G.S. 1983. Limited-dependent and qualitative variables. In *Frontiers in Econometrics*. New York: Cambridge University Press.

Majid, I., J.A. Sinden, and A. Randall. 1983. Benefit evaluation of increments to existing systems of public facilities. *Land Economics* 59:377-392.

Mäler, K.-G. 1974. *Environmental economics: a theoretical inquiry*. Baltimore: Johns Hopkins University Press.

————. 1977. A note on the use of property values in estimating marginal willingness to pay for environmental quality. *Journal of Environmental Economics and Management* 4:355-369.

————. 1981. *A note on the possibility of calculating demand for a public good from information on individual behavior*. Stockholm School of Economics, Research paper no. 6209. Stockholm, Sweden.

————. 1985. Welfare economics and the environment. In *Handbook of Natural Resources and Energy Economics*, eds. A.V. Kneese and J.L. Sweeney, vol. I. Amsterdam: North-Holland Publishing Company.

Malinvaud, E. 1966. *Statistical methods of econometrics*. Amsterdam: North-Holland Publishing Company.

Manuel, E.H., Jr., et al. 1981. *Benefits analysis of alternative national ambient air quality standards for particulate matter*, vol. 2. Report for USEPA contract no. 68-02-3392. Princeton, NJ: Mathtech, Inc.

Marglin, S.A. 1963. The social rate of discount and the optimal rate of investment. *Quarterly Journal of Economics* 77:95-111.

Marin, A., and G. Psacharopoulos. 1982. The reward for risk in the labor market: evidence from the United Kingdom and a reconciliation with other studies. *Journal of Political Economy* 90:827-853.

Marshall, J. 1976. Moral hazard. *American Economic Review* 66:880-890.

Marwell, G., and R.E. Ames. 1981. Economists free ride, does anyone else? Experiments on the provision of public goods, IV. *Journal of Public Economics* 15:295-310.

Math-Tech, Inc. 1982. *Benefits analysis of alternative secondary national ambient air quality standards for sulfur dioxide and total suspended particulates*, vol II. Report to U.S. Environmental Protection Agency. Research Triangle Park, NC: Office of Air Quality Planning and Standards, U.S. EPA.

McConnell, K.E. 1975. Some problems in estimating the demand for outdoor recreation. *American Journal of Agricultural Economics* 57:330-334.

————. 1977. Congestion and willingness to pay: a study of beach use. *Land Economics* 53:185-195.

McConnell, K.E., and T.T. Phipps. 1987. Identification of preference parameters in hedonic models: consumer demands with nonlinear budgets. *Journal of Urban Economics* 22:35-52.

McConnell, K.E., and I.E. Strand. 1981. Measuring the cost of time in recreation demand analysis. *American Journal of Agricultural Economics* 63:153-156.

McConnell, K.E., I.E. Strand, and N.E. Bockstael. 1990. Habit formation and the demand for recreation. In *Advances in Applied Microeconomics*, ed. V.K. Smith, vol. 5, pp. 217-235. Greenwich, CT: JAI Press.

McFadden, D. 1974. Conditional logit analysis of qualitative choice behavior. In *Frontiers in econometrics*, ed. P. Zarembka, pp. 105-142. New York: Academic Press.

————. 1978. Modelling the choice of residential location. In *Spatial interaction theory and planning models*, ed. A. Karlqvist et al., pp. 75-96. Amsterdam: North-Holland Publishing Company.

————. 1981. Econometric models of probabilistic choice. In *Structural analysis of discrete data*, eds. C.F. Manski and D. McFadden, pp. 198-272. Cambridge, MA: MIT Press.

————. 1984. Econometric analysis of qualitative response models. In *Handbook of Econometrics*, eds. Z. Griliches and M.D. Intriligator, vol. 2, chapter 24, pp. 1395-1457. Amsterdam: North-Holland Publishing Company.

McGuckin, J.T., and R.A. Young. 1981. On the economics of desalination of brackish household water supplies. *Journal of Environmental Economics and Management* 8:79-91.

Mendelsohn, R. 1984. Estimating the structural equations of implicit markets and household production functions. *Review of Economics and Statistics* 66:673-677.

————. 1985. Identifying structural equations with single market data. *Review of Economics and Statistics* 67:525-529.

————. 1987. Measuring the aesthetic damages from hazardous wastes. Working paper, Yale University, New Haven, CT.

Merrill, T.W. 1985. Trespass, nuisance, and the costs of determining property rights. *Journal of Legal Studies* 14:13-48.

Meyer, J. 1987. Two-moment decision models and expected utility maximization. *American Economic Review* 77:421-430.

Michaels, R.G., and V.K. Smith. 1990. Market segmentation and valuing amenities with hedonic models: the case of hazardous waste sites. *Journal of Urban Economics* 28:223-242.

Michaels, R.G., V.K. Smith, and D. Harrison, Jr. 1987. Market segmentation and valuing amenities with hedonic models: the case of hazardous waste sites. Working paper, Department of Economics, North Carolina State University, Raleigh, NC.

Miller, D.M. 1984. Reducing transformation bias in curve fitting. *The American Statistician* 38:124-126.

Miller, J.R., and M.J. Hay. 1984. Estimating sub-state values of fishing and hunting. In *Transactions of the 49th North American wildlife and natural resources conference,* Boston, MA.

Milon, J.W. 1988. A nested demand shares model of artificial marine habitat choice by sport anglers. *Marine Resource Economics* 5:191-214.

Miranowski, J.A., and B.D. Hammes. 1984. Implicit prices of soil characteristics for farmland in Iowa. *American Journal of Agricultural Economics* 66:745-749.

Mishan, E.J. 1971. Evaluation of life and limb: a theoretical approach. *Journal of Political Economy* 79:687-705.

————. 1981. The value of trying to value a life. *Journal of Public Economics* 15:133-137.

Mitchell, R.C., and R.T. Carson. 1981. An experiment in determining willingness to pay for national water quality improvements. Draft report for the U.S. Environmental Protection Agency, Washington, DC.

————. 1986a. Some comments on the state of the arts report. In *Valuing environmental goods,* eds. R. Cummings, D.S. Brookshire, and W.D. Schulze. Totawa, NJ: Rowman and Allanheld Publishers.

————. 1986b. Valuing drinking water risk reductions using the contingent valuation method: a methodological study of risks from THM and Giardia. Draft report to the U.S. Environmental Protection Agency, Washington, DC.

————. 1988. Contingent valuation and the legal arena. Paper presented at the Resources for the Future conference on Assessing Natural Resource Damages, June, Washington, DC.

————. 1989. *Using surveys to value public goods.* Baltimore: John Hopkins University for Resources for the Future.

Mjelde, J.W., R.M. Adams, B.L. Dixon, and P. Garcia. 1984. Using farmers' actions to measure crop loss due to air pollution. *Journal of the Air Pollution Control Association* 34:360-365.

Moore, M.J., and W.K. Viscusi. 1988. Doubling the estimated value of life: results using new occupational fatality data. *Journal of Policy Analysis and Management* 7:476-490.

Moser, D.A., and M. Dunning. 1986. *A guide for using the contingent valuation methodology in recreation studies.* Fort Belvoir, VA: Institute for Water Resources, Army Corps of Engineers.

Mulligan, P.J. 1978. Willingness to pay for decreased risk from nuclear plant accidents. Working paper no. 43, Center for the Study of Environmental Policy, Pennsylvania State University, University Park, PA.

Munley, V.G., and V.K. Smith. 1976. Learning-by-doing and experience: the case of whitewater recreation. *Land Economics* 52:545-553.

————. 1977. A note on learning-by-doing and willingness-to-travel. *Annals of Regional Science* 11:61-70.

Murdoch, J.C., and M.A. Thayer. 1988. Hedonic price estimation of variable urban air quality. *Journal of Environmental Economics and Management* 15:143-146.

Murray, M.P. 1983. Mythical demands and mythical supplies for proper estimation of Rosen's hedonic price model. *Journal of Urban Economics* 14:327-337.

Muth, R. 1966. Household production and consumer demand functions. *Econometrica* 34:699-708.

Naraganan, R.L., and B. Lancaster. 1973. Household maintenance costs and particulate air pollution. *Clean Air* 7:10-13.

Naylor, T.H., J.L. Balintfy, D.S. Burdick, and K. Chu. 1968. *Computer simulation techniques.* New York: John Wiley & Sons, Inc.

Needleman, L. 1976. Valuing other people's lives. *Manchester School of Economics and Social Studies* 44:309-342.

Neil, J.R. 1986. Bounds on willingness to pay for non-traded goods: a possibility theorem. *Journal of Public Economics* 30:267-272.

————. 1988. Another theorem on using market demands to determine willingness to pay for non-traded goods. *Journal of Environmental Economics and Management* 15:224-232.

Nelson, J.P. 1978. Residential choice, hedonic prices, and the demand for urban air quality. *Journal of Urban Economics* 5:357-369.

————. 1982. Highway noise and property values. *Journal of Transport Economics and Policy* 16:117-138.

Niskanen, W.A., and S.H. Hanke. 1977. Land prices substantially underestimate the value of environmental quality. *Review of Economics and Statistics* 59:375-377.

Nriagu, J.O. 1978. *Sulfur in the environment,* part 2. New York: John Wiley & Sons, Inc.

O'Connor, J.J. 1913. *The economic cost of the smoke nuisance to Pittsburgh.* Mellon Institute of Industrial Research and School of Specific Industries, Smoke Investigation Bulletin no. 4, Pittsburgh, PA.

Ohsfeldt, R.L., and B.A. Smith. 1985. Estimating the demand for heterogeneous goods. *Review of Economics and Statistics* 67:165-171.

Olson, C.A. 1981. An analysis of wage differentials received by workers on dangerous jobs. *Journal of Human Resources* 16:167-185.

Oster, S. 1977. Survey results on the benefits of water pollution abatement in the Merrimack River basin. *Water Resources Research* 13:882-884.

Palmquist, R.B. 1980a. Alternative techniques for developing real estate price indexes. *Review of Economics and Statistics* 62:442-448.

————. 1980b. *Impact of highway improvements on property values in Washington state.* Report no. PB80- 170 715. Olympia, WA: Washington State Department of Transportation.

————. 1982a. The effects of air pollutants on residential property values. Working paper, North Carolina State University, Raleigh, NC.

————. 1982b. Measuring environmental effects on property values without hedonic regressions. *Journal of Urban Economics* ll:333-347.

————. 1983. Estimating the demand for air quality from property values studies: further results. Working paper, North Carolina State University, Raleigh, NC.

————. 1984. Estimating the demand for the characteristics of housing. *Review of Economics and Statistics* 66:394-404.

————. 1988. Welfare measurement for environmental improvements using the hedonic model: the case of nonparametric marginal prices. *Journal of Environmental Economics and Management* 15:297-312.

————. 1989. Land as a differentiated factor of production: a hedonic model and its implications for welfare measurement. *Land Economics* 65:23-28.

————. 1990. A note on transactions costs, moving costs, and benefit measurement. Working paper, North Carolina State University, Raleigh, NC.

————. Valuing localized externalities. *Journal of Urban Economics,* forthcoming.

Palmquist, R.B., and L.E. Danielson. 1989. A hedonic study of the effects of erosion control and drainage on farmland values. *American Journal of Agricultural Economics* 71:55-62.

Parsons, G.R. 1986. An almost ideal demand system for housing attribute. *Southern Economic Journal* 53:347-363.

Peltzman, S. 1975. The effects of automobile safety regulation. *Journal of Political Economy* 83:677-725.

Phlips, L. 1974. *Applied consumption analysis.* Amsterdam: North-Holland Publishing Company.

Pines, D., and Y. Weiss. 1976. Land improvement projects and land values. *Journal of Urban Economics* 3:1-13.

Plott, C.R. 1982. Industrial organization theory and experimental economics. *Journal of Economic Literature* 20:1485-1527.

Polinsky, A.M., and D.L. Rubinfeld. 1977. Property values and the benefits of environmental improvements: theory and measurement. In *Public economics and the quality of life,* eds. L. Wingo and A. Evans, pp. 154-180. Baltimore: Johns Hopkins University Press.

Polinsky, A.M., D.L. Rubinfeld, and S. Shavell. 1974. Economic benefits of air quality improvements as estimated from market data. In *Air quality and automotive emissions,* NAS/NAE, vol. 4, chap. 4, sections 3-5. Washington: National Academy of Science/National Academy of Engineers.

Polinsky, A.M., and S. Shavell. 1976. Amenties and property values in a model of an urban area. *Journal of Public Economics* 5:119-129.

————. 1978. Amenities and property values in a model of an urban area: a reply. *Journal of Public Economics* 9:111-112.

Pollak, R.A. 1969. Conditional demand functions and consumption theory. *Quarterly Journal of Economics* 83:60-78.

Pollak, R.A., and M. Wachter. 1975. The relevance of the household production function and its implications for the allocation of time. *Journal of Political Economy* 83:255-277.

Pollak, R.A., and T.J. Wales. 1981. Demographic variables in demand analysis. *Econometrica* 49:1533-1551.

Pope, R.D. 1980. The generalized envelope theorem and price uncertainty. *International Economic Review* 21:75-86.

Quigley, J.M. 1976. Housing demand in the short run: an analysis of polytomous choice. *Explorations in Economic Research* 3:76-102.

————. 1982. Non-linear budget constraints and consumer demand: an application to public programs. *Journal of Urban Economics* 12:177-201.

————. 1985. Consumer choice of dwelling, neighborhood and public services. *Regional Science and Urban Economics* 15:41-63.

Rae, D.A. 1983. The value to visitors of improving visibility at Mesa Verde and Great Smokey National Parks. In *Managing air quality and scenic resources at national parks and wilderness areas,* eds. R.D. Rowe and L.G. Chestnut, pp. 217-234. Boulder, CO: Westview Press.

Randall, A. 1986. The possibility of satisfactory benefit estimation with contingent markets. In *Valuing environmental goods,* eds. R.G. Cummings, D.S. Brookshire, and W.D. Schulze, pp. 114-122. Totowa, NJ: Rowman and Allanheld.

Randall, A., G.C. Bloomquist, J.P. Hoehn, and J.R. Stoll. 1985. National aggregate benefits of air and water pollution control. Interim report to the US Environmental Protection Agency, Washington, DC.

Randall, A., J.P. Hoehn, and C. Sorg Swanson. 1990. *Estimating the recreational, visual, habitat and quality of life benefits of Tongass National Forest.* General Technical Report RM-192. Fort Collins, CO: USDA Forest Service.

Randall, A., O. Grunewald, A. Pagoulators, R. Ausness, and S. Johnson. 1978. Reclaiming coal surface mines in central Appalachia: a case study of the benefits and costs. *Land Economics* 54:427-489.

Randall, A., B.C. Ives, and C. Eastman. 1974. Bidding games for valuation of aesthetic environmental improvements. *Journal of Environmental Economics and Management* 1:132-149.

Randall, A., and J.R. Stoll. 1980. Consumer's surplus in commodity space. *American Economic Review* 70:449-455.

————. 1983. Existence value in a total valuation framework. In *Managing air quality and scenic resources at national parks and wilderness areas,* eds. R.D. Rowe and L.G. Chestnut. Boulder, CO: Westview Press.

Rao, P., and R.L. Miller. 1971. *Applied econometrics.* Belmont, CA: Wadsworth.

Ratti, R.A., and A. Ullah. 1976. Uncertainty in production and the competitive firm. *Southern Economic Journal* 42:703-710.

Repetto, R. 1987. The policy implications of non-convex environmental damages: a smog control case study. *Journal of Environmental Economics and Management* 14:13-29.

Rich, C.L. 1977. Is random digit dialing really necessary? *Journal of Marketing Research* 14:300-305.

Ridker, R.G. 1967. *Economic costs of air pollution.* New York: Praeger.

Ridker, R.G., and J.A. Henning. 1967. The determinants of residential property values with special reference to air pollution. *Review of Economics and Statistics* 49:246-257.

Riesman, D. 1958. Some observations on the interviewing in the teacher apprehension study.

In *The academic mind: social scientists in a time of crisis,* eds. P. Lazorsfeld and W. Thielens, Jr. Glencoe, IL: Free Press.

Roback, J. 1982. Wages, rents, and the quality of life. *Journal of Political Economy* 90:1257-1278.

————. 1988. Wages, rents, and amenities: differences among workers and regions. *Economic Inquiry* 26:23-41.

Roberts, K.J., M.E. Thompson, and P.W. Pawlyk. 1985. Contingent valuation of recreational diving at petroleum rigs, Gulf of Mexico. *Transactions of the American Fisheries Society* 114:214-219.

Rosen, S. 1974. Hedonic prices and implicit markets: product differentiation in pure competition. *Journal of Political Economy* 82:34-55.

————. 1979. Wage-based indexes of urban quality of life. In *Current issues in urban economics,* eds. P. Mieszkowski and M. Straszheim. pp. 74-104. Baltimore: Johns Hopkins University Press.

Rowe, R.D., and L.G. Chestnut. 1985. Oxidants and asthmatics in Los Angeles: a benefits analysis. Report no. EPA-230-07-85-010. Washington: U.S. Environmental Protection Agency Office of Policy Analysis. (Addendum, March 1986.)

————. 1986. Addendum to *Oxidants and asthmatics in Los Angeles: a benefits analysis.* USEPA Report no. 230-09-86-017. Washington: U.S. Environmental Protection Agency.

————. 1989. New national park visibility value estimates. Paper presented at the International Specialty Conference: Visibility and Fine Particles, Air and Waste Management Association, October, 1989, Estes Park, CO.

Rowe, R.D., R.C. d'Arge, and D.S. Brookshire. 1980. An experiment on the economic value of visibility. *Journal of Environmental Economics and Management* 7:1-19.

Russell, F.A. Rollo. 1899. London fog and smoke. Paper presented at the Building Trades Exposition of 1899. (As referenced above in *The economic cost of the smoke nuisance to Pittsburgh,* by J.J. O'Connor.)

Sagoff, M. 1988. *The economy of the earth: philosophy, law, and the environment.* New York: Cambridge University Press.

Samples, K.C., J.A. Dixon, and M.M. Gowen. 1986. Information disclosure and endangered species valuation. *Land Economics* 62:306-312.

Samuelson, P.A. 1952. Spatial price equilibrium and linear programming. *American Economic Review* 42:283-291.

————. 1953. Prices of factors and goods in general equilibrium. *Review of Economic Studies* 21:1-20.

Sandmo, A. 1971. On the theory of the competitive firm under price uncertainty. *American Economic Review* 41:65-73.

Schelling, T.C. 1968. The life you save may be your own. In *Problems in public expenditure analysis,* ed. S.B. Chase, Jr. Washington: The Brookings Institution.

Schmalensee, R. 1972. Option demand and consumer's surplus: valuing price changes under uncertainty. *American Economic Review* 62:813-824.

Schnare, A.B., and R.J. Struyk. 1976. Segementation in urban housing markets. *Journal of Urban Economics* 3:146-166.

Schulze, W.D. 1988. Use of direct methods for valuing natural resource damages. Paper prepared for the Conference on Natural Resource Damage Assessment, June 15-17, 1988. Washington, DC.

Schulze, W.D., D.S. Brookshire, E.G. Walther, K.K. McFarland, M.A. Thayer, R.L. Whitworth, S. Ben-David, W. Malm, and J. Molenar. 1983. The economic benefits of preserving visibility in the national parklands of the southwest. *Natural Resources Journal* 23:149-173.

Schulze, W.D., R.G. Cummings, D.S. Brookshire, M.A. Thayer, R.L. Whitworth, and M. Rahmatian. 1983. Experimental approaches to valuing environmental commodities: vol. 2. Draft report for *Methods development in measuring benefits of environmental improvements,* USEPA grant no. CR 808-893-01, U.S. Environmental Protection Agency, Washington, DC.

Schulze, W.D., R.C. d'Arge, and D. Brookshire. 1981. Valuing environmental commodities: some recent experiments. *Land Economics* 57: 151-172.

Schulze, W., G. McClelland, B. Hurd, and J. Smith. 1986. Improving accuracy and reducing costs of environmental benefit assessments: Volume IV, A case study of hazardous waste sites: perspectives from economics and psychology. Draft report for USEPA cooperative agreement no. CR812054-02-1, US Environmental Protection Agency, Washington, D.C.

Schuman, H., and G. Kalton. 1985. Survey methods. In *Handbook of social psychology,* eds. G. Lindzey and E. Aronson, vol. 1. New York: Random House.

Scotchmer, S. 1985. Hedonic prices and cost/benefit analysis. *Journal of Economic Theory* 37:55-75.

————. 1986. The short-run and long-run benefits of environmental improvement. *Journal of Public Economics* 30:61-81.

Sellar, C., J.R. Stoll, and J.-P. Chavas. 1985. Validation of empirical measures of welfare change: a comparison of nonmarket techniques. *Land Economics* 61:156-175.

Sen, A.K. 1983. Development: which way now? *Economic Journal* 93:745-762.

Shapiro, P., and T. Smith. 1981. Preference for nonmarketed goods revealed through market demands. In *Advances in applied micro-economics,* ed. V.K. Smith, vol. I, pp. 105-122. Greenwich, CT: JAI Press.

Shaw, D. 1988. On-site samples regression: problems of non-negative integers, truncation, and endogenous stratification. *Journal of Econometrics* 37:211-224.

Shechter, M. 1989. Valuation of morbidity reduction due to air pollution abatement: direct and indirect measurements. Paper presented at the Association of Environmental and Resource Economists Conference on Estimating and Valuing Morbidity in a Policy Context, June 8-9, 1989, Research Triangle Park, NC.

Shepard, D.S., and R.J. Zeckhauser. 1982. Life cycle consumption and willingness to pay for increased survival. In *The value of life and safety,* ed. M.W. Jones-Lee, pp. 95-141. Amsterdam: North-Holland Publishing Company.

————. 1984. Survival vs. consumption. *Management Science* 30:423-439.

Shibata, H., and J.S. Winrich. 1983. Control of pollution when the offended defend themselves. *Economica* 50:425-438.

Shogren, J.S., and T.D. Crocker. The effects of self-protection on ex ante economic value. Working paper, University of Wyoming, Department of Economics, Laramie, WY.

Shortle, J.S., J.W. Dunn, and M. Phillips. 1986. *Economic assessment of crop damage due to air pollution: the role of quality effects.* Staff paper 118. Department of Agricultural Economics, Pennyslvania State University, State College, PA.

Silvey, S.D. 1980. *Optimal design.* London: Chapman and Hart.

Simmons, P.J. 1984. Multivariate risk premia with a stochastic objective, *Economic Journal,* supplement 94:124-132.

Slovic, P., B. Fischhoff, and S. Lichtenstein. 1979. Rating the risks. *Environment* 21:14-20, 36-39.

————. 1980. Facts versus fears: understanding perceived risk. In *Societal risk assessment: how safe is enough?,* eds. R.C. Schwing and W.A. Albers, pp. 181-216. New York: Plenum Press.

————. 1982. Response mode, framing, and information-processing effects in risk assessment. In *Question framing and response consistency,* ed. R.M. Hogarth. San Francisco: Jossey-Bass.

Small, K.A. 1975. Air pollution and property values: further comment. *Review of Economics and Statistics* 57:105-107.

Small, K.A., and H.S. Rosen. 1981. Applied welfare economics with discrete choice models. *Econometrica* 49:105-130.

Smith, R.S. 1976. *The occupational safety and health act.* Washington: American Enterprise Institute for Public Policy Research.

Smith, T.W. 1983. The hidden 25 percent: an analysis of nonresponse on the 1980 general social survey. *Public Opinion Quarterly* 47:386-404.

Smith, V.K. 1974. *Technical change, relative prices and environmental resource evaluation.* Baltimore: Johns Hopkins University Press.

————. 1979. Indirect revelation of the demand for public goods: an overview and critique. *Scottish Journal of Political Economy* 26:183-189.

————. 1983. The role of site and job characteristics in hedonic wage models. *Journal of Urban Economics* 13:296-321.

————. 1984. A bound for option value. *Land Economics* 60:292-296.

————. 1987a. Nonuse values in benefit cost analysis. *Southern Economic Journal* 54:19-26.

————. 1987b. Uncertainty, benefit-cost analysis, and the treatment of option value. *Journal of Environmental Economics and Management* 14:283-292.

————. 1988. Selection and recreation demand. *American Journal of Agricultural Economics* 70:29-36.

————. 1990. Taking stock of progress with travel cost recreation demand methods: theory and implementation. *Marine Resource Economics,* in press.

Smith, V.K., and W.H. Desvousges. 1985. The generalized travel cost model and water quality benefits: a reconsideration. *Southern Economic Journal* 52:371-381.

————. 1986a. Averting behavior: does it exist? *Economic Letters* 20:291-296.

————. 1986b. *Measuring water quality benefits.* Boston: Kluwer-Nijhoff Publishing.

————. 1987. An empirical analysis of the economic value of risk changes. *Journal of Political Economy* 95:89-114.

————. 1988. The valuation of environmental risks and hazardous waste policy. *Land Economics* 64:211-219.

————. 1989. Subtle behavior, nonresponse and mitigation. Working paper, North Carolina State University, Raleigh, NC.

Smith, V.K., W.H. Desvousges, A. Fisher, and F.R. Johnson. 1987. *Communicating radon risk effectively: a mid-course evaluation,* U.S. EPA-230-07-87-029. Washington: Office of Policy Analysis, U.S. Environmental Protection Agency.

Smith, V.K., W.H. Desvousges, and A.M. Freeman III. 1985. Valuing changes in hazardous waste risks: a contingent valuation approach. Draft report to the United States Environmental Protection Agency. Research Triangle Park, NC: Research Triangle Institute.

Smith, V.K., W.H. Desvousges, and M.P. McGivney. 1983. Estimating water quality benefits: an econometric analysis. *Southern Economic Journal* 50:422-437.

Smith, V.K., and T.A. Deyak. 1975. Measuring the impact of air pollution on property values. *Journal of Regional Science* 15:277-288.

Smith, V.K., and Y. Kaoru. 1987. The hedonic travel cost model: a view from the trenches. *Land Economics* 63:179-192.

————. 1988. Signals or noise: explaining the variation in recreation benefit estimates. Unpublished paper, November, North Carolina State University, Raleigh, NC.

————. 1989. Are we learning from partial equilibrium demand models? Working paper, May, North Carolina State University, Raleigh, NC.

————. 1990. Signals or noise? Explaining the variation in recreation benefit estimates. *American Journal of Agricultural Economics* 72:419-433.

Smith, V.K., R.B. Palmquist, and P. Jackus. 1989. A non-parametric hedonic travel cost model for valuing estuarine quality. Working paper, North Carolina State University, Raleigh, NC.

Sonstelie, J.C., and P.R. Portney. 1980. Gross rents and market values: testing the implications of Tiebout's hypothesis. *Journal of Urban Economics* 7:102-118.

Sorg, C., and D.S. Brookshire. 1984. Valuing increments and decrements in wildlife resources — further evidence. Report to the U.S. Forest Service Rocky Mountain Forest and House Experiment Station, Fort Collins, CO.

Sorg, C., and J.B. Loomis. 1984. *Empirical estimates of amenity forest values: a comparative review.* General Technical Report RM-107. Rocky Mountain Forest and Range Experiment Station, U.S. Forest Service, Fort Collins, CO.

Sorg, C.F., J.B. Loomis, D.M. Donnelly, G.L. Peterson, and L.J. Nelson. 1985. Net economic value of cold and warm water fishing in Idaho. Resources Bulletin RM-11. Rocky Mountain Forest and Range Experiment Station, United States Forest Service, Fort Collins, CO.

Sorg, C.F., and L.J. Nelson. 1986. Net economic value of elk hunting in Idaho. Resources bulletin RM-12. Rocky Mountain Forest and Range Experiment Station, United States Forest Service, Fort Collins, CO.

Spitzer, J.J. 1982. A primer on Box-Cox estimation. *Review of Economics and Statistics* 64:307-313.

Stankey, G. 1972. A strategy for the definition and management of wilderness quality. In *Natural environments: studies in theoretical and applied analysis,* ed. J.V. Krutilla, pp. 88-114. Baltimore: John Hopkins University.

Stephens, S.A., and J.W. Hall. 1983. Measuring local policy options: question order and question wording effects. Paper presented at the Annual Conference of the American Association for Public Opinion Research. May 1983, Buck Hill Falls, PA.

Stinchcombe, A., C. Jones, and P. Sheatsley. 1981. Nonresponse bias for attitude questions. *Public Opinion Quarterly* 45:359-375.

Stoll, J.R., and L.A. Johnson. 1984. Concepts of value, nonmarket valuation, and the case of the whooping crane. *Transactions of the forty-ninth North American wildlife and natural resources conference* 49:382-393.

Stoll, J.R., and L.A. Johnson. 1985. Concepts of value, nonmarket valuation, and the case of the whooping crane. Texas Agricultural Experiment Station Article no. 19360, Department of Agricultural Economics, Texas A & M University, College Station, TX.

Straszheim, M.R. 1973. Estimation of the demand for urban housing services from household interview data. *Review of Economics and Statistics* 55:1-8.

————. 1974. Hedonic estimation of housing market prices: a further comment. *Review of Economics and Statistics* 56:404-406.

Strotz, R.H. 1968. The use of land rent changes to measure the welfare benefits of land improvement. In *The new economics of regulated industries: rate-making in a dynamic economy,* ed. J.E. Haring, pp. 174-186. Los Angeles: Economics Research Center, Occidental College.

Sudman, S. 1976. *Applied sampling.* New York: Academic Press.

Sudman, S., and N.M. Bradburn. 1982. *Asking questions: a practical guide to questionnaire design.* San Francisco, CA: Jossey-Bass.

Sussman, F.G. 1984. A note on the willingness to pay approach to the valuation of longevity. *Journal of Environmental Economics and Management* 11:84-89.

Sutherland, R.J., and R.G. Walsh. 1985. Effect of distance on the preservation value of water quality. *Land Economics* 61:281-291.

Takayama, T., and G.G. Judge. 1971. *Spatial and temporal price and allocation models.* Amsterdam: North- Holland Publishing Company.

Thaler, R. 1982. Precommitment and the value of a life. In *The value of life and safety,* ed. M.W. Jones-Lee. Amsterdam: North-Holland Publishing Company.

Thaler, R., and S. Rosen. 1976. The value of life savings. In *Household production and consumption,* ed. N. Terleckyj. New York: Columbia University Press.

Thayer, M.A. 1981. Contingent valuation techniques for assessing environmental impacts: further evidence. *Journal of Environmental Economics and Management* 8:27-44.

Thompson, W.A., C.S. Holling, D. Kira, C.C. Huang, and I. Vertinsky. 1979. Evaluation of alternative forest system management policies: the case of the spruce budworm in New Brunswick. *Journal of Environmental Economics and Management* 6:51-68.

Thurstone, L. 1927. A Law of Comparative Judgement. *Psychological Review* 34:273-286.

Tinbergen, Jan. 1956. On the theory of income distribution. *Weltwirtschaftliches Archiv* 77:155-174.

Tobin, J. 1958. Estimation of relationships for limited dependent variables. *Econometrica* 26:24-36.

Tolley, G.S., and L. Babcock. 1986. Valuation of reductions in human health symptoms and risks. University of Chicago, final report to the Office of Policy Analysis, U.S. Environmental Protection Agency.

Tolley, G.S., L. Babcock, M. Berger, A. Bilotti, G. Blomquist, R. Fabian, G. Fishelson, C. Kahn, A. Kelly, D. Kenkel, R. Krumm, T. Miller, R. Ohsfeldt, S. Rosen, W. Webb, W. Wilson, and M. Zelder. 1986. Valuation of reductions in human health symptoms and risks. Report for USEPA grant no. CR-811053-01-0, US Environmental Protection Agency, Washington, D.C.

Tolley, G.S., and R. Fabian. 1988. *The economic value of visibility.* Mount Pleasant, MI: Blackstone Books.

Tolley, G.S., A. Randall, G. Blomquist, R. Fabian, G. Fishelson, A. Frankel, J.P. Hoehn, R. Krumm, and E. Mensah. 1986. *Establishing and valuing the effects of improved visibility in the eastern United States.* Report for USEPA contract no. 807768-01-0, U.S. Environmental Protection Agency, Washington, DC.

Tschirhart, J., and T.D. Crocker. 1987. Economic valuation of ecosystems. *Transactions of the American Fisheries Society* 116:369-478.

Tull, D.S., and D.I. Hawkins. 1984. *Marketing research: measurement and method,* 3rd ed. New York: Macmillian.

U.S. Department of Interior. 1986. Final rule for natural resource damage assessments under the Comprehensive Environmental Response, Compensation, and Liability Act of 1980 (CERCLA). *Federal Register* 51(148):27674-27753.

U.S. Environmental Protection Agency. 1986a. *Air quality criteria for lead.* Research Triangle Park, NC: Environmental Criteria and Assessment Office.

————. 1986b. Review of the national ambient air quality standards for lead: assessment of the scientific and technical information. Draft paper, Office of Air Quality Planning and Standards, Research Triangle Park, NC.

————. 1987. *Summary of selected new information on effects of ozone on health and welfare.* Corvallis, OR: U.S. Environmental Protection Agency.

U.S. Joint Economic Committee. 1969. *The analysis and evaluation of public expenditures: the PPB system.* 91st Congress, 1st Session.

Ulph, A. 1982. The role of *ex ante* and *ex post* decisions in the valuation of life. *Journal of Public Economics* 18:265-276.

Usher, D. 1973. An imputation to the measure of economic growth for changes in life expectancy. In *The measurement of economic and social performance: studies in income and wealth,* ed. M. Moss, vol. 38. New York: National Bureau of Economic Research.

Varian, H.R. 1984. *Microeconomic analysis,* 2nd ed. New York: W.W. Norton and Co.

Vartia, Y.D. 1983. Efficient methods of measuring welfare changes and compensated income in terms of orderly demand functions. *Econometrica* 51: 79-98.

Vaughan, W.J. 1988. *Empirical issues in the estimation of hedonic rent or property value equations and their use in prediction.* Research report no. OEO/WP-02/88, Inter-American Development Bank, Washington, DC.

Vaughan, W.J., and C.S. Russell. 1982. *Fresh water recreational fishing: the national benefits of water pollution control.* Washington: Resources for the Future.

Viscusi, W.K.. 1978. Labor market valuations of life and limb: empirical evidence and policy implications. *Public Policy* 26:359-386.

————. 1979. *Employment hazards: an investigation of market performance.* Cambridge, MA: Harvard University Press.

————. 1981. Occupational safety and health regulation: its impact and policy alternatives. In *Research in Public Policy Analysis and Management,* ed. J. Crecine, vol. 2, pp. 281-291. Greenwich, CT: JAI Press.

Viscusi, W.K., W.A. Magat, and J. Huber. 1989. Pricing environmental health risks: survey assessments of risk-risk and risk-dollar trade-offs. In *AERE Workshop Proceedings: Estimating and Valuing Morbidity in a Policy Context.*

Viscusi, W.K., and M.J. Moore. 1989. Rates of time preference and valuations of the duration of life. *Journal of Public Economics* 38:297-313.

Viscusi, W.K., and C.J. O'Connor. 1984. Adaptive responses to chemical labeling: are workers bayesian decision makers? *American Economic Review* 74:942-956.

Walsh, R.G., R.K. Ericson, J.R. McKean, and R.A. Young. 1978. *Recreation benefits of water quality: Rocky Mountain National Park, South Platte River Basin, Colorado.* Environmental Resources Center Technical Report no. 12. Fort Collins, CO: Colorado State University.

Walsh, R.G., and L.O. Gilliam. 1982. Benefits of wilderness expansion with excess demand for Indian Peaks. *Western Journal of Agricultural Economics* 7:1-12.

Walsh, R.G., J.B. Loomis, and R.A. Gillman. 1984. Valuing option, existence, and bequest demands for wilderness. *Land Economics* 60:14-29.

Walsh, R.G., N.P. Miller, and L.O. Gilliam. 1983. Congestion and willingness to pay for expansion of skiing capacity. *Land Economics* 59:195-210.

Water Resources Council. 1983. *Principles and guidelines for water and related land resources implementation studies.* Washington: United States Government Printing Office.

Watson, W.D., and J.A. Jaksch. 1982. Air pollution: household soiling and consumer welfare losses. *Journal of Environmental Economics and Management* 9:248-262.

Wegge, T.W., R. Carson, and W.M. Hanemann. 1988. Site quality and the demand for sportfishing in southcentral Alaska. Paper presented at Sportfishing Institute Symposium, March 14-15, 1988, Charleston, SC.

Wegge, T.W., W.M. Hanemann, and I.E. Strand. 1985. An economic analysis of recreational fishing in Southern California. Report submitted to National Marine Fisheries Service, La Jolla, CA.

Weisbrod, B.A. 1964. Collective-consumption services of individual-consumption goods. *Quarterly Journal of Economics* 78:471-477.

Weitzman, M.B. 1988. Consumer's surplus as an exact approximation when prices are appropriately deflated. *Quarterly Journal of Economics* 103:543-553.

Wellman, J.D., E.G. Hawk, J.W. Roggenbuck, and G.J. Buhyoff. 1980. Mailed questionnaire surveys and the reluctant respondent: an empirical examination of differences between early and late respondents. *Journal of Leisure Research* 12:164-172.

Wicksell, K. 1967. A new principle of just taxation. In *Classics in the theory of public finance*, eds. R.A. Musgrave and A.T. Peacock. New York: St. Martins.

Williams, A. 1979. A note on 'trying to value a life'. *Journal of Public Economics* 10:257-258.

Willig, R.D. 1976. Consumer's surplus without apology. *American Economic Review* 66:589-597.

————. 1978. Incremental consumer's surplus and hedonic price adjustment. *Journal of Economic Theory* 17:227-253.

Witte, A.D., H.J. Sumka, and H. Erekson. 1979. An estimate of a structural hedonic price model of the housing market: an application of Rosen's theory of implicit markets. *Econometrica* 47:1151-1173.

Wolak, F.A., and C.D. Kolstad. Homogeneous input demand under price uncertainty. *American Economic Review,* forthcoming.

Wood, S.E., and A.H. Trice. 1958. Measurement of recreation benefits. *Land Economics* 34:195-207.

AUTHOR INDEX

SUBJECT INDEX